Mechanical Science and Technology Update (MSTU 2017)

Journal of Physics: Conference Series Volume 858

Omsk, Russia
27 – 28 February 2017

ISBN: 978-1-5108-4246-5
ISSN: 1742-6588

Printed from e-media with permission by:

Curran Associates, Inc.
57 Morehouse Lane
Red Hook, NY 12571

Some format issues inherent in the e-media version may also appear in this print version.

Printed by Curran Associates, Inc. (2017)

For permission requests, please contact the Institute of Physics
at the address below.

Institute of Physics
Dirac House, Temple Back
Bristol BS1 6BE UK

Phone: 44 1 17 929 7481
Fax: 44 1 17 920 0979

techtracking@iop.org

Additional copies of this publication are available from:

Curran Associates, Inc.
57 Morehouse Lane
Red Hook, NY 12571 USA
Phone: 845-758-0400
Fax: 845-758-2633
Email: curran@proceedings.com
Web: www.proceedings.com

Table of contents

Volume 858

Mechanical Science and Technology Update
27–28 February 2017, Omsk, Russia

Accepted papers received: 26 May 2017
Published online: 7 June 2017

Preface

Papers

IOP Conf. Series: Journal of Physics: Conf. Series **858** (2017) 011001 doi:10.1088/1742-6596/858/1/011001

Mechanical Science and Technology Update (MSTU-2017)

27-28 February, 2017, Omsk State Technical University, Omsk, Russia

Alexander V. Myshlyavtsev, professor, doctor of chemical sciences

The results of theoretical and experimental studies in the field of mechanical engineering and technology are presented in the conference materials.

Particular attention is paid to the development of theory and methods of research processes affecting the technical condition of engineering objects; designing new machines, aggregates and processes, new control methods; evaluation of the technical condition of mechanical engineering objects, physical and chemical deformation transformations in production; processing and operation of a variety of materials , including computer modeling; analyzing, testing and evaluation of physical, mechanical and performance properties of materials on the samples and products, surface coatings and nano-materials, in order to determine the parameters of the equipment, machines, mechanisms to ensure the implementation of the set of technological operations and increase productivity, quality and environmental friendliness.

IOP Conf. Series: Journal of Physics: Conf. Series **858** (2017) 011002 doi:10.1088/1742-6596/858/1/011002

Peer review statement

All papers published in this volume of *Journal of Physics: Conference Series* have been peer reviewed through processes administered by the proceedings Editors. Reviews were conducted by expert referees to the professional and scientific standards expected of a proceedings journal published by IOP Publishing.

Geometric modeling of multilayer structures with the application of curves and surfaces Bezier

T V Ayusheev[1], S V Pavlova[1], R N Bulychev[1], K L Panchuk[2], A A Lyashkov[2], V Yu Yurkov[2]

[1]East Siberian State University of Technology and Management, Klyuchevskaya str., house 40, building 1, Ulan-Ude 670013, Russia
[2]Omsk State Technical University, Mira ave., 11, Omsk 644050, Russia

E-mail: Panchuk_kl@mail.ru

Abstract. In the article is reported a method of constructing the model of three-dimensional parametric solids free-form and their description using Bezier curves and surfaces. There are shown the equations of solid patch that bounded curves and surfaces cubic Bezier. The experiment was carried on an octagonal carcass of curved solid patch. Computer experiment showed that the method works well and correctly determines the position of the external and internal surfaces of both the initial condition, and under deformation of the solid shape's with the aid of control points. This method makes easyly controlling the shape and completely describes the behavior of the intermediate surfaces inside of the simulated object's solid. The proposed method can be used for numerical simulation of multilayer structures.

1. Introduction

Methods of geometric modeling of three-dimensional parametric solids, including multi-layer structures, make it possible to optimizing the design and development of a wide variety of industrial products different that often functionally different. For example, the volumetric geometric modeling method was developed for the design of parts manufactured from composite material by winding [1]. Capabilities and features of this method are also outlined for the designing of light industry products, which have a complex anthropomorphic form and the multi-layer structure [2, 3].

There are different approaches for simulating these solids. There are superficial solid model representation bodies through flat faces, modeling bodies with the help of constructive geometry solid methods [4]. Each of these methods has its advantages and disadvantages. The general approach is described the solid by sum total boundary portions, faces and edges that are defined parametrically. Each patch is constructed from a set of matching with each other freeform surfaces, each of which contains full information about its borders and relations with neigh borings. Such a description is called a representation of the solid with the help of borders. It allows performing a number of operations over the bodies while maintaining a consistent way of "interior structure". Presentation of solids through boundaries allows modeling objects of arbitrary shape and complexity, including multilayer structures.

In the works [1, 5] were used a generalized linear interpolation for describe the three-dimensional solid patch. This method of a solid description enables constructing a composite solid out of the patches of this type. Though at the boundaries of patches of the solid will be only continuous but not smooth. Thus, in result composite solid is formed with order zero smoothness. Essential for most practical applications a continuous gradient is achieved by a more complicated way. In this way, the solid patch is determined not only by calculating the boundary surfaces, but also through the settlement boundary of the slopes in the directions that are transversally to the boundary surfaces. For this purpose, in [6] was posed and solved the problem of constructing the equation of the solid patch, which takes into account the specified slopes across the boundary surfaces. In this paper, the equation of patches obtained by applying the generalized Hermite interpolation. For a description of the boundary curves used parametric cubic splines. The obtained equation makes it possible to form a compound solid of the first order of smoothness. That kind of method is useful if a shape of the modeling object is known. Although the conversion of an algebraic equation Hermite equation in the form of the curve allows you to work with curve on an intuitive level, Hermitian form does not fully meet the requirements of application systems developers. It is not easy to predict the curve shape in a magnitude of the tangent vectors. Therefore, in some applications, such as in the designing of light industry products, where the physical model is often absent or present starting imperfections of product shape, it is preferable the modeling of curves and surfaces in the form of Bezier.

In this article is considered the problem of constructing a three-dimensional equation of the solid patch by a cubic Bezier curves and surfaces. The proposed method allows describing the composite solid of the first order of smoothness.

2. Problem formulation

Suppose we know the shape of the solid surface of the modeled object. Mesh of curves that defining the surface of the solid, was built. Grid of curves divides the surface into rectangular patches that are bounded *u, v, w* curves, as shown in figure 1. Let the length of these parametric curves varies from 0 to 1. Then $\mathbf{r}(u,v,w)$, $0 < u,v,w < 1$, represents the interior of the solid patch, and r (i, v, w), r (u, j, w), r (u, v, k), i, j, k = 0, 1 determine its known boundary surfaces. It is necessary to determine the function $\mathbf{r}(u,v,w)$, provided that $u = i$, $v = j$, $w = k$ represents the desired boundary surface of the solid patch.

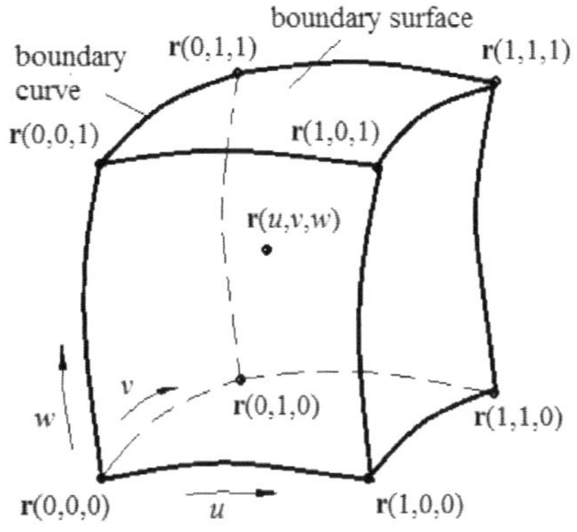

Figure 1. A patch of the solid

3. Theory

A patch of the solid is an octagonal curved carcass consisting of 8 anchor points, 12 boundary curves, and 6 boundary surfaces (figure 2).

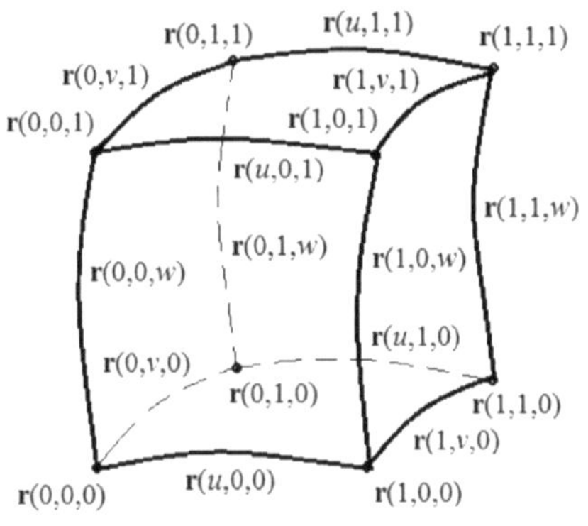

Figure 2. The boundary curves of the solid patch

For a description of the boundary curves we used a parametric cubic Bezier curve (figure 3). The cubic Bezier curve equation in matrix form is (7):

$$\mathbf{r}(t) = \mathbf{F}(t)\mathbf{MP}, \tag{1}$$

where $\mathbf{F}(t) = (t^3 \quad t^2 \quad t \quad 1)$ – row matrix of parameter *t*;

$\mathbf{P} = (\mathbf{P}_0 \quad \mathbf{P}_1 \quad \mathbf{P}_2 \quad \mathbf{P}_3)^{\mathrm{T}}$ – column matrix of guided curve points;

$\mathbf{M} = \begin{pmatrix} -1 & 3 & -3 & 1 \\ 3 & -6 & 3 & 0 \\ -3 & 3 & 0 & 0 \\ 1 & 0 & 0 & 0 \end{pmatrix}$ – matrix of coefficients cubic curve Bezier.

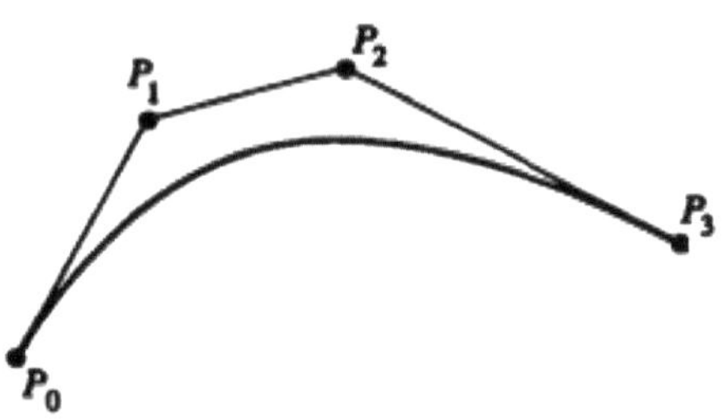

Figure 3. Cubic curve Bezier

Using equation (1), we write the equation of the boundary curves of the solid patch:

$$\mathbf{r}(u,j,k) = \mathbf{F}(u)\mathbf{MP}_{ujk}, \mathbf{r}(i,v,k) = \mathbf{F}(v)\mathbf{MP}_{ivk}, \\ \mathbf{r}(i,j,w) = \mathbf{F}(w)\mathbf{MP}_{ijw}, i,j,k \in \{0,1\}, \tag{2}$$

where $\mathbf{P}_{ujk}$, $\mathbf{P}_{ivk}$, $\mathbf{P}_{ijw}$ – control points along curves *u*, *v*, *w*.

The equations of the boundary surfaces of the solid patchs are of the form:

$$\mathbf{r}(i,v,w) = \mathbf{F}(v)\mathbf{MP}_{ivw}\mathbf{M}^{\mathrm{T}}\mathbf{F}(w)^{\mathrm{T}}, \mathbf{r}(u,j,w) = \mathbf{F}(u)\mathbf{MP}_{ujw}\mathbf{M}^{\mathrm{T}}\mathbf{F}(w)^{\mathrm{T}},$$

$$\mathbf{r}(u,v,k) = \mathbf{F}(u)\mathbf{M}\mathbf{P}_{uvk}\mathbf{M}^{\mathrm{T}}\mathbf{F}(v)^{\mathrm{T}}, i,j,k \in \{0,1\}, \tag{3}$$

where $\mathbf{P}_{ivw}$, $\mathbf{P}_{ujw}$, $\mathbf{P}_{uvk}$ – control points of boundary surfaces.

Equation solid patch in matrix form is

$$\mathbf{r}(u,v,w) = \mathbf{F}(u)\mathbf{M}\mathbf{P}_{uv0}\mathbf{M}^{\mathrm{T}}\mathbf{F}(v)^{\mathrm{T}}\mathbf{F}(w)\mathbf{M}^{(0)} + \mathbf{F}(u)\mathbf{M}\mathbf{P}_{uv1}\mathbf{M}^{\mathrm{T}}\mathbf{F}(v)^{\mathrm{T}}\mathbf{F}(w)\mathbf{M}^{(1)} + \mathbf{F}(u)\mathbf{M}\mathbf{P}_{uv2}\mathbf{M}^{\mathrm{T}}\mathbf{F}(v)^{\mathrm{T}}\mathbf{F}(w)\mathbf{M}^{(2)} + \mathbf{F}(u)\mathbf{M}\mathbf{P}_{uv3}\mathbf{M}^{\mathrm{T}}\mathbf{F}(v)^{\mathrm{T}}\mathbf{F}(w)\mathbf{M}^{(3)}, \tag{4}$$

where P_{uv0}, $\mathbf{P}_{uv1}$, $\mathbf{P}_{uv2}$, $\mathbf{P}_{uv3}$h – control points of surfaces.

4. Illustrative example

As an experiment was described solid's patch on an octagonal curved carcass. Computer experiment showed that the method works well and correctly determines the position of the external and internal surfaces of both the initial condition, and when the deformation forms using control points.

Computational experiments were carried out using mathematical MathCAD application. Changes the position of the nodal points of the solid patch and defined by the intermediate patch of the surface of the solid with parameter values of *=0,25; 0,5; 0,75; 1.* The experimental results are shown in figure4.

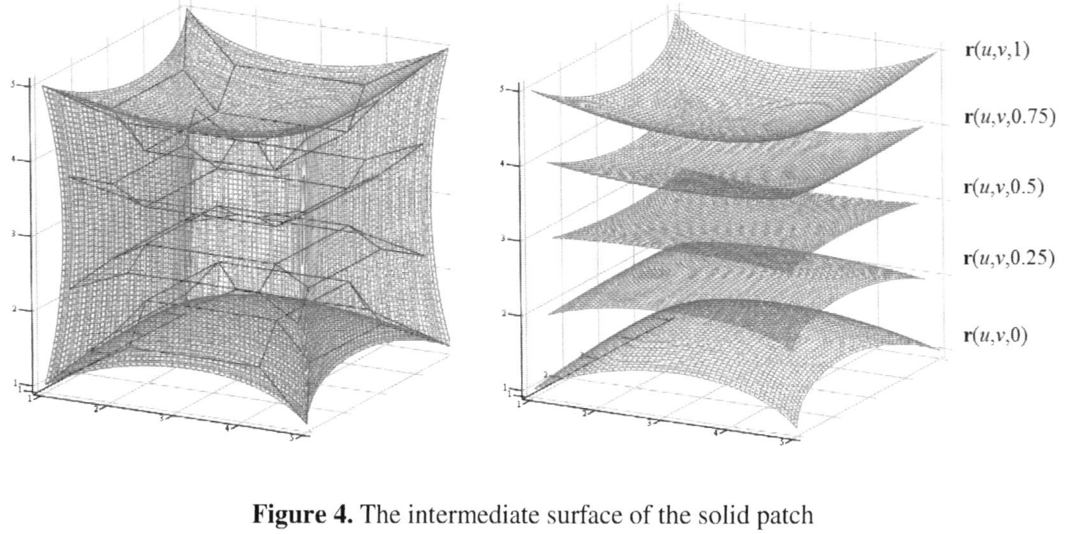

Figure 4. The intermediate surface of the solid patch

5. Results

Computational experiments demonstrated the correctness of the obtained formula (4) for describe the patch of solid. Sequential substitutions of specific numerical values in the above formula satisfy the required conditions of the task. Manage form of simulated object can use the control points of the boundary surfaces Bezier. The boundary surfaces are changed by changing the carcass points. Thus the resulting structure is completely describes the behavior of the intermediate solid surface, including the middle surface. This is a very important result for certain applications where it is necessary to determine the median surface of the thick-walled multi-layer shell with variable thickness with the change in the shape of the boundary surfaces exposed to different loads.

6. Conclusions

This model setting the shape and structure of the multilayer product design will optimize the tasks of modeling the volume of products in various branches of industrial production. This will shorten the

IOP Conf. Series: Journal of Physics: Conf. Series **858** (2017) 012001 doi:10.1088/1742-6596/858/1/012001

product design, without causing deterioration of the quality of the designed object. Results of the experiments showed that the proposed method does not require the overriding the solid carcass when determining the position and shape of the intermediate surfaces. The shape of the solid surface is easily controlled with the aid of the control points. Using a parametric cubic Bezier curves and surfaces provides a composite with first order of smoothness.

References

[1] Ayusheev T V 2005 *Information Technol.* **6** pp 35–40.
[2] Pavlova S V and Ayusheev TV 2014 Geometric modeling method of multi-layer constructions of light industry products *Bulletin ESSTU* **1** pp 13–18.
[3] Pavlova S V and Ayusheev T V 2007 Description of the virtual product model in the fashion industry from the perspective of geometric modeling of the form *Bulletin ESSTU* **3** pp 32–35.
[4] Golovanov N N Geometric modeling 2002 *Moscow* 472
[5] R. Durikovic S and Czanner 2004 Modeling with three types of Coons Bodies *International Journal of Modeling & Simulation (IASTED)* **24(2)** pp 97–101
[5] Ayusheev T V 2005 Method of construction of solid bodies using generalized Hermite interpolation *Information Technology* **6** pp 27-32.
[5] Ivanov V P and Batrakov A S 1995 Three-dimensional computer graphics *Moscow* 224

IOP Conf. Series: Journal of Physics: Conf. Series **858** (2017) 012002 doi:10.1088/1742-6596/858/1/012002

Power and energy ratios in mechanical CVT drive control

P D Balakin[1], L O Stripling[1]

[1]Omsk State Technical University, Mira ave. 11, Omsk, 644050, Russia

Abstract. Being based on the principle of providing the systems with adaptation property to the real parameters and operational condition, the mechanical system capable to control automatically the components of convertible power is offered and this allows providing stationary operation of the vehicle engine in the terms of variable external loading. This is achieved by drive control integrated in the power transmission, which implements an additional degree of freedom and operates on the basis of the laws of motion, with the energy of the main power flow by changing automatically the kinematic characteristics of the power transmission, this system being named CVT. The power and energy ratios found allow performing the necessary design calculations of the sections and the links of the mechanical CVT scheme.

1. Introduction

The implementation of the mechanical design principle by adapting to the real-world settings of the mechanical systems and operating mode at design phase allows, for example, creating power transmission with the auto control of transforming power components [1].

In traditional systems [2-6] in the context of variable external loading, this function is performed by an operator or an automatic control system (ACS) based on the elements of automation and with self-contained power supply. In vehicles equipped with ACS, there are nodes and components operating in the same site that are based on the use of physically heterogeneous environments or fields, which is always problematic and complicates the system and reduces its reliability.

In the proposed system design, there is an internal adapting link that implements the degree of freedom added to the main section movement. Additional section motion is used to control the characteristics of the system, including purpose autochange of its velocity transfer function.

The adapting link activity involves the use of laws of motion and the energy adapting movement is performed by the variable component of the force characteristic of the main force flow. The additional motion is implemented by mechanical drive control integrated into power transmission that takes the power flow, compares it with a nominal value, and forwards the formed difference to change automatically the kinematic characteristics of the mechanical system.

The principle of designing mechanical systems with adaptive properties has been theory-based and is contained in numerous publications of the scientific and pedagogical staff of the Machine Department of the Omsk State Technical University, for example, [1]. However, the solutions of applied tasks to create adaptive systems of specific application are relevant, because even during the design phase of such systems, it is necessary to know the levels of sections force loading and the links of both the main chain and the adapting mechanical drive control.

The study subject is power transmission based on the friction CVT with automatically changeable velocity transfer function.

2. Problem Statement

Now we clarify the structure and operation of the mechanical system based on the engineering solution [7]. Figure 1 shows the CVT driven unit node. CVT works as follows. When the resistance on the driven shaft increases, the semi-pulleys 4 and 6 and the nuts 3, respectively, receive relative angle motion that due to multidirectional thread in the screw joint causes semi-pulley axial approaching and pushing the belt to the periphery of active semi-pulley surfaces, thus increasing the kinematic size of the driven unit.

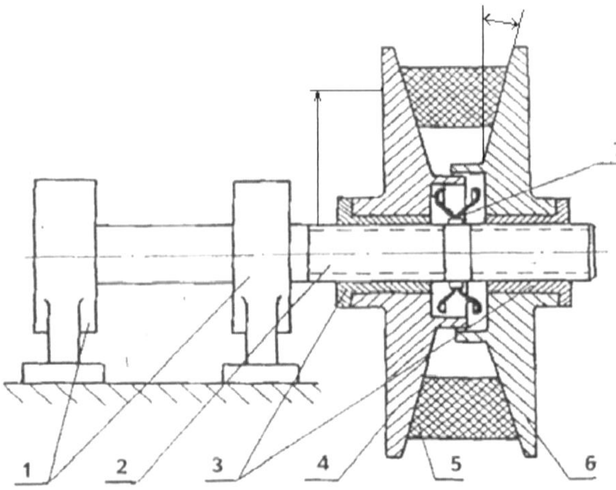

Figure 1. V-belt CVT with helical block control [5]:

1-the support of the driven shaft; 2-helical surface with right and left threads; 3-nuts (left and right);

4 and 6 – semi-pulleys; 5 – belt; 7 – springs.

The axial rapprochement of the semi-pulleys produces the deformation of the elastic element 7, elastic force of which is used to reverse the evolution of the detail node while reducing the external load of the driven shaft.

The driving unit can be formed using similar design elements, but in the first approaching we state the driving unit having the unchanged kinematic size, with belt tension in evolution being provided by the tension looper, for example by [8].

The intermediate can be performed either by a cone synthetic or rubber-cord belt or by a flexible metallic solid or section belt.

The drive control, consisting of semi-pulley, screw joint and elastic elements, is integrated into the main scheme. In addition to the main movement the adapting one occurs due to the energy of the transformed main force flow generated by the operating member of the machine.

Our task is to define power and energy ratios in the sections of the given CVT depending on the variable characteristics of external force load, as it is needed to evaluate its useful evolution parameters and calculate drive elements.

3. Theory

The automatic control by components of convertible power allows a stationary, economical running of the Power plant (engine) in the context of a variable external force load.

The denoted M_1, ω_1 and M_2, ω_2 are the power elements and the speed of engine shafts and operating member of the machine. Without taking into account the losses variational relation is:

$$M_1\omega_1 = M_2\omega_2, \tag{1}$$

where

$$\omega_2 = \frac{M_1\omega_1}{M_2}. \tag{2}$$

If, under the conditions of the problem, M_1 = *const* and ω_1 = *const*, then the mechanical system output velocity ω_2 at variable M_2 is connected with it by hyperbolic law, but the automatic CVT transfer function Π^ω

$$\Pi^\omega = U_{1,2} = \frac{\omega_1}{\omega_2} \tag{3}$$

is linear relating to M_2, i.e. a multiple change of M_2 requires a multiple modification of Π^ω that is provided by adapting evolution of the automatic CVT kinematic characteristic.

Figure 2 shows the design model of the driven unit with integral drive control.

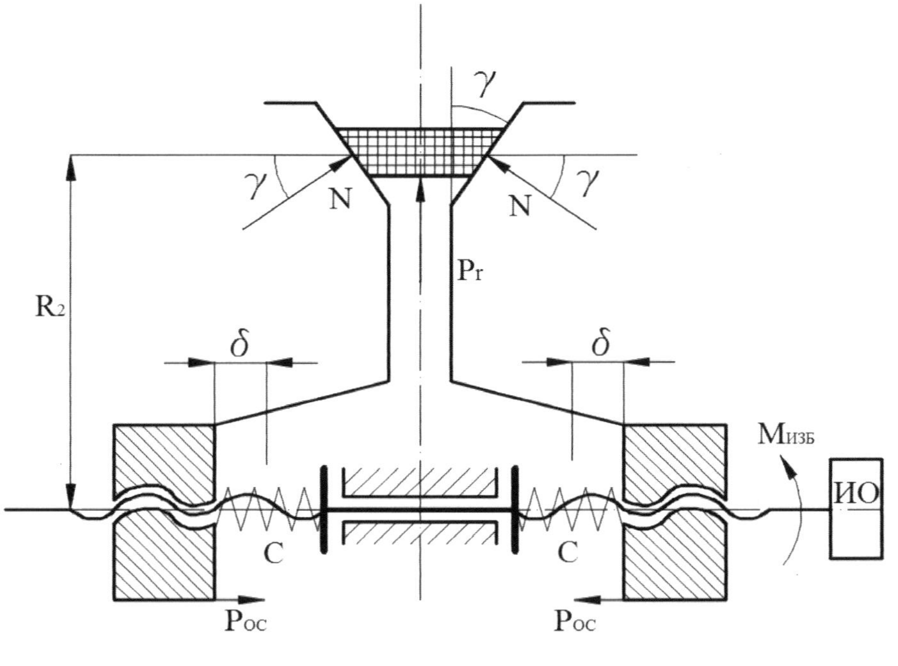

Figure 2. The design model of the driven unit with integral drive control.

In figure 2 R_2 is a changeable kinematic size of auto CVT driven unit; M_{ex} $(=M_{изб})$ is a variable part of force external load from the operating member (OM); δ - is the axial movement of the nuts due to axial force P_{ax} $(=P_{oc})$, derived by M_{ex} ;c is the stiffness of the elastic elements, other symbols are clear.

The known force ratio in the screw joint is:

$$P_{oc} = \frac{2M_{изб}}{d_{cp}tg(\beta+\theta)}, \tag{4}$$

where d_{cp} is a pitch diameter of the helicoidal surface; θ *is arctgf'*, with $f'=f/cos\alpha$, where f is the friction coefficient of active screw joint surfaces, and a is the thread angle of helicoidal surface; β is the thread angle of helical line.
The screw joint is proposed to be executed by ball screw one with sliding friction replaced by rolling friction, therefore, the friction is minimized. Besides, the joint is to be nonlocking with reverse movement, so $\beta \gg \theta$ and in the design calculation the value θ can be deleted.

Figure 3b shows

$$tg\beta = \frac{T}{\pi d_{cp}}, \tag{5}$$

where T is a helix lead in the screw joint.

We substitute (5) into (4), dividing M_{ex} into every nut evenly, and obtain axial force values P_{ax}, acting on each semi-pulley:

$$P_{oc} = \frac{2\frac{M_{u3\delta}}{2}\pi d_{cp}}{d_{cp}T} = \frac{\pi M_{u3\delta}}{T}. \tag{6}$$

The axial force operation A on each screw joint point is:
$A=P_{ax}\delta$, and in general $A=2P_{ax}\delta$ taking into account (6)

$$A = \frac{2\pi M_{u3\delta}\delta}{T}. \tag{7}$$

Axial force P_{ax} on the semi-pulleys generates radial force P_r (Fig.2), pushing out the belt to the cone pulley periphery on semi-pulley approaching;

$$P_r = 2P_{oc}tg\gamma, \tag{8}$$

where γ is the angle of active semi-pulley cone surface (Fig.2), taking into account (6) we obtain:

$$P_r = 2\frac{\pi M_{u3\delta}}{T}tg\gamma. \tag{9}$$

Distributed across the active semi-pulley cone surface, buoyancy P_r is to ensure force S_o overcoming for belt pre-tensioning, i.e. $P_r \geq 2S_o$, taking into account (9)

$$\frac{\pi M_{u3\delta}}{T}tg\gamma \geq S_o. \tag{10}$$

Belt pre-tensioning defines belt-pulley friction sufficient to transmit both the constant and variable parts of the power flow. If the aggregate value of the transformed moment is stated M, then

$$S_1 - S_2 = \frac{M}{R_2} \tag{11}$$

$S_1+S_2=2S_o$, where S_1 and S_2 are the effort in pulling and idling branches respectfully.
With values M and R_2 the ratios (11) enable to define S_o and maintain its value by automatic tension looper [8].

Axial force P_{ax} generates potential strain energy of integral elastic elements. During semi-pulley approaching equal to 2δ energy relation is:

$$P_{oc} \geq 2c\delta. \tag{12}$$

IOP Conf. Series: Journal of Physics: Conf. Series **858** (2017) 012002 doi:10.1088/1742-6596/858/1/012002

Under the axial force P_{ax} the evolution of the driven unit design is accompanied with additional irreversible friction losses very significant in the belt-semi-pulley contact. Friction force $F_{fr}=fN$, where f is the friction coefficient of belt material and active pulley surface, N is a normal force in the belt-pulley contact, defined as (Fig.3a): $N=P_{ax}/cos\gamma$, or taking into account (6) $N=\pi M_{ex}/Tcos\gamma$, hence

$$F_{mp} = \frac{f\pi M_{u3б}}{T\cos\gamma}. \quad (13)$$

The operation of buoyancy P_r to overcome pretension forces S_o on the movement $\pi\Delta R_2$, $\Delta R_2 = \frac{\delta}{tg\gamma}$ is equal to (figure 3c)

$$A_r = \pi S_o \frac{\delta}{tg\gamma}. \quad (14)$$

The work of the integral elastic components under semi-pulley approaching is equal to:

$$A^*_{осев} = c\delta^2. \quad (15)$$

The friction work with evolution is:

$$A_{mp} = F_{mp}\frac{\delta}{\sin\gamma} или$$

$$A^*_{mp} = \frac{\pi f M_{u3б}\delta}{T\cos\gamma\sin\gamma}. \quad (16)$$

The energy balance equation is as follows:

$$\frac{2\pi M_{u3б}}{T} = \frac{\pi S_o\delta}{tg\gamma} + c\delta^2 + \frac{\pi f M_{u3б}\delta}{T\cos\gamma\sin\gamma} \quad (17)$$

after transformations we express δ with convertible M_{ex} and taken drive control settings:

$$\delta = \frac{M_{u3б}\left[\frac{2\pi}{T} - \frac{\pi f}{T\cos\gamma\sin\gamma}\right] - \frac{\pi S_o}{tg\gamma}}{c}. \quad (18)$$

Changing discretely c, in (18) with known f, T, γ, S_o we obtain dependencies «δ» with the variable value of the power flow, generated by operating process in the operating member of the machine.

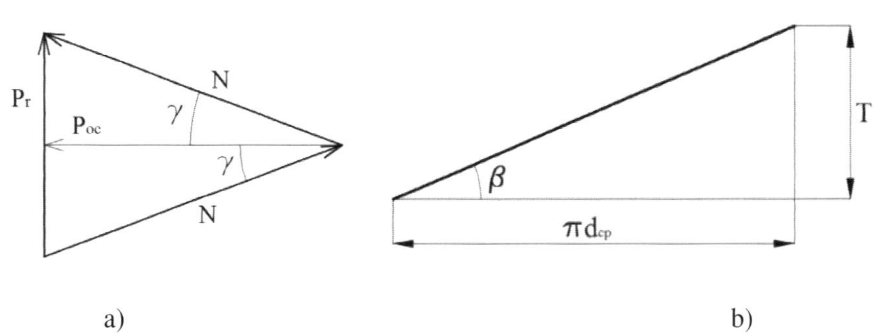

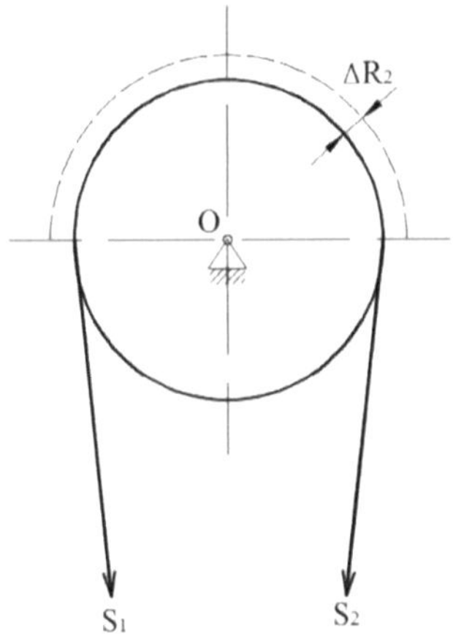

c)

Figure 3. To the calculation of power and energy ratios in the drive control.

4. Result discussion

According to (18) the obtained value of semi-pulley approaching δ, depending on M ex and accepted constants is easily transformed into the change of kinematic size $\Delta R_2 = \frac{\delta}{tg\gamma}$ that defines the transfer function of the auto CVT velocity. Therefore, power and energy ratios in the drive control enable to perform the task of synthesizing auto CVT scheme in a given range of the transfer function change, and consequently, the range of transformed power component auto control.

It is obvious that the regulatory range can be extended by embedding the drive control into the design of the auto CVT driving unit.

It is clear that friction losses in the evolution of the main sections, especially with the relative belt movements on the active semi-pulley surfaces, should be clarified to take into account the need for transmission and the constant friction of the force moment. In the second approximation during choosing elastic element stiffness *c*, the general semi-pulley approaching should be divided into the constant and variable parts. The former is defined by constant power moment component, the latter - by the variable one.

The power and energy ratios are dependent on both the kinematic diagram of the auto CVT as a whole and the scheme of the integral drive control.

5. Results and summary

- The power and energy ratios in the drive control of the mechanical auto CVT based on V-belt diagram with expanding semi-pulleys have been obtained.
- It has been given the principled feasibility of CVT structure capable to regulate the components of the transformed power that operates using the laws of mechanics through the energy of the main force flow in a variable load case.
- It has been shown that the drive control is able to ensure linear dependence of CVT velocity transfer function on the variable external force load.
- On the base of power and energy ratios one can perform the needed amount of unit design calculations and CVT elements of the offered scheme.

Reference

IOP Conf. Series: Journal of Physics: Conf. Series **858** (2017) 012002 doi:10.1088/1742-6596/858/1/012002

[1] Balakin P D 2016 Elements of the theory of real mechanical systems: Monograph (Omsk, Russia) 272

[2] Yildiz A and Bottiglione F Carbone G 2015 An experimental study on the shifting dynamics of the chain CVT *Proc. of The Third Intl. Conf. on Advances in Mechanical and Robotics Engineering (AMRE-2015)*

[3] Brown A W Rooij J Van and Frank A A 2004 The Design of an Inline GCI Chain CVT For Large Vehicles, *International Continuously Variable and Hybrid Transmission Congress, UC Davis* **2004-40-0054**

[4] Srivastava N and Haque I 2009 A review on belt and chain continuously variable transmissions (CVT) dynamics and control *Mechanism and Machine Theory* **44(1)** pp 19–41

[5] Bottiglione F, De Pinto S, Mantriota G, Sorniotti A 2014 Energy Comsumption of a Battery Electric Vehicle with Infinitely Variable Transmission *Energies* **7(12)** pp 8317–8337

[6] Mangialardi L and Mantriota G 1992 The Advantages of USING Continuously Variable Transmission in Wind Power Systems *Renewable Energy* **2(3)** pp 201–209

[7] Patent RU No. 2122670. MCI 6 F 16 Mr. 9/18. Automatic V-belt CVT/P. D. Balakin, V. V. Bienko (Russia) // Bull. No. **33-1998**

[8] Patent RU No. 2122670. MCI 6 F 16 H 7/08. Tension device for transmission with flexible link / P.D. Balakin, V. V. Bienko (Russia) Bull. No **33-1998**

Molybdenum protective coatings adhesion to steel substrate

A I Blesman[1], D V Postnikov[1], D A Polonyankin[1], A A Teplouhov[1], A V Tyukin[2] and E A Tkachenko[1]

[1]Omsk State Technical University, 11, Mira Ave., Omsk, 644050, Russia
[2]Siberian State Automobile and Highway University, 5, Mira Ave., Omsk, 644080, Russia

E-mail: dvpostnikov@omgtu.ru

Abstract. Protection of the critical parts, components and assemblies from corrosion is an urgent engineering problem and many other industries. Protective coatings' forming on surface of metal products is a promising way of corrosionprevention. The adhesion force is one of the main characteristics of coatings' durability. The paper presents theoretical and experimental adhesion force assessment for coatings formed by molybdenum magnetron sputtering ontoa steel substrate. Validity and reliability of results obtained by simulation and sclerometry method allow applying the developed model for adhesion force evaluation in binary «steel–coating» systems.

1. Introduction

Corrosion resistance of metals has a direct impact on the product's performance. A corrosion losses in countries with developed industry is up to 5% of gross national product (GNP) [1]. Protective coatings' formation on metal surface relates to methods of corrosion control [2–5]. High adhesion between coating and substrate is one of the major criteria for coatings' material selection. At high coating strength and low adhesion to the substrate coatings' detachment and disconnection occurs. At low coating strength the bubbles with a gap appear which dramatically increases the rate of metal oxidation.

Definitely, other problems related to the protective coatings application can be indicated, however, and they are reduced to wrong coatings' material selection. Qualitativemethods of adhesion evaluation include mechanical (scratch test, a test of abrasion resistance, tensile strength) as well as non–mechanical methods (XRD) [6].XRD is usedfor stress state investigation of polycrystalline coatings (e.g.molybdenumcoatings)deposited by magnetron sputtering [7]. Application of this method allows carrying out an estimation of residual stresses that have a direct impact on adhesion of molybdenum coating to the substrate [8–16]. Among the quantitative methods of adhesion evaluation allocate sclerometry, nanoindentation [17] and scribing as the most widespread in applied research of coatings'adhesive resistant [18].Scientific and practical interest of thin molybdenum coatings formation is causedby its heat resistance, high mechanical strength and unique electrical properties.Molybdenum is used as a material for the manufacture of contact groups and protective layers in integrated circuits [19]and thin–film solar cells [20, 21], superconducting thin films [22] and optical interference coatings [23]. Magnetron sputtering of refractory metals including molybdenum is used for wear–resistant anticorrosive coatings formation in various fields of industry [24, 25].

2. Materials and methods

J24056 grade steel is used as a coating's substrate. Products made of this steel operating under high temperatures conditions and corrosive environments that promote corrosive wear. The study assesses experimentaladhesion force (σ) of molybdenum coating to substrate (defined as the interaction force between coating and substrate per unit area, Pa) obtained by sclerometry in comparing with calculation results of developed model. Theoretical calculations can be used for service life evaluation of coated products operating at elevated temperatures in corrosive environments.

This paper describes the adhesion force by sclerometry. At adhesion force determination by this method the coating breaking occurs to baring a substrate with it following detachment. It must be an accurate measurement horizontal load, crack width, the film thickness and hardness to quantify the adhesion value. Horizontal force F was measured during coating's scratching at such vertical load when the substrate remained clean on indenter track. It was assumed that horizontal force F (1) is the sum of three components. The first component, F_1 is the force that occurs when coating is scratching (2). The second component, F_2 (3) is the force that provides full coating's scratching from the substrate. The third component, F_3 is the force acting on indenter while it moves across substrate without coating:

$$F = F_1 + F_2 + F_3 \tag{1}$$

$$F_1 = dhH \tag{2}$$

Here d is the track width, h is the film thickness, H is the hardness of coating.

$$F_2 = 1/4\ (\sigma\pi d^2) \tag{3}$$

Here σ is the adhesion force, determined by equation:

$$\sigma = 4(F - F_3 + dhH)\ /(\pi d^2) \tag{4}$$

Experimental installation for adhesion determination has been collected on the basis of Vickers' hardness tester «PMT–3M». A horizontal force is applied to the rotating platform and the microhardness was determined using a strain gauge. The following analysis of indenter track was performed by using scanning electron microscopy (JEOL JCM–5700). Determination of adhesion properties of the coating was carried out with a vertical force 0.5 N applied to indenter.

3. Results and discussion

Experimental scheme for the adhesion force research is shown on figure 1.The indenter track and the width scratch for randomly selected area is shown on figure 2a. The initial parameters'values are the track width (d=23 mm), coating hardness (Hv=450), coating thickness (h=1 mm) and the difference between horizontal forces (F_1–F_3=0.45 N). Side micrograph of the diamond indenter is shown on figure2b. Top micrograph of the diamond indenter is shown on figure 2c.

The adhesion force value calculated by equation (1) gives the result equal to σ=760 MPa. Coincidence of adhesion force evaluation for a magnetron sputtered «steel–molybdenum coating» system with values, obtained earlier by authors, leads to conclusion about reliability of the developed measurement techniques.

IOP Conf. Series: Journal of Physics: Conf. Series **858** (2017) 012003 doi:10.1088/1742-6596/858/1/012003

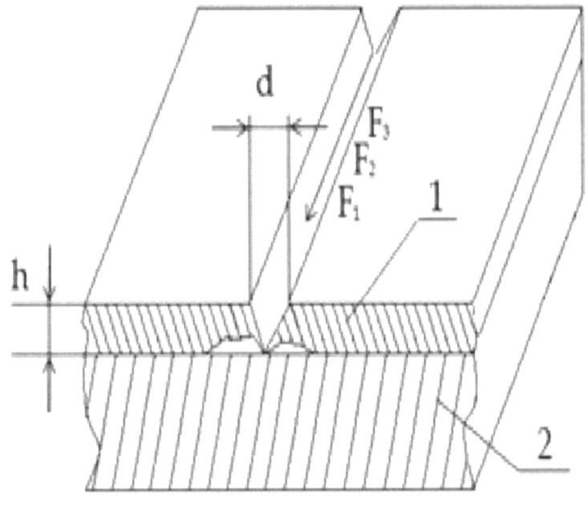

Figure 1. Cross–sectional scheme of the scratch (1–coating; 2–substrate).

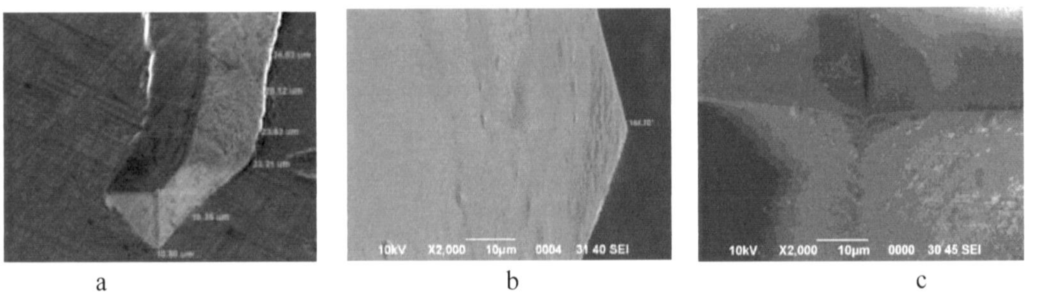

a b c

Figure 2.SEM images: a – track of indenter on the surface of molybdenum coating; b – side micrograph of the diamond indenter; c – top micrograph of the diamond indenter.

An example of adhesion force calculation for molybdenum coating obtained by magnetron sputtering onto the surface of J24056 grade steel is considered to verify received results. For the adhesion force calculation,it is applied an approach described for magnetron sputtered coatings in reference [26]. The adhesion force could be calculated by equation:

$$\sigma = kNf \tag{5}$$

Here k is the dimensionless coefficient that takes into account the reduction of possible bonds number, depending on the presence of structure defects; f is the strength of a single metallic atoms' bond near the border of «steel–coating» transition layer; N is the atoms' distribution through the layer depth with thickness in order of the lattice constant.Transition layer is a substitutional solid solution, in this case it is appropriate to describe its contribution to adhesion force by each material with equation (6):

$$\sigma = \sigma_1 + \sigma_2 = kN_1f_1 + kN_2f_2 = k(N_1f_1 + N_2f_2) \tag{6}$$

Here f_1, f_2 is the strength of a single metallic atom bond of each material (N); N_1, N_2 is the each material atoms' distribution of through the layer depth with thickness in order of the lattice constant and one square meter area. With regard to the number of atoms in crystalline lattice of the target layer it was found the depth distribution $N_1(x)$, $N_2(x)$ of transition layer:

$$N_1 = C_1N_0; \quad N_2 = C_2N_0 \tag{7}$$

The single bond force of the atom was calculated using the material melting point (T). An average kinetic energy of random atoms' motion at such temperature goes into overcome work of the interatomic attraction forces thus this process can be represented by equation:

$$3/2\ k_B T = nfd \tag{8}$$

Here k_B is the Boltzmann constant; n is the number of overcoming atomic bonds in crystal lattice and d is the lattice constant.
Thus for calculation of the atoms' single bond force can be used the following expression:

$$f = (3/2\ k_B T) = (2nd) \tag{9}$$

The adhesion force σcan also be estimated through the adhesion energy:

$$W_a = (W_{p1}W_{p2}(K_1 + K_2)^2)/(W_{p1}K_1^2 + W_{p2}K_2^2) \tag{10}$$

Here W_{p1} and W_{p2} is the surface energy of contacting materials (reference values) that can be also calculated by method described in [26]; K_1 and K_2 are the coefficients calculated by equation:

$$K_1 = \frac{\upsilon_1}{(1 - \upsilon_1)}; \quad K_1 = \frac{\upsilon_1}{(1 - \upsilon_1)} \tag{11}$$

Here υ_1 and υ_2 are the Poisson's coefficients of contacting materials. Thus the estimated formula for adhesion force σ has a form:

$$\sigma = (W_a E)^{1/2} /(\pi h)^{1/2} \tag{12}$$

Here E is the elasticity modulus of surface material.The impurity content dependence from the depth of transition layer in «molybdenum–steel» system calculated on the basis of mass transfer model [27, 28]. Interatomic forces calculated for iron and molybdenum correspondently received the following values: f_1=3.76·10^{-11} N and f_2=5.98·10^{-11} N. The adhesion force is also influenced by purity of product's surface, by presence of impurities and defects in it, and can also be influenced by various other factors. Adhesion force measurements show that as a result of these factors a contact area of coating is 45%...50% less than surface area on which the coating is applied to by magnetron sputtering [18]. Because of this reason it is used a coefficient k=0.5for adhesion force modeling calculations.
The coating's adhesion force σ calculated by equation (5) based on iron and molybdenum atoms' distribution through the depth of transition layer is presented on figure 3. There is a slight discrepancy on figure 3 between calculated values of the adhesion force σ compared with the reference: 760 and 820 MN/m^2, so we can assume that calculation results are consistent with published data. Maximum of σ corresponds to the highest concentration of implanted impurity, so when attempting detach the coating, the greatest probability of detachment corresponds to the surface edge or to the depth of 1.02 μm where the transition layer ends.
It should be noted that the estimated value of adhesion force calculated through adhesion energy substantially higher than the experimental data. According to estimation the adhesion force on steel surfaces for coatings with 1 μm thickness can reach about 800 N/m^2, at the same time adhesion of molybdenum coating to the surface of pure iron attains more than 820 N/m^2 [26]. This overestimation of calculated results apparently due to the fact that evaluation model considered perfectly smooth, clean surface without defects.

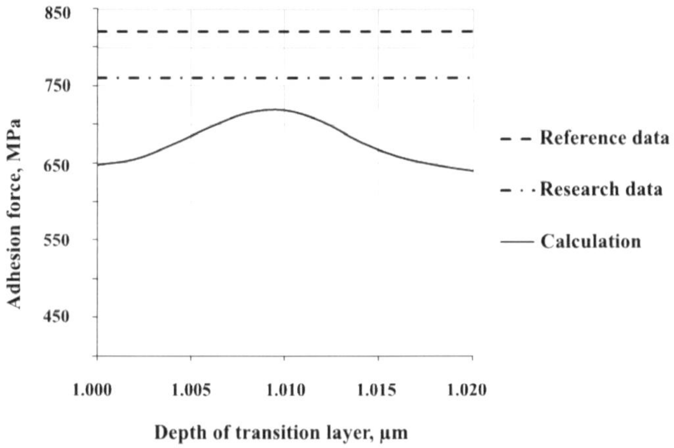

Figure 3. Adhesion force dependence from the depth of «steel–molybdenum» transition layer at 1µm coating's thickness.

The obtained estimationslead to an error of more than 100% due to fact that it is not possible to accurately determine the interaction potential between atoms, so approximate potentials are used: Morse and others.The values of adhesion forceσ will be close to the estimated values if coefficient k=1 will be used in the proposed model (ideal case).

4. Conclusion

This investigation assessestheadhesion force between molybdenum coating deposited by magnetron sputtering and J24056 grade steel substrate, which is amounted of 760 MPa from the experimental data.The developed model is used for adhesion force evaluation of thin–film coatings and calculation results show that the minimum adhesion force is equal to 650 MPa.

The results of adhesion force calculation are in satisfactory agreement with reference and verified with experimental data. Thus, the proposed model can be used in the adhesion force evaluation for the binary «steel–coating» systems.

Acknowledgments

The work was carried out under the research project № 17053B in Omsk State Technical University (OmSTU).

References

[1] Thompson N G, Yunovich M and Dunmire D 2007*Corros. Rev.***25(3–4)**pp 247–262

[2] SeoJ H, SeoB H, LeeS H, LeeI K, JeonJ H, ChoeH and LeeK W 2007*SID Symposium Digest of Technical Papers***38(1)**pp 381–384

[3] HerranenM, BauerA D, CarlssonJ O and BunshahR F 1997*Surf. Coat. Technol.***96(2–3)**pp 245–254

[4] KimH S, KimB O and SeoJ H 2015 *J. Nanosci. Nanotechnol.***15(11)**pp 8743–8748

[5] KuoS Y, JengM J, ChangL B, LinW T, HuS C, LuY T and WuC W 2009 *34thIEEE Photovoltaic Specialists Conference (PVSC)*pp 83–86

[6] RickerbyD 1996Measurement of coating adhesion *Metallurgical and ceramic protective coatings*ed K H Stern(Springer Nature) pp306–333

[7] Kishan Singh I C, Ilango S, Polaki S R, Dash S and Tyagi A K2014*Mater. Res. Express***1(3)**036401

[8] Andritschky M and Teixeira V1992*Vacuum***43(5–7)**pp 455–458

[9] Depla D and BraeckmanB R 2016 *Thin Solid Films***604**pp 90–93
[10] MagnfältD, FillonA, BoydR D, HelmerssonU, SarakinosK and AbadiasG 2016*J. Appl. Phys.***119(5)** 055305
[11] Blesman A I, Polonyankin D A and Postnikov D V 2015*Procedia Eng.***113**pp 413–417
[12] YoderK B, ElmustafaA A, LinJ C, HoffmanR A and StoneD S 2003*J. Phys. D: Appl. Phys.***36(7)**pp 884–895
[13] Chason E 2012*Thin Solid Films***526**pp 1–14
[14] ChasonE, KarlsonM, ColinJ J, MagnfältD, Sarakinos K and AbadiasG. 2016 *J. Appl. Phys.***119(14)** 145307
[15] GuisbiersG, OverscheldeO V, WauteletM, LeclèreP and LazzaroniR. 2007 *J. Phys. D: Appl. Phys.***40(4)**pp 1077–1079
[16] GuisbiersG, StrehleS and WauteletM 2005*Microelectron. Eng.***82(3–4)** 665–669
[17] Lu M and HuangH 2015Interfacial Adhesion of Film/Substrate System Characterized by Nanoindentation*Advances in Materials Science and Engineering*ed S Zhang (London: CRC Press) pp 315–376
[18] Seong J and Frankel G S2012*Corrosion***68(3)** 032501–1–032501–4
[19] KlabundeF, Lo hmannM, Bla singJ and Dru sedauT 1996*J. Appl. Phys.***80(11)** 6266
[20] AshrafeeT, AryalK, RajanG, KarkiS, RanjanV, RockettA, CollinsR W, AyalaO and Marsillac S 2015 *IEEE 42nd Photovoltaic Specialist Conference (PVSC)*pp 1–5
[21] LiuX, HaoX, GreenM A, Huang S and ConibeerG 2012*COMMAD*pp 61–63
[22] AndreoneA, BaroneA, ChiaraA Di, MascoloG, PalmieriV, PelusoG and U. Scotti di Uccio1989 *IEEE Trans. Magn.***25(2)**pp 1972–1975
[23] Jakubiak R, Murphy N, Sun L, Grant J and Jones J 2013*Optical Interference Coatings*FC.5
[24] Sidky P S and HockingM G 1999 *Br. Corros. J.***34(3)**pp 171–183
[25] BeilisI I, KoulikY, YankelevichY, Arbilly D and BoxmanR L 2015*IEEE Trans.Plasma Sci.***43(8)**pp 2323–2328
[26] Kostyuk G I and Voliak E A 2004 *XXIst International Symposium on Discharges and Electrical Insulation in Vacuum (ISDEIV)*pp 517–520
[27] BlesmanA I, PostnikovD V and PolonyankinD A 2015 *IOP Conf. Ser.: Mater. Sci. Eng.***81**pp 012-031
[28] Blesman A I, Postnikov D V Polonyankin D A and Rogachev EA2014 *Dynamics of systems, mechanisms and machines (Dynamics 2014)*

Gas dynamic model of electrothermal thrusters of small spacecraft and possibility of applying microwave heating of a working

V N Blinov, VV Shalay, I S Vavilov, V V Kositsin, V I Ruban, A I Lykyanchik, P S Yachmenev, A S Vlasov

Omsk State Technical University, 11, Mira ave., Omsk, 644050, Russia

E-mail: Petrov@mail.ru

Abstract. This paper is devoted to development and approbation of the gas dynamic model of ammonia thruster with low power consumption and ultra small thrust for picosatellite weighing up to 5 kg and possibility of applying microwave heating of a working fluid. It is shown, that simplest electrothermal thruster consisting of propellant tank, solenoid valve, expension cavity and heating chamber can provide ultra small trust due to gas dynamic processes and small heat supply. The results of the study set tasks for further design of small spacecrafts microwave generators.

1. Introduction

Twenty-first century with its microminiaturization, electronics, nanotechnology and the aspiration for cheaper without quality loss begets rigid boundary value problems, including intensive commercial, military and scientific exploration of space. In recent years, in Russia has increased significantly the interest of States to the space industry. This is due to the natural processes of restoration science intensive high-tech industries, the launch of the first missile from a new cosmodrome Vostochny, the need to create your navigation system, etc. These processes are especially noticeable in the field of designing and manufacturing small spacecraft.

Space ceases to be the fate of elected and universities and laboratories all over the world are intensively develop their satellite destinations. An example of such adaptation to the new realities of technical progress is creating such small spacecrafts like a CubeSat. At the relative cheapness of manufacture of such equipment and their components can be displayed in near-Earth space group and breeding method of orbits of functioning. Jet micro-engines with different thrust creation are used for exploring and correction of putting into orbit errors.

Ammonia trust systems with ohmic heating element have been designed in PA "Polet" in Omsk and were supplied to customer worldwide. Developments were transferred to OmSTU futher. This systems showed its reliability and posibility to perfom the task, but its have been designed for heavy satellites. Time set the task to minimize dimensions, mass, energy consumption of satellite while maintaining reliability. Respectively, should be developed and go on the market engines with less consumption of working fluid, less weight, less power, less prices and high reliability.

Also should also consider other ways of transmission of energy to working body, have as less as possible intermediaries in this way, which ultimately increases the efficiency of propulsion. One such

IOP Conf. Series: Journal of Physics: Conf. Series **858** (2017) 012004 doi:10.1088/1742-6596/858/1/012004

way is microwave heating. It is distinguished by three-dimensional heat transfer and lack of contact with the working fluid.

2. Formulation of the problem

Objective in this research is gas-dynamic calculation of working fluid flow tract of small power elementary electrothermal thruster, which used on ultra small spacecrafts like CubeSat for creating micro-thrust and the use microwave enegergy for working fluid heating.

3. Theory

The scheme of the model electrothermal truster is shown in Figure 1.

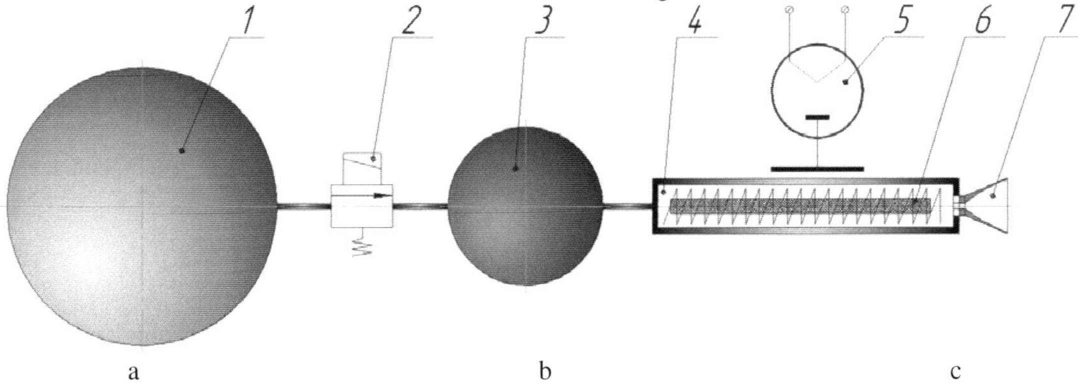

Figure 1. Scheme elementary electrothermal micromotor:
1– high-pressure tank with working fluid; 2–valve solenoid; 3 – expansion cavity; 4– the working fluid heating chamber ; 5 –magnetron; 6 –radar absorbing material; 7 – Nozzle.

Ammonia has been chosen as working fluid. The choice of ammonia as a working body is caused, primarily, a low boiling point and high pressure on the saturation line. Ammonia in critical condition in tank 1(fig. 1), this condition ensures a constant pressure at the outlet of the tank at a constant temperature inside the tank, regardless of the volume of the working fluid. Using ammonia as the working fluid to simplify truster construction without using expulsion system.

Thermodynamic parameters of ammonia in the tank 1 further denoted by index "1". Output Parameters of expension cavity 2 denoted by index "2".

Consider the expiration of ammonia from the high preassure tank to expension cavity. Low temperature gas obeys the equation of state Van-Der-Waals with high accuracy under high pressure. Expiration process is accompanied by the Joule-Thomson effect, which leads to a drop in the temperature of the gas at adiabatic throttling.

In General, a change in temperature when adiabatic throttling can be expressed as the ratio [1]:

$$\frac{\Delta T}{\Delta P}=\frac{\dfrac{b\cdot R\cdot T}{(V-b)^2}-\dfrac{2\cdot a}{V^2}}{c_p\cdot\left(\dfrac{\partial P}{\partial V}\right)_T},$$

where a,b - Van-der-Waals constants; c_p -isobaric heat capacity; R -Universal gas constant; P, T, V - the pressure, temperature and volume, respectively.

Calculating the derivative $\left(\dfrac{\partial P}{\partial V}\right)_T$ from Clapeyron equation we get:

$$\frac{\Delta T}{\Delta P} = \frac{\frac{2 \cdot a}{R \cdot T} - b}{c_p} \qquad (1)$$

Ammonia Van-der-Waals Constants:

$$a_{NH_3} = \frac{27}{64} \cdot \frac{R^2 \cdot T_{cr}^2}{P_{cr}}, \; b_{NH_3} = \frac{R \cdot T_{cr}}{8 \cdot P_{cr}},$$

where $P_{cr} = 113.2 \cdot 10^5$ Pa; $T_{cr} = 405.55$ K - ammonia critical point parameters [2].
Thermodynamic parameters of ammonia in the tank P_1, T_1, c_{p1}.
Thermodynamic parameters of ammonia in the expansion cavity P_2, T_2, c_{p2}.
Then from equation (1) we have:

$$T_2 = \left(\frac{\frac{2 \cdot a_{NH_3}}{R \cdot T_1} - b_{NH_3}}{c_{p1}} \right) \cdot (P_2 - P_1) + T_1 \qquad (2)$$

Pressure and heat capacity of ammonia in the tank depend on temperature of the tank, which, in turn, can take different values depending on the light, the work of the on-board equipment. Temperature range of engine construction is in the range -10° C ... + 400°C in space (e.g., telemetry processing small spacecraft QL series showed the construction temperature range 0°C ... 100° C).
According to [2], it is possible plot the dependence of ammonia pressure in the tank on its temperature at saturation line. Relation is shown in Fig. 2. Gaseous ammonia area is highlighted.

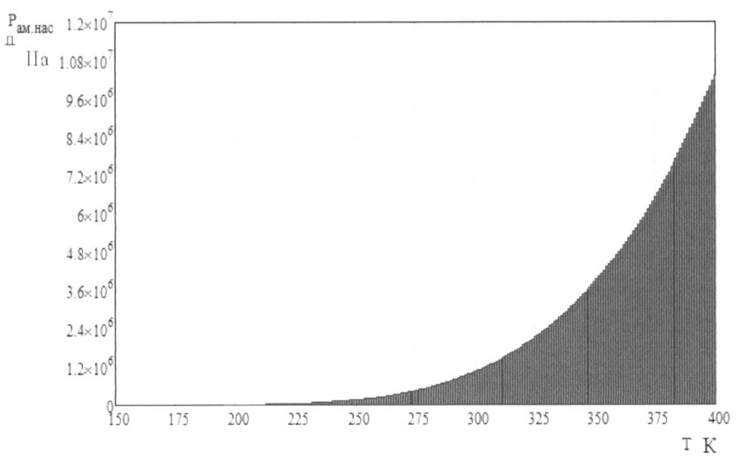

Figure 2. Dependence of ammonia pressure on temperature at saturation line.

Expansion cavity pressure is value assigned and is chosen according to the required microengine trust. In this way dependence (2) allows to get ammonia temperature value in expansion cavity. It is possible to judge in which state of aggregation is working fluid based on obtained parameters and ratio is shown in Fig. 2.
Working fluid falls into heating chamber 4 after cavity 3 (Fig. 1), where gas spead increased by supplying heat. The heating value is calculated from the condition guaranteed obtaining gas fraction of ammonia in the Chamber.
The temperature in the Chamber can be determined from the ratio:

$$Q = m \cdot c_{p1} \cdot (T_3 - T_2)$$
$$Q = m \cdot r_3 \tag{3}$$

where Q - heat is spent on evaporation; m - ammonia mass; r_3 - specific heat of of vaporization of ammonia at the thermodynamic parameters of the gas in the expansion chamber.

Expressing the temperature T_3 of the system (3), we get: $T_3 = \frac{r_3}{c'_{p2}} + T_2$. Here is c'_{p2} the heat capacity of liquid ammonia, if the parameters of the working body in expansion cavity presuppose the existence of a liquid phase. If the ammonia in the cavity gaseous can be taken $c'_{p2} = c_{p1}$.

Further, movement of gas is considered as a one-dimensional flowing with heating. Heat supply process makes thermal resistance, full pressure falls [3].

Gas velocity is determined from the Bernoulli equation at existing pressure drop and heating value. In case of the expiration from expansion cavity are:

$$V_2 = \sqrt{\frac{2 \cdot P_2}{\rho_2 \cdot \left(2 \cdot \frac{T_3}{T_2} - 1\right)}} \tag{4}$$

where ρ_2 is the density of the gas at the inlet in heating Chamber (density expansion cavity). This value can be determined from equation of State of Van der Waals:

$$\left(P_2 + \frac{a_{NH_3} \cdot \rho_2^2}{M^2}\right) \cdot \left(\frac{M}{\rho_2} - b_{NH_3}\right) = T_2 \cdot R$$

where is the M molecular mass.

Ammonia isentrope index in expension cavity can be determined from the ratio:

$$k_2 = \frac{c''_{p2}}{c''_{p2} - R_{NH_3}} \tag{5}$$

where c''_{p2} is the heat capacity of the gas phase; $R_{NH_3} = 488.162$ kJ/(kg · K) -ammonia gas constant.

Thus, determining the temperature by the expression (3) and isentrope index by the expression (5), it is can be calculated the critical sound velocity at the outlet from expansion cavity:

$$a_{cr2} = \sqrt{\frac{2 \cdot k_2 \cdot R_{NH_3} \cdot T_2}{k_2 + 1}}.$$

Define relative velocity $\lambda_2 = \frac{V_2}{a_{cr2}}$.

We will use the gas-dynamic function

$$q(\lambda) = \left(\frac{k+1}{2}\right)^{\frac{1}{k-1}} \cdot \lambda \cdot \left(1 - \frac{k-1}{k+1} \cdot \lambda^2\right)^{\frac{1}{k-1}} \tag{6}$$

For expiration from expansion cavity according to (6) we have:

$$q(\lambda_2) = \left(\frac{k_2+1}{2}\right)^{\frac{1}{k_2-1}} \cdot \lambda_2 \cdot \left(1 - \frac{k_2-1}{k_2+1} \cdot \lambda_2^2\right)^{\frac{1}{k_2-1}}$$

Gas-dynamic function

$$z(\lambda_2) = \lambda_2 + \frac{1}{\lambda_2} \tag{7}$$

Gas flows into the heating Chamber after expension cavity, which generally represents a vortex tube. For the convenience of research consider the heating chamber as two consecutive elements: local resistance and cylindrical heating pipe (fig. 3).

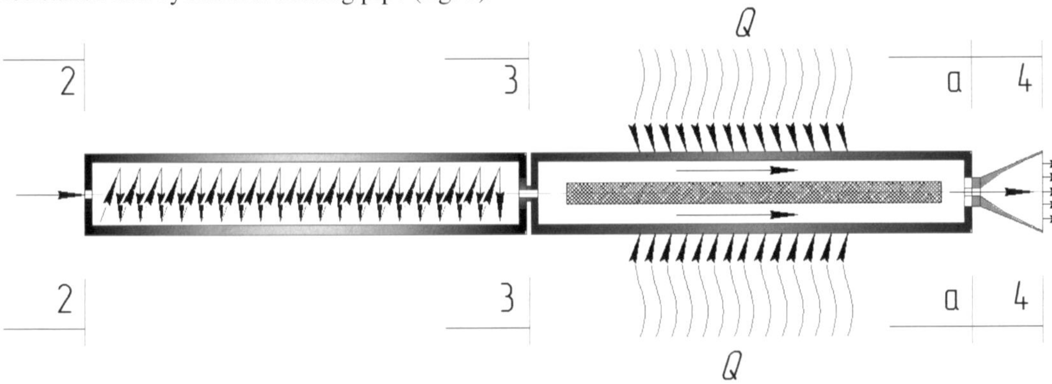

Figure 3. Heating chamber physical model: section 2-2 – input gas from expansion cavity; section 3-3 – input gas into heating pipe; a-a – nozzle critical section; section 4-4 – nozzle cross-section; Q – working fluid heating

Relative velocity in critical cross-section a-a at the thermal crisis (achieving sound speed) $\lambda_a = 1$. Then according to expression (7) we have

$$z(\lambda_a) = \lambda_a + \frac{1}{\lambda_a} = 1 + \frac{1}{1} = 2.$$

Brake temperature in nozzle (on a plot of 3-3-a-a) is equal $T_3 = T_2$.
Preasure loss in engine tract (plot 2-2-3-3 Fig. 3) can be determined by Darcy-Weisbach formule [4] for smooth-wall tube:

$$\Delta P_{TR} = 32 \cdot \nu_2 \cdot \frac{l_{TR}}{d_{TR}^2} \cdot \rho_2 \cdot V_2,$$

where ν_2 is the kinematic viscosity of the gas at the inlet into heating Chamber; l_{TP}, d_{TP} -length and diameter pipeline respectively.
Then the heating chamber pressure:

$$P_3 = P_2 - \Delta P_{TR}.$$

According to the obtained values of the thermodynamic parameters of the gas in the heating chamber P_3 and T_3 the reference data [2] define isobaric heat capacity of ammonia c''_{p3}. Similarly, the definition of ammonia isentrope index in the expansion cavity heating chamber isentrope index is defined by the expression (5):

$$k_3 = \frac{c''_{p3}}{c''_{p3} - R_{NH_3}}.$$

The critical velocity in the heating Chamber is equal to:

$$a_{cr3} = \sqrt{\frac{2 \cdot k_3 \cdot R_{NH_3} \cdot T_3}{k_3 + 1}}.$$

Total pressure losses in the half-thermal nozzle at the thermal crisis ($\lambda_a = 1$) can be determined by [3]:

$$P_4 = P_3 \cdot \left(\frac{1+\lambda_2}{2}\right)^2 \cdot \left[\frac{1 - \frac{k_2 - 1}{k_2 + 1} \cdot \lambda_2^2}{\frac{2}{k_2 + 1}}\right]^{\frac{1}{k_2 - 1}}$$

or (8)

$$\sigma_{cr} = \frac{P_4}{P_3} = \left(\frac{1+\lambda_2}{2}\right)^2 \cdot \left[\frac{1 - \frac{k_2 - 1}{k_2 + 1} \cdot \lambda_2^2}{\frac{2}{k_2 + 1}}\right]^{\frac{1}{k_2 - 1}}$$

Gas consumption through supersonic nozzle is defined by the expression [3]:

$$G_a = \frac{P_3 \cdot F_a}{\sqrt{T_3}} \cdot \left[\frac{2}{k_3 + 1}\right]^{\frac{k_3 + 1}{2 \cdot (k_3 - 1)}} \cdot \sqrt{\frac{k_3}{R_{NH_3}}} \tag{9}$$

Here F_a - is the critical cross-section square of nozzles (specified by the Designer).

To determine the cross-section of the supply channel of heating chamber use the condition of equality of ammonia rates in the studied cross sections of the engine tract. We have:

$$F_\kappa = \frac{G_a}{q(\lambda_2) \cdot B_{1G} \cdot \frac{P_2}{\sqrt{T_2}}} \tag{10}$$

where $B_{1G} = \sqrt{k_2 \cdot \left(\frac{2}{k_2 + 1}\right)^{\frac{k_2 + 1}{k_2 - 1}}} \cdot \frac{1}{\sqrt{R_{NH_3}}}$ is a constant [5].

From the expression (10) is define the geometric parameters of the flow cross section of the supply channel. Section may be round, rectangular or triangular.

The engine trust is determined by the formula:

$$P_C = P_3 \cdot P_\kappa \cdot \left[\sigma_{cr} \cdot \frac{f(\lambda_4)}{q(\lambda_4)} - f(\lambda_2)\right] \tag{11}$$

$$P_C = P_a \cdot F_a \cdot \left(\lambda_a + \frac{1}{\lambda_a}\right) \cdot \left(\frac{2}{k_3 + 1}\right)^{\frac{1}{k_3 - 1}} \tag{12}$$

where: $f(\lambda_4) = (\lambda_4^2 + 1) \cdot \left(1 - \frac{k_3 - 1}{k_3 + 1} \cdot \lambda_4^2\right)^{\frac{1}{k_3 - 1}}$ is the gas-dynamic function;

$q(\lambda_4) = \left(\frac{k_3 + 1}{2}\right)^{\frac{1}{k_3 - 1}} \cdot \lambda_4 \cdot \left(1 - \frac{k_3 - 1}{k_3 + 1} \cdot \lambda_4^2\right)^{\frac{1}{k_3 - 1}}$ - gas-dynamic function.

Trust value increases with growth λ_4 at the expiration of the gas in the vacuum ($P_{vac} = 0$). In this case, the gas-dynamic pressure function equals $\pi(\lambda_4) = \frac{P_{vac}}{P_4} = 0$. It is possible to determine the relative velocity at the nozzle exit:

$$\pi(\lambda_4) = \frac{P_{vac}}{P_4} = \left(1 - \frac{k_3 - 1}{k_3 + 1} \cdot \lambda_4^2\right)^{\frac{k_3}{k_3 - 1}} = 0.$$

Pressure in a critical section at $\lambda_a = 1$ can be determinem from the equation [3]:

$$\frac{P_3}{P_a} = \left(\frac{k_3 + 1}{2}\right)^{\frac{k_3}{k_3 - 1}}.$$

According to the speed of response of valve (Figure 1) can be judged on the extent of the expansion of the cavity:

$$V_{rp} = \frac{G_a \cdot t}{\rho_2},$$

where t is the time of opening and closing the valve section.
The power needed to heat the propellant with rate G_a to the temperature T_3 is equal to:

$$P = G_a \cdot c''_{p3} \cdot (T_3 - T_2) \tag{13}$$

At the same time, the specific power of electromagnetic vibrations emitted per unit volume of radioabsorbing material is equal to [6]:

$$P_{sp} = 0.278 \cdot 10^{-12} \cdot f \cdot \varepsilon' \cdot tg\delta \cdot |E|^2 \tag{14}$$

where: $\varepsilon' = \frac{\varepsilon_a'}{\varepsilon_0} \varepsilon_a'$ - is the real part of the absolute dielectric permittivity of medium; $\varepsilon_0 = 8.85418 \cdot 10^{-12} \frac{\Phi}{m}$ -Electric constant; f - microwave frequency (for microwave heating installation are distinguish the following frequencies: 433 Mhz ± 0.2%, 915 Mhz ± 2.73%; 2450 Mhz ± 2.04%, 5800 Mhz ± 1.29%, 22125 Mhz ± 0.56% [6]); $tg\delta = \frac{\varepsilon_a''}{\varepsilon_a'} + \frac{\sigma}{\omega \cdot \varepsilon_a'}$ -dielectric loss tangent; ε_a'' -minimal part of the absolute dielectric permittivity of medium; σ -the specific conductivity of the medium; $\omega = 2 \cdot \pi \cdot f$ -circular frequency; E -intensity of the electric field.
Dielectric loss tangent in the General case $tg\delta$ □ 1, and the amplitude of the electric field in dielectric fades. For metallic conductors (and graphite) $tg\delta$ □ 1, due to the high conductivity [7]. The result is a more intense heat at a low intensity electrical field.
An important feature of the conductors used in microwave technology is the depth of the a screen-layer. It depends on the magnetic permeability of medium. In this case, graphite, like diamagnetik, has an advantage over metal because it possesses more depth screen-layer.

4. Research Results

The research of elementary electrothermal micro engine was carried out according to a mathematical model.
Input data for calculation are shown in Table I.

IOP Conf. Series: Journal of Physics: Conf. Series **858** (2017) 012004 doi:10.1088/1742-6596/858/1/012004

Table 1. Input data

Parameter	Value	U. measurement
Working fluid	ammonia	-
Engine case temperature	280	K
Critical diameter nozzle cross-section	$0.8 \cdot 10^{-3}$	m
The inner diameter of the working fluid piplines	$2 \cdot 10^{-3}$	m
The total length of pipelines	0.8	m
Pressure in the expansion cavity	$1 \cdot 10^{4}$	Pa
The time of opening-closing of the valve	2	s
Frequency of microwave radiation	$2.45 \cdot 10^{9}$	Hz

Solving equation (1)...(14) were obtained design values of of thermodynamic parameters of engine and its power characteristics. The main results of the research are presented in Table II.

Table 2. Research Results

Parameter	Designation	Value	U. measurement
Van der Waals constants	a_{NH_3}	$4.237 \cdot 10^{5}$	$N \cdot m^4 / k - mol^2$
	b_{NH_3}	0.037	$m^3 / k - mol$
The gas temperature in the expansion cavity	T_2	213.326	K
The density of the gas in the expansion cavity	ρ_2	0.0961	kg/m^3
The temperature in the heating Chamber	T_3	533.239	K
Gas pressure in the heating Chamber	P_3	$9.95 \cdot 10^{3}$	Pa
Critical velocity at the inlet in the heating Chamber	a_{cr2}	344.049	m/s
Critical velocity at the inlet in the nozzle	a_{cr3}	540.982	m/s

Nozzle gas consumption	G_a	$6.385 \cdot 10^{-6}$	kg/s
Engine trust	P_C	$3.496 \cdot 10^{-3}$	N
The power to heat the working fluid	P	5.477	W
Expansion cavity volume	V_{RP}	$1.328 \cdot 10^{-4}$	m^3

As for microwave heating, the resulting gas-dynamic calculation of engine power performance allow to approach the issue of choice of radar absorbing material. According to the formula (14) at the volume of material $V_{RP} = 1 \cdot 10^{-6}$ m^3 complex value can be determined:

$$\varepsilon' \cdot tg\delta \cdot |E|^2 = \frac{P}{0.278 \cdot 10^{-12} \cdot f \cdot V_{RP}} = \\ = \frac{5.477}{0.278 \cdot 10^{-12} \cdot 2.45 \cdot 10^9 \cdot 1 \cdot 10^{-6}} = 8.042 \cdot 10^9 \quad (15)$$

5. Results and Discussion

The results of this research showed that the simplest electrothermal thruster for picosatellites has low power consumption and a simple functional dependence of the thermodynamic parameters of the design temperature of the apparatus. Having plotted the engine characteristics for all possible temperatures we can get the field of possible engine thrust and delineate the scope of applicability. Regulation of engine thrust may be controlled by changing of pressure in the expansion chamber, which in turn is achieved by the opening time of the solenoid. Working fluid heating quantity in the chamber is dependent magnitude, but it can be used as a control feedback. Calculation was carried out for case not-constant isentrope index of engine tract, which introduces additional bonds in the construction of mathematical models.

Selecting the method of heat transmission to working body is also relevant. In this research microwave method of power transmission to working fluid is offered. This method differs by fast volumetric heating of the working fluid or surface circumfluous by working fluid.

As in this research as the working fluid ammonia was selected, which is not a conductor and is transparent to microwave radiation, the method considered heat transmission by radar absorbing material (e.g., graphite, copper). It should be noted the possibility of additives (for example, water, which is infinitely soluble in ammonia), which will enhance the conductivity of the working fluid and, accordingly, will disappear in need of radioabsorbing mediators.

6. Conclusion

Research leads to the following conclusions:

- ammonia Consumption is 6.385 mg/s;
- Engine trust is 3.496 mN;
- The temperature in the heating chamber 533.329 k;
- Working fluid heating power 5.477 W.

Diamagnetic and ferromagnetic material with high thermal conductivity and a deep screen-layer suitable as radar absorbing inverter microwave energy into thermal energy according to the formula (15).

Findings allow to start designing microwave generators for electrothermal engines with ultra small thrust.

From the calculation follows that the electrothermal thruster can be used to create a corrective micro-thrust for ultrasmall spacecrafts like CubeSat. These apparatus are low-powered, so such propulsion system of this satellites should to consume no more than 10 Watts. Another feature of this class of spacecrafts is that they used in corrective propulsion systems and shall not has high thrust. This is due to the minimization of disturbing moment of the expiration of gas through a jet nozzle.
In addition to the heat in the engine there is only one electricity consumer is a solenoid valve. Currently, production is able to provide developers with low-power small-sized high-speed low flow valves with the consumption up to 2 W.
It can be seen that the total forecasted consumption power of the engine does not exceed 8 W.

7. Acknowledgments

The researches carried in the framework of the financial support of the Russian Foundation for Basic Research under the Contract number 31 16-38-60089 \ 15 from 12.02.2015, the (research Gr.46-15 number, reg. Number A161160202100195 AAAA).

References

[1] Sivukhin V N 2005General physics course*Thermodynamics and Molecular Physics* (Moscow, Phismathlit) 544
[2] Golubev I F 1978 Ammonia thermal properties (Moscow, Standards publishing House) 264
[3] Abramovich G N 1991 Applied Gas Dynamics (Moscow, Nauka Publishers)600
[4] Idelchik I E 1992 Hydraulic resistance manual 672
[5] Davidson V E 1965Fundamentals of gas dynamics in tasks (Moscow, Higher School) 208
[6] Arkhangelskii I S 2011Handbook of Microwave electrothermics (Saratov) 560
[7] Didenko, A. N. Microwave power:Theory and practice (Moscow, Nauka Publishers) 446

Wire electrical discharge machining of E110 zirconium alloy

N V Bobkov, A A Fedorov, A I Blesman, D V Postnikov and D A Polonyankin

Omsk State Technical University, 11 Mira ave., Omsk, 644050, Russia

E–mail: fedot83@gmail.com

Abstract. The paper deals with the results of experimental research carried out by scanning electron microscopy (SEM) and X–ray diffraction analysis (XRD) to define, how the modes of wire electrical discharge machining (WEDM) influence on the elemental and the phase composition of E110 zirconium alloy's surface layer.Investigation of the phase composition allowed us to determine the main α and δ phase's distribution through the depth of zirconium surface layer, in common with phases of oxygen, copper, zirconium, and niobium specific compounds. It was also established the maximum depth of the defect level containing amorphous phase for all of WEDM modes, and proposed the grinding and polishing as potential mechanical methods of its removal.

1. Introduction

Zirconium alloys are widely used in nuclear power engineering, zirconium is also used to create multifunctional coatings, including biocompatible ones; this makes it really important to solve the problem of efficient treatment and waste minimizing in the course of manufacturing the heavy–duty products using expensive zirconium alloys. Zirconium as other rare metals is demanded in engineering technology besides traditional alloys (steel, cast iron, heat–resistant, aluminum and brass alloy). Zirconium targets are used to create multifunctional coatings by magnetron sputtering. Zirconium alloys are also applied in manufacture of fuel elements, assemblies and spacer grids and nuclear reactors. The production cost of machine engineering products with its output rate, shaping and quality has always been one of the main criteria of their production efficiency. That is why we need to pay special attention to efficient waste minimizing in the process of rare metal parts manufacture, which is very cost–intensive. For example, we can lessen the cutting width, thus reducing the chip amount. Applying of wire EDM method helped to achieve significant results in waste minimizing, because the cutting depth in case of WEDM is only one and a half times bigger than the wire electrode diameter.
It is possible to process any electrically–conductive metal by using EDM wire cutting. The machinery manufacturers provide machinery units having a set of modes for treatment of most commonly used materials, such as steel, copper, aluminum, hard alloy, and graphite (ISO 63). In case, when there is a need to treat the metal, which is not included into the unit's data base, e.g. zirconium, we have to define the most efficient treatment mode experimentally.Despite the fact that EDM treatment of metals, compared to other treatment methods, has a lot of advantages, diffusion of the electrode material and dielectric medium into the surface layer of the product under treatment negatively affects characteristics and properties of the treated surface. The authors [1] particularly mention a high hydrogen absorption degree of the surface of Zr–2,5Nb zirconium alloy during its EDM treatment in a hydrocarbon dielectric. The requirements to the zirconium target surface are especially high in case of forming biocompatible coatings using the method of magnetron sputtering [2, 3]. In the course of

treatment, particles of the electrode material and decomposition products are carried to the surface layer of the product and form a defect layer. Therefore, the aim of the current research is evaluation of depth and phase composition of a defect layer, which becomes one of the main quality criteria for wire electrical discharge machining [4].

2. Materials and methods

In the purpose of zirconium surface layer investigation after EDM treatmentit was prepared E110 zirconium alloy samples in the form of disks with diameter of 30 mm and width of 3 mm. The surface was treated using 1, 2, 3, and 4 cuts to control phase changes due to different number of cuts. The parameters of the WEDM modes of zirconium samples were experimentallydefinedand presentedin table 1.The treatment was performed using EDM machine unit Sodick VZ300L with brass wire electrode (weight content of copper and zinc was 65% and 35% respectively) and distilled water as a dielectric fluid. The basic elements content in the E110 alloy is as follows: Zr=99%, Nb=1% (by weight).

Microsections were made as follows: mounting into acrylic compound, grinding using paper SiC P320, P600, P800, P1500, P2500, polishing with diamond suspensions 6, 3 and 1 μm, final polishing with polishing suspension with the grain size of 0.05 μm.

Table 1. Parameters of the WEDM modes of E110 zirconium alloy.

Serial cut number	Average voltage, V	Average amperage, A	Offset (H), mm	Pulse duration (On)	Pause duration (Off)
1	60	2.3	0.106	008	014
2	55	2.1	0.031	002	011
3	45	1.3	0.011	001	001
4	40	1.0	0.005	001	001

Experimentally defined parameters of the modes provide stable treatment without wire breakages. Electron microscope investigations were carried out by using *JEOL JCM–5700* microscope with an energy dispersive X–ray spectrometer *JED–2300* in a high–vacuum mode. The signal type was secondary electrons (SEI). The SpotSize parameter was chosen 60, the value of accelerating voltage – 20 kV, magnification upto 5000×.

X–ray phase analysis was performed using X–ray diffractometer Shimadzu XRD–7000 by the sliding beam method [5] using Kα–radiation of a copper tube with Θ angles from 1° to 10°. The X–rays patterns were interpreted by using PowderCell 2.4 software.

3. Results and discussion

Interpretation of X–rays patterns allowed us to receive the numeric data for phase depending on a tube angle Θ.For example the phase composition of a surface layer after 4th cut is shown in table 2. We can clearly see that there are two main phases of zirconium: hexagonal and cubical, compounds of copper and niobium ($CuNbO_3$), amorphous phase, in the second cut there is pure copper. Moreover, there are two phases of zirconium dioxide: tetragonal and monoclinic.

Table 2. The phase composition of a surface layer after 4th cutdepending on a tube angle Θ.

Θ	Zr hexagonal	Zr cubic	ZrO_2 tetragonal	ZrO_2 monoclinic	$CuNbO_3$	Amorphous phase
1	55.80	6.40	8.80	0.20	1.80	27.00
2	54.10	11.90	6.80	1.20	4.00	22.00
3	47.40	14.10	8.00	1.90	4.60	24.00

4	45.00	16.20	10.60	1.50	6.70	20.00
5	46.90	19.40	10.20	3.40	5.10	15.00
6	55.30	17.30	9.60	1.90	5.90	10.00
7	54.00	16.40	11.10	5.20	3.30	10.00
8	59.20	15.50	5.50	2.10	7.70	10.00
9	68.50	16.50	6.10	0.80	8.10	0.00
10	71.50	16.70	5.00	1.50	5.30	0.00

The diffusion depth of wire electrode material was evaluated on cross–sections by energy–dispersive spectroscopy (EDS). For example dotted analysis of E110 zirconium alloy sample after 4th WEDM cut is presented on figure 1.Cross–sectioncontains the points where the energy dispersive analysis was performed.We moved from the surface deep into the material, until impurity matter disappeared, i.e. until zirconium concentration achieved 100%. Maximum depth of the copperdiffused layer amount13.0 μm for the 1st cut,6.7 μm for the 2nd cut,3.8 μm for the 3rd cut and 3.0 μm for the 4th cut.

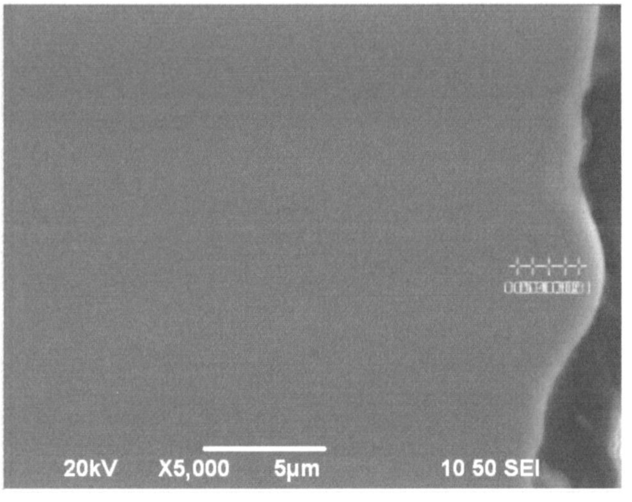

Figure 1. Cross–section of E110 zirconium alloy after 4th WEDM cut. SEM, x5000.

Technical requirements under cathodes manufacture for magnetron sputtering setup [6, 7] do not allow any foreign matter present in the surface layer, because in the course of sputtering first the foreign matter is sputtered, and only after that – zirconium; this results in different thickness of zirconium layer in different parts of the product [8]. It is obvious that during WEDM treatment the workpiece's surface and electrode are exposed to high temperatures and pressure. It causes formation of different phases complying with minimum potential energy principle. The receive data show that different phases form in a pre–surface layer, thus adversely affecting usage of different parts, applied for manufacture of fuel elements, cathodes and other products [9, 10].

4. Conclusion

Taking into account the above–said, we can argue that the WEDM treatment is an efficient method of precise zirconium machining. Maximum depth of the copperdiffused layer amount13.0 μm for the 1st cut, 6.7 μm for the 2nd cut, 3.8 μm for the 3rd cut and 3.0 μm for the 4th cut.

Though the received data, concerning depth of the electrode material penetration into the surface layer of zirconium target, we can make a conclusion that it is necessary to oversize for finishing grinding or

IOP Conf. Series: Journal of Physics: Conf. Series **858** (2017) 012005 doi:10.1088/1742-6596/858/1/012005

polishing before using the products as intended. Further research of the mentioned zirconium samplesfor defining exact grinding or polishingmodesare necessary to be performed.

Acknowledgments

The work was carried out under the research projects№17033B, № 17053B and №17067B in Omsk State Technical University (OmSTU).

References

[1] Chow C K, Brady G R, Urbanic V F and Coleman C E 1998 *J. of Nucl. Mat.***257(1)**pp 35–43
[2] Lyon D, Chevalier J, Gremillard L and Cam C D 2011 *Compr. Biomater.***20**pp 95–108
[3] Rabiee S M and Azizian M 2012*Int. J. Appl. Ceram. Tech.***10(1)**pp 33–39
[4] Huang H and Yan J 2015 *Appl. Surf. Sci.***355**pp 1306–1315
[5] Tkachenko E A, Postnikov D V, Blesman A I and Polonyankin D A 2016 *IOP Conf. Ser.: Mater. Sci. Eng.* **110** 012009
[6] Kelly P J, O'Brien Jand Arnell R D 2004 *Vacuum***74(1)**pp 1–10
[7] O'Brien J and Arnell R D 1996 *Surf. and Coat. Tech.***86–87**pp 200–206
[8] Pamu D, Sudheendran K, Krishna M G and Raju K C J 2008 *J. of Vacuum Sci. Tech. A***26(2)**pp 185–192
[9] Lassmann K, Schubert A, van de Laar J and Van Uffelen P 2015 *Ann. of Nuclear Energy***81** pp 332–335
[10] Wang L, Hu X and Nie X 2013 *Surf. and Coat. Tech.***221**pp 150–157

Regularities of the working equipment elements mass reduction to the hydraulic power cylinder piston for the bucket boom machines size standard

I V Boyarkina, V N Tarasov

Siberian State Motor Road University, 5, Mira ave., Omsk, 644050, Russia

E-mail: iriboyarkina@yandex.ru

Abstract. The working equipment mass reduction is performed proceeding from the condition of the kinetic energy equality of the reduced mass and kinetic energy of the working equipment. Analytic connections of the boom hydraulic cylinder piston kinematic characteristics with the kinematic characteristics of the working equipment rotation are determined for the establishing of this dependence. Dependences of the working equipment reduced masses are obtained for different boom positions: transport, horizontal and maximum raised position of the working equipment. As opposed to similar researches in mechanical engineering the results of this research consist in the establishing of reduced masses dependences on the machine weight, its lifting capacity, boom positions for the machines size standard. Relative characteristics of the reduced mass depending on the lifting capacity and weight of the machine are obtained.

1. Introduction

Modern process boom hydraulic excavators, loaders and other machines at the world market are represented by samples of the standard series, power of which ranges from several kW to 2000-3000 kW. The total weight of super heavy samples of machines reaches 250-300 tons and this is not the limit.

Machines of such class are in demand at the market and have heavy unbalanced boom working equipment, which inertial and stiffness properties are understudied.

In paper [1] the authors Z. Miaofen, S. Shaohui, G. Youping, Z. Dada establish connections of mechanical and hydraulic control models of the loader working process. Multiparametric optimization of the engine operation mode is performed by the methods of numerical integration.

In paper [2] the authors Z. Zhihong, W. Yunxin, M. Changxun consider the dynamic simulation of the boom in horizontal and other positions. Mechanical parameters, reducing dynamic loads are established.

In paper [3] the authors S. Kang, J. Park, S. Kim, B. Lee, Y. Kim, P. Kim, H. J. Kim consider the excavator working equipment control by means of the controller, implementing efficient control processes of a boom, bucket, arm, simulating actions of an experienced operator.

In paper [4] the authors H. Xie, G. Zhang established that when the automobile crane boom was in the upper position then the pressure in the hydraulic system was increased and vibration was increased at the expense of the reduced mass increasing and power arm decreasing.

IOP Conf. Series: Journal of Physics: Conf. Series **858** (2017) 012006 doi:10.1088/1742-6596/858/1/012006

The reduced masses of the attached working equipment of heavy hydraulic machines have large values and they are significantly changed in different working positions. That's why it's expedient to establish the reduced masses influence on the quality of transient processes of acceleration, piston speed and pressure changing in hydraulic cylinders of the working equipment [5, 6, 7].

For bucket machines the main parameters of the loading equipment are nominal lifting capacity Q, nominal bucket capacity V_{Γ}, geometric and mass characteristics of the working equipment [7].

Paper [5] examines the task of the scraper bucket mass reduction to the bucket digging edge. The formula of the reduced mass m_{Π} represents the function from the elements mass of the working equipment. Paper [5] gives the graph of the reduced mass dependence at the cutting edge of the bulldozer DET-250 on the hydraulic cylinder rod movement $S_{\text{Ш}}$. It's concluded that the reduced mass m_{Π} is changed insignificantly during the rod movement of such machines as scrapers, bulldozers.

Paper [8] examines the tasks of the mass reduction on impact of two objects, rotating on two parallel axes. The reduced masses of each rotating object to the impact point are performed according to the formulas:

$$m_{1\Pi} = J_1 / r_1^2 \; ; \; m_{2\Pi} = J_2 / r_2^2 \, ,$$

where J_1, J_2 – inertia moments of rotating objects; r_1, r_2 - distances from rotation axes to the objects impact point.

Paper [8] examines actuation devices control systems. The inertia moment of moving masses at small steering angles is determined by the formula

$$J_{\Pi} = ml^2 + J_p, \tag{1}$$

where m – moving mass; l – distance from the mass gravity center to the rotation axis; J_p – inertia moment of the reduced mass relative to the own gravity center .

Essentially, formula (1) is the theorem on the object inertia moment for parallel axes.

This paper examines boom machines, which boom swing angle changes for more than 90°, these machines include cranes, excavators, loaders. In scientific and technical literature there are no data about the mass reduction methods for this class of machines. This is a constraint of the dynamic characteristics improvement of such machines due to the absence of methods and numerical data relative to the boom machine inertia moments.

2. Task definition

The article sets the task to obtain the general rated function of the boom bucket machine working equipment elements mass reduction to the hydraulic power cylinder piston. The problem peculiarity consists in a significant change of the reduced masses values depending on a boom swing angle.

Figure 1 shows the working equipment of a front loader in two positions: transport position with the bucket at the bottom over the bearing surface and horizontal position of the boom when the overturning moment created by the gravity forces of the load and bucket is minimum one.

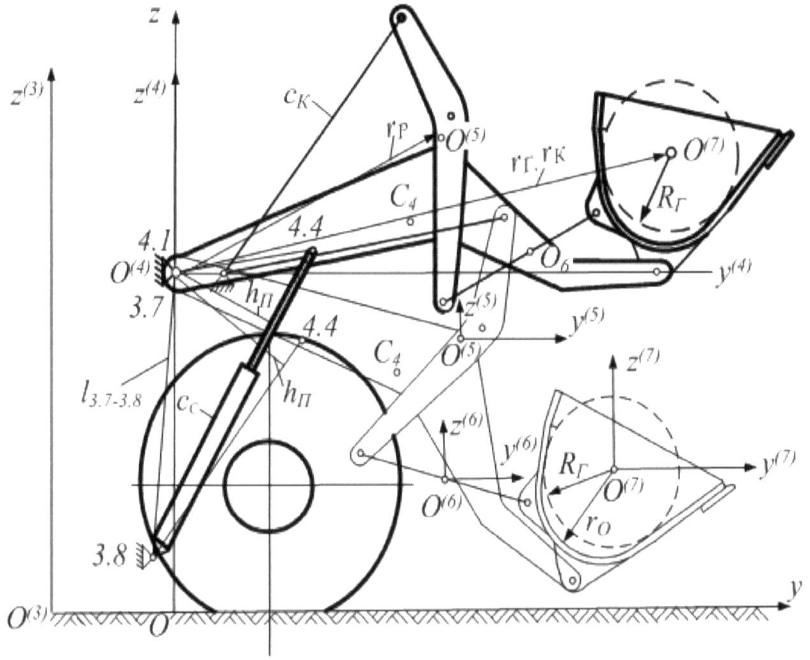

Figure 1. Design diagram of the boom bucket machine working equipment.

Figure 1 shows $h_П$ – arm of the hydraulic power cylinder; r_K, $r_Г$, r_P – accordingly radius vectors of the elements mass centers of the working equipment; c_C – length of the boom hydraulic cylinder.

For the determination of the bucket mass inertia moment (of the bottom and walls) we can use the thin-walled cylinder inertia moment formula, having taken the cylinder radius equal to the radius of the bucket bottom r_o (see Figure 1). Then the bucket inertia moment relative to the own mass center can be calculated according to formula [7]

$$J_{K.Xc} \approx m_K r_o^2, \tag{2}$$

where m_K – mass of the bucket.

For the calculation of the load inertia moment in the bucket $J_{Г.Xc}$ relative to the own mass center we'll perform the conventional replacement of the cargo in the bucket by an equivalent ground cylinder, which radius is calculated according to the formula:

$$R_Г = \sqrt{\frac{V_Г}{\pi B_o}}, \tag{3}$$

where $V_Г$ – nominal capacity of the bucket; B_o – linear dimension of the ground cylinder, equal to the internal dimension of the bucket width.

For small and medium-size machines the bucket bottom radius r_o differs from the ground cylinder radius $R_Г$ by 2…3%, for super heavy machines $R_Г > r_o$ by 12 … 20%.

The lever inertia moment $J_{P.Xc}$ relative to the own gravity center can be determined by mechanics formula [7]

$$J_{P.Xc} = \frac{m_p L_p^2}{12} \quad , \tag{4}$$

where m_p – lever mass; L_p – lever arm.

3. Theory

The mass reduction of the working equipment to the hydraulic power cylinder piston is performed proceeding from the equality condition of the working equipment kinetic energy and kinetic energy of the mass reduced to the piston.

The general formula of the kinetic energy of the reduced mass and kinetic energy of the working equipment elements is as follows:

$$\frac{m_П V_1^2}{2} = \frac{(J_{K.Xc} + J_{Г.Xc})\omega_C^2}{2} + \frac{J_{P.Xc}\omega_C^2}{2} + \frac{(m_K + m_Г)(r_K \omega_C)^2}{2} + \frac{m_P (r_P \omega_C)^2}{2} + \frac{(m_C L_C^2/3)\omega_C^2}{2}, \tag{5}$$

where $m_П$ – working equipment mass reduced to the piston; V_1 – piston speed; m_C, L_C – boom mass and length accordingly; ω_C – boom angular speed; r_K, r_P – radius vectors, drawn from the boom rotation axis to the gravity centers of the bucket and lever.

The first two components of expression [5] determine the working equipment kinetic energy during relative rotation around the own gravity centers. Three other components characterize the kinetic energy of the working equipment mass movable rotation together with the boom.

Taking into consideration that the boom angular speed is determined by operation formula [7]

$$\omega_C = \frac{V_1}{h_П},$$

we'll receive the final form of the formula for the working equipment mass reduced to the piston

$$m_П = \frac{1}{h_П^2}(J_{K.X.C} + J_{Г.X.C} + J_{P.X.C} + (m_K + m_Г)r_K^2 + (m_K + m_Г)r_K^2 + m_P r_P^2 + m_C L_C^2/3). \tag{6}$$

The important value in the reduced mass formula is the arm $h_П$ of the hydraulic power cylinder, which can be determined in any position of the boom according to the formula for the determination of the triangle altitude $h_П$ with vertices *3.7*, *3.8*, *4.4* by known lengths of the sides of this triangle l_1, l_2, c_c, where $l_1 = l_{3.7-3.8}$; $l_2 = l_{3.7-4.4}$; $c_c = l_{3.8-4.4}$.

The theorem of the triangle vertex heights was obtained for the first time by the authors of this article in papers [9, 10]

$$h_4^2 = \left(\frac{l_1 \cdot l_2}{c_c}\right)^2 - \left(\frac{l_1^2 + l_2^2 - c_c^2}{2c_c}\right)^2. \tag{7}$$

4. Research results

The general formula is obtained for the determination of the reduced mass $m_П$ of boom machines for different boom positions: lower, horizontal and higher.

Table 1 represents the calculation results of the reduced masses $m_П$ for the modern size standard of front loaders.

IOP Conf. Series: Journal of Physics: Conf. Series **858** (2017) 012006 doi:10.1088/1742-6596/858/1/012006

Table 1. Results of the working equipment mass reduction to the hydraulic power cylinder piston for the loaders size standard

Lifting capacity Q, t	Reduced mass m_{Π}, kg		
	Transport boom position with cargo / without cargo	Horizontal boom position with cargo / without cargo	Upper boom position with cargo / without cargo
2.2	76409 / 18552	76584 / 17633	382022 / 82971
3.3	78432 / 18554	121669 / 27536	771725 / 161711
3.8	209514 / 84297	176201 / 70554	692063 / 277310
6.6	183093 / 73944	261122 / 103685	1433033 / 550669
7.3	210524 / 92251	318438 / 142322	1341941 / 571617
15.0	405808 / 170782	499917 / 209762	1679183 / 697663
30.0	572611 / 213574	933802 / 375873	3891260 / 1378189
75.0	1313186 / 489803	2427603 / 890688	10971278 / 4115359

The reduced mass values are shown for three positions of the working equipment with cargo in the bucket and without cargo.

For the transport position of the working equipment Table I shows the reduced mass for a super heavy loader with the lifting capacity Q=75 tons, which value is equal to m_{Π}=1313186 kg.

For the horizontal boom position the reduced mass of this loader m_{Π}=2427603 kg, i.e. it has increased by 1.85 times. And finally, for the upper boom position the working equipment reduced mass for this loader with the lifting capacity Q=75 tons is equal to m_{Π}=10971278 kg, i.e. it has increased by 8.35 times in comparison with the transport mode.

For the size standard of frontal loaders Figure 2 shows dependencies of the reduced weight m_{Π} on the lifting capacity Q for three positions of the working equipment with cargo in the bucket. The regularity is established which lies in the fact that the reduced mass m_{Π} for this type of machines exceeds many times the operating mass $m_{Э}$ of the whole machine.

Dependencies shown in Figure 2 have regression equations: for transport mode $m_{\Pi} = 18{,}28Q$; for horizontal boom position $m_{\Pi} = 32{,}4Q$; for upper boom position $m_{\Pi} = 143{,}96Q$.

These results prove the proper use of huge reduced masses for analytical calculations in differential equations of the machine working equipment dynamics.

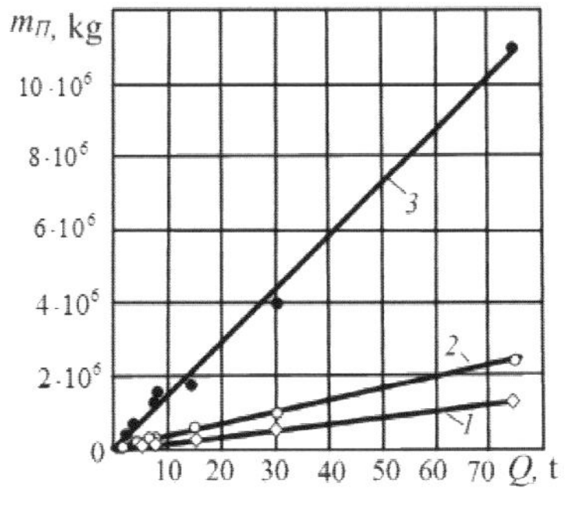

Figure 2. Dependence of the reduced mass $m_{П}$ on the lifting capacity for the front loaders size standard in different boom positions: 1 – boom is in transport position; 2 – boom is in horizontal position; 3 – boom is in upper end position.

Table 2 represents relative values of the reduced mass and operating mass $m_{П} / m_{Э}$, as well as relative value of the reduced mass and lifting capacity $m_{П} / Q$, where $m_{Э}$ – operating mass without a cargo in the bucket.

Table 2. Relative parameters of reduced masses of loaders size standard

Lifting capacity Q, tons	Operating mass $m_{Э}$, tons	Ratio $m_{П} / m_{Э}$ for three boom positions			Ratio $m_{П} / Q$ for three boom positions		
		Transport position	Horizontal position	Upper position	Transport position	Horizontal position	Upper position
2.2	7.5	10.187	10.21	50.936	34.731	34.810	173.646
3.3	11.1	7.0559	10.96	69.520	23.767	36.869	233.856
7.5	26.0	8.097	12.20	51.595	28.069	42.456	178.865
15.0	74.0	5.4839	6.76	22.69	27.053	33.327	111.945
30.0	139.0	4.1195	6.761	27.990	19.087	31.126	129.708
75.0	245.0	5.3599	9.91	44.780	17.509	32.368	146.283

5. Discussion of results

Dynamic processes of the working equipment are described by differential equations of hydraulic power cylinders.

Figure 3 shows the transient processes of acceleration and piston speed of the hydraulic power cylinder of the super heavy loader Le Torneau L-2350 with different reduced masses, obtained in the result of working equipment differential equation solution [5, 7]

$$\ddot{V} + 2n\dot{V} + \omega^2 V = \frac{K_{Г}}{m_{П}} x, \tag{8}$$

where V – hydraulic power cylinder piston speed; n – resistance coefficient, $n = \frac{\vartheta_{Г}}{2m_{П}}$; $\vartheta_{Г}$ – viscous drag coefficient; ω – rotational frequency of natural oscillations, $\omega = \sqrt{C_{Г}/m_{П}}$; $C_{Г}$ – stiffness coefficient of hydraulic power cylinders; $K_{Г}$ – working fluid flow coefficient of hydraulic power cylinders, $K_{Г} = \frac{C_{Г} V_{con}}{x}$; V_{con} – steady-state value of the piston speed after the electric spool opening.

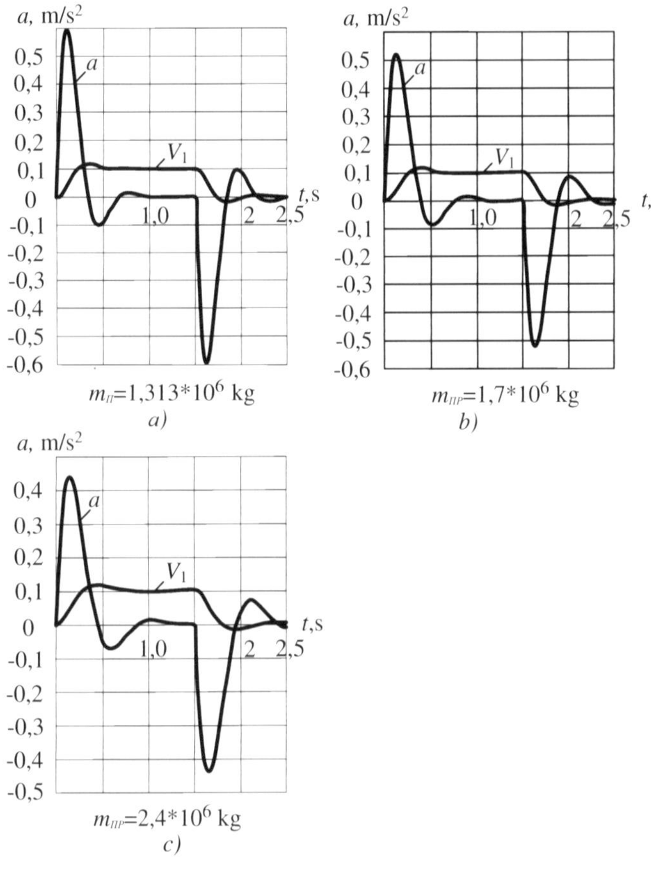

Figure 3. Transient processes of the hydraulic power cylinder piston movement for different reduced masses of the loader Le Torneau L-2350.

The reduced masses varied within the real variation range. With the nominal lifting capacity Q=75 tons for the loader Le Torneau L-2350 with a cargo in the bucket in the transport mode the reduced mass is equal to $m_{П}$ =1313186 kg.

Transient processes shown in Figure 3, performed with different reduced masses within the indicated range, allow to make important conclusions. The graphs show that the first acceleration amplitude $a = a_{\max}$ is decreased with the mass increase practically in inverse proportion to the reduced mass changing. At the same time the transient process of the piston speed changing depends little on the reduced mass changing within the given range.

Obtained results allow to determine dynamic pressures in hydraulic power cylinders excited by inertia forces.

For the loader Le Torneau L-2350 the dynamic inertia force on the piston for a maximum mass is determined by the formula $\varPhi_{\partial} = m_{П} a_{\max}$.

The dynamic pressure in hydraulic power cylinders, corresponding to this force

$$p_{\partial} = \frac{\varPhi_{\partial}}{2(\pi D_{Ц}^2/4)}.$$

The following parameters: acceleration a =0.42 m/s^2, dynamic inertia force on the piston of hydraulic power cylinder $\varPhi_{\partial}$ =4607820 N; pressure in hydraulic power cylinders p_{∂} =18.33 MPa correspond to the reduced mass $m_{П}$ =10971000 kg and cylinder diameter $D_{Ц}$ =0.4 m.

Peak maximum pressure acts during the time interval *t*=0.3 s (see Figure 3,*a*).

6. Summary and conclusion

1. The proper use in engineering of the reduced masses of the boom bucket machine working equipment is proved, which have order of values, for example, $m_{П} = 1{,}1 \cdot 10^{7}$ kg with the operating mass of the machine itself $m_{Э} = 2{,}4 \cdot 10^{5}$ kg.

2. The reduced mass of the bucket boom machine working equipment is the main parameter in the differential equation which forms oscillation frequency, period of oscillations, dynamic inertia forces and pressure in the hydraulic system.

References

[1] Miaofen Z, Shaohui S, Youping G, and Dada Z. 2010 Research on the Multi-domain Modeling and Optimizing Method for Loader Executing System *Intelligent System Design and Engineering Application* (ISDEA-2010) **2** pp 854–857

[2] Zhihong Z, Yunxin W and Changxun M 2011 Simulation and Optimization of the Driving Forces of Hydraulic Cylinders for Boom of Truck Mounted Concrete Pump *Intelligent Computation Technology and Automation (ICICTA-2011)* **1** pp 915–919

[3] Kang S, Park J, Kim S, Lee B, Kim Y, Kim P and Kim H J 2014 Path tracking for a hydraulic excavator utilizing proportional-derivative and linear quadratic control *IEEE Conference on Control Applications (CCA-2014)* 2014 pp 808–813.

[4] Xie H and Zhang G 2013 Research on Characteristics of the Piloted Follow-up Load Control Valve in Automobile Crane Luffing System *Fifth International Conference on Measuring Technology and Mechatronics Automation* pp 887–892

[5] Tarasov V N 1975 Dynamics of control systems of working processes of earth-moving machines (monography) (Omsk) 182

[6] Tarasov V N and Boyarkina I V 2006 Theory of impact in the construction and mechanical engineering industries 336

[7] Tarasov V N and Boyarkina I V 2006 Theory of impact in the construction and mechanical engineering industries 336

[8] Tarasov V N and Boyarkina I V 2010 Theorem of triangle altitude *Bulletin of the Siberian Motor Road Academy* **3(17)** pp 69–72

[9] Tarasov V N and Boyarkina 2012 Theorem of triangle altitudes in mathematics and technology *Materiály VIII mezinárodní vědecko-praktická konference "Dnyvědy–2012" pp* 3–9.

IOP Conf. Series: Journal of Physics: Conf. Series **858** (2017) 012007 doi:10.1088/1742-6596/858/1/012007

Dynamics of vibration isolation system with rubber-cord-pneumatic spring with damping throttle

Yu A Burian, M V Silkov

Omsk State Technical University, 11, Mira ave., Omsk, 644050, Russia

E-mail: silkovmv@mail.ru

Abstract. The study refers to the important area of applied mechanics; it is the theory of vibration isolation of vibroactive facilities. The design and the issues of mathematical modeling of pneumatic spring perspective design made on the basis of rubber-cord shell with additional volume connected with its primary volume by means of throttle passageway are considered in the text. Damping at the overflow of air through the hole limits the amplitude of oscillation at resonance. But in contrast to conventional systems with viscous damping it does not increase transmission ratio at high frequencies. The mathematical model of suspension allowing selecting options to reduce the power transmission ratio on the foundation, especially in the high frequency range is obtained

1. Introduction

The actual problem for vibration insulation of various objects is reducing the power transmission ratioon thefoundation in a wide range of frequencies. Pneumatic springs based on rubber-cord shells [1–4] with a throttling passageway3 between main 1 and supplementary 2 cavities are prospective for this purpose (Figure 1). Damping at the overflow of air through this passageway limits the amplitude of oscillation at resonance. But in contrast to conventional systems with viscous damping, it does not increase the transmission factor at high frequencies. It is due to air flowingthrough the passageway between the cavities of pneumatic spring is carried out to a certain flow fluctuations rate, and if the resistance in the damper is high enough tomakea gate which disables further cavity. Thus, the efficiency of such a system is much higher at high frequencies.

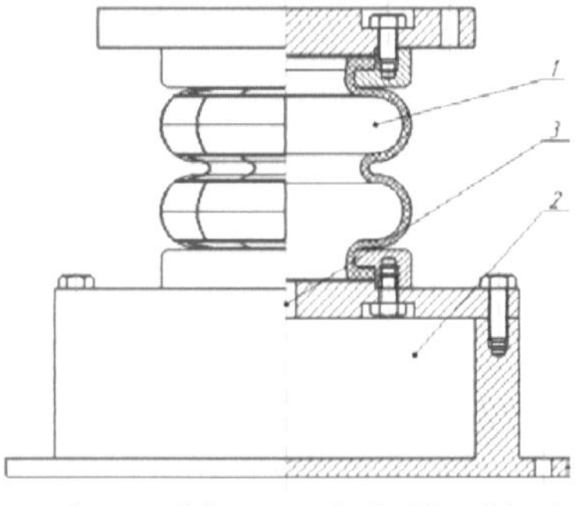

Figure 1. Pneumatic springs based on rubber-cord shell with throttling passagewaybetween main and supplementary cavities

2. Setting a problem

To quantify this effect compared to a conventional viscous damping it is necessary to create a mathematical model of such pneumatic suspension and carry out the necessary calculations.

3. Theory

Thermodynamic processes in the pneumatic springs will be regarded as quasistationary and proceeding at steady expiration modes [1]. The process of gas compression is considered adiabatic and it is usually true for frequencies greater than 3 Htz.

Gas enthalpy i changes due to changes in internal energy u and gas expansion or compression work A, i.e.

$$\frac{di}{dt} = \frac{du}{dt} + \frac{dA}{dt} = Gc_p T_0, \tag{1}$$

where G is mass flow of gas;
c_p, c_v are specific heat of gas at constant pressure and volume;
T_0 is absolute gas temperature.

Besides $\frac{dA}{dt} = p\frac{dV}{dt}$ end $\frac{du}{dt} = Gc_v T_0$ if we assume the gas to be ideal, then using the gas to be ideal, then using the gas equation and having differentiated it with respect to time, $GT_0 = \frac{1}{R}\frac{d}{dt}(pV)$ we will obtain $\frac{du}{dt} = \frac{c_v}{R}\frac{d}{dt}(pV)$.

Thus, the rate of change of enthalpy in the main chamber 1 is

$$Gc_p T_0 = \frac{c_v}{R}(p_1 V_1) + p_1\frac{dV_1}{dt}, \tag{2}$$

where p_1, V_1 are pressure and volume in chamber 1;
R is gas constant.

Given that $c_p - c_v = R$, $\frac{c_p}{c_v} = k$, where k = 1.4 is adiabatic index from (1) after some transformations we will get

$$G = \frac{1}{RT_0}\left(\frac{V_1}{k}\cdot\frac{dp_1}{dt} + p_1\frac{dV_1}{dt}\right). \tag{3}$$

Since the volume of the supplementary chamber 2 (Figure 1) does not change, and G is the same, then from (3) one can derive

$$G = \frac{V_2}{kRT_0}\frac{dp_2}{dt}, \tag{4}$$

where p_2, V_2 are pressure and volume of supplementary chamber.

Assuming that the connecting passageway between the chambers is a capillary (the length is much greater than the diameter) the flow through it can be determined from Poiseuille formula [5]

$$G = \frac{\pi d^4 \rho (p_1 - p_2)}{128\mu\ell}, \tag{5}$$

where d, ℓ are diameter and length of capillary;
ρ is air density;
μ is dynamic viscosity of air.

If we express the density from the gas equation, $\rho = \frac{p}{RT_0}$ and take $p = \frac{p_1 + p_2}{2}$ as average pressure, from expression (5) one can write

$$G = \frac{B}{2RT_0}\left(p_1^2 - p_2^2\right), \quad B = \frac{\pi d^4}{128\mu\ell}. \tag{6}$$

There can be carried out linearization of expressions (3), (4) and (6) in the close to parameters corresponding to their steady-state values in the close to the equilibrium position of the pneumatic spring $G = G_0$, $V_1 = V_{10}$, $p_1 = p_2 = p_0$. At the same time one can exclude from the obtained expressions ΔG and Δp_2 and combine them into a single expression

$$\left[V_{10}\left(\frac{sV_2}{kBp_0} + 1\right) + V_2\right]\Delta p_1 = kp_0\left(\frac{sV_2}{kBp_0} + 1\right)\Delta V_1, \tag{7}$$

where s is operator of differentiation.

In case we divide this expression by Δx, where x is pneumatic spring absolute movement of masses, and consider that $\frac{dV_1}{dx} = F_0$ is effective area of pneumatic spring and $\frac{\Delta p_1}{\Delta dx} = \frac{dp}{dx}$ then expression (7) can be written as

$$\frac{dp_1}{dt} = \frac{kp_0F_0}{V_{10} + \dfrac{V_2}{1 + \dfrac{V_2}{kBp_0}\cdot s}}. \tag{8}$$

Stiffness c of pneumatic spring can be expressed as follows:

$$c = \frac{d(p_1F_0)}{dx} = F_0\frac{dp_1}{dx} + p_1\frac{dF_0}{dx}. \tag{9}$$

If we take $F_0 = \text{const}$, from (8) и (9) we will obtain

$$c = \frac{kp_0F_0^2}{V_{10}\left(1 + \dfrac{\dfrac{V_2}{V_{10}}}{1 + \dfrac{V_2}{kBp_0}\cdot S}\right)} = \frac{c_{10}}{1 + \dfrac{N_1}{1 + B_1S}}, \tag{10}$$

where $c_{10} = \frac{kp_0F_0^2}{V_{10}}$ is static stiffness of the spring without considering the supplementary cavity $B_1 = \frac{V_2}{kBp_0}$, $N_1 = \frac{V_2}{V_{10}}$.

IOP Conf. Series: Journal of Physics: Conf. Series **858** (2017) 012007 doi:10.1088/1742-6596/858/1/012007

If one makes the substitution $s = \omega j$ in (10) then at low frequency vibrations (ω approaches 0) it follows from (10) that

$$c_0 = \frac{c_{10}}{1+N_1} = \frac{kp_0F_0^2}{V_{10}+V_2},$$

i.e. with overflow of air through the capillary the pressure values in both chambers have time to equalize and stiffness of the spring is determined with their total volume. At high frequencies (ωapproaches infinity)equation(10) implies $c_\infty = c_{10}$, which means that the pressure of the main chamber has no time to be transferred to the additional stiffness and is determined only with the volume of the main cavity. At intermediate frequencies stiffness takes intermediate values. Furthermore one can observe oscillation damping effect at low frequencies due to overflow of the gas

Equation of spring mass motion m can be written as $(ms^2 + c)\, x = P$, where P is harmonic forcing process acting in the x direction. Then the force transmitted to the foundation $P_f = cx$, and the transfer function characterizing the efficiency of vibration protection, change to

$$W(s) = \frac{P_f}{P} = \frac{c}{ms+c}, \tag{11}$$

where c is dynamic stiffness determined from the expression (10).

Having substituted (11) into (10) and expressed the module of the transfer function (power transfer coefficient on the foundation - K_t), we obtain

$$K_t = \sqrt{\frac{(B_1\omega)^2+1}{\left[1-\left(\frac{\omega}{\omega_0}\right)^2\right]^2 - (B_1\omega)^2\left[1-N_0\left(\frac{\omega}{\omega_0}\right)^2\right]}}, \tag{12}$$

where $\omega_0^2 = \frac{c_0}{m}$; $c_0 = \frac{c_{10}}{1+N_1} = \frac{kp_0F^2}{V_{10}+V_2}$; $N_0 = \frac{V_{10}}{V_0} = 1+N_1$.

We are going to consider the weight fluctuations on the same pneumatic spring without additional volume V_2 but with a conventional viscous damper creating a resistance force $P_{res} = B_2\dot{x}$. Then we can write $(ms^2 + B_2\,s + c_{10})x = P$, and for the force on the foundation one can get $P_f = (c_{10} + B_2 s)x$. In this case, the force transmission ratio on the foundation is determined this way

$$K_t' = \sqrt{\frac{\left(\frac{B_2}{c_{10}}\omega\right)^2+1}{\left[1-\left(\frac{\omega}{\omega_{10}}\right)^2\right]^2 + \left(\frac{B_2}{c_{10}}\omega\right)^2}}; \qquad \omega_{10}^2 = \frac{c_{10}}{m};\ c_{10} = \frac{kp_0F_0^2}{V_{10}}. \tag{13}$$

4. The results of the numerical experiment and discussion

To compare the effect of vibration isolation of these two schemes with expressions (12) and (13) the calculations were carried out. The following inputs for a pneumatic spring with rubber-cord shell I-08 [7]: m= 35kg, $V_{10} = 0.7\cdot10^{-3}$ m^3, $V_2 = 1.4\cdot10^{-3}$ m^3, $F_0 = 72\cdot10^{-4}$ m^2, $p_0 = 1.54\cdot10^5$ Pa; for a throttle passageway $d = 1{\cdot}10^{-3}$ m, $\ell = 2{\cdot}10^{-2}$ m, $\mu = 18.2{\cdot}10^{-6}$ Pa·s re provided as an example. B_2 resistance coefficient for the second scheme was chosen so that the value of K_t at resonance for both schemes was similar and could be compared on vibration isolation efficiency at the super resonance zone of frequency changes.

The results of the calculations (see Figure 2 and particularly Figure 3) demonstrate that the scheme with a pneumatic spring and the throttle passageway at high frequencies allows reducing the K_t value in about 4 times compared to the scheme with a conventional viscous damper (curves 1 and 2). Furthermore both schemes indicate the same efficiency when passing through the resonance zone at low frequencies Furthermore there is an optimum diameter of the throttle passageway, ensuring the most damping while passing through resonance zone (curve 1 in Figure 2). By increasing the diameter of the passageway (from optimum) resistance value is reduced, which causes an increase in vibration amplitude at resonance (curve 3 in Figure 2).With further increase in the diameter the damping passageway tends to zero, and turns into pneumatic spring with the static volume $V_{10}+V_2$. By decreasing the diameter (from optimum) damping is also reduced as less amount of air has time to flow from chamber 1 to chamber 2 and back. In addition to that there is a growth of the oscillation amplitude at resonance (curve 4 in Figure 2). Following further decrease in the diameter the damping goes to zero and the support turns into a pneumatic spring with static volume V_{10}.

At high frequencies, the small deviations from the optimum diameter of the throttling passageway have very little effect on the values of K_t (Figure 3, curves 1, 3, 4). This is due to the fact that at high frequencies of the passageway turns into a gate which disables supplementary cavity. Damping does not occur, and that is necessary to improve the vibration isolation in this frequency range.

One provides on Fig.1 an example of a pneumatic spring with rubber-coated balloon-type shell of small size. But this scheme with the throttle pasageway can also be implemented for other types of shells [1, 7]. It is particularly suitable for a pneumatic spring with a diaphragm shell, which involves the construction of two volumes (Figure 4). Calculations carried out for this springwith m = 480 kg, $V_{10} = 6{\cdot}10^{-3}$ m^3, $V_2 = 9{\cdot}10^{-3}$ m^3, $F_0 = 6{\cdot}10^{-2}$ m^2, $p_0 = 1.8{\cdot}10^5$m^2 showed that optimal values for the throttle passageway is $d = 1.8{\cdot}10^{-3}$ m, when $\ell = 3{\cdot}10^{-2}$ m.

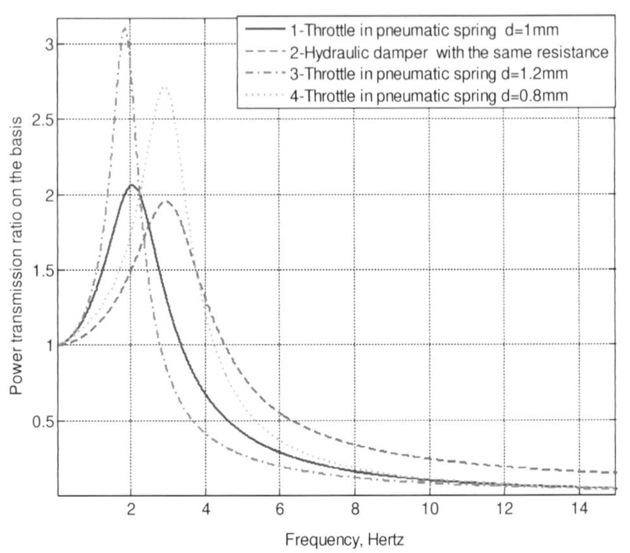

Figure 2. Dependence of the power transmission ratio on the basis of frequency for throttling passageway with pneumatic springs between the primary and secondary volume of comparison with the same spring but without additional volume of a hydraulic damper with the same resistance (in the frequency range up to 15 Hz)

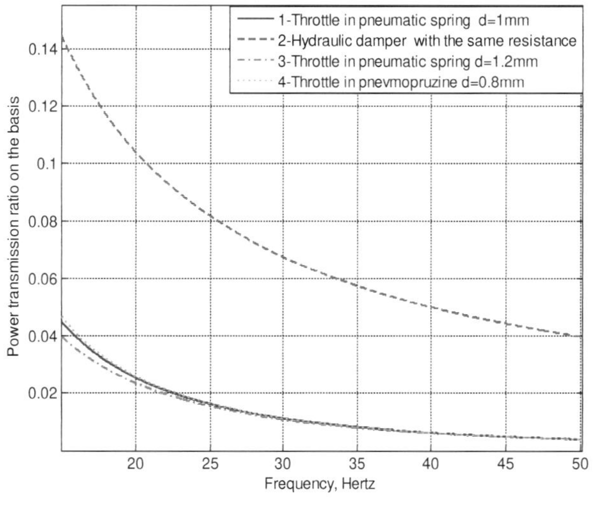

Figure 3. The same for the frequency range above 15 Hz.

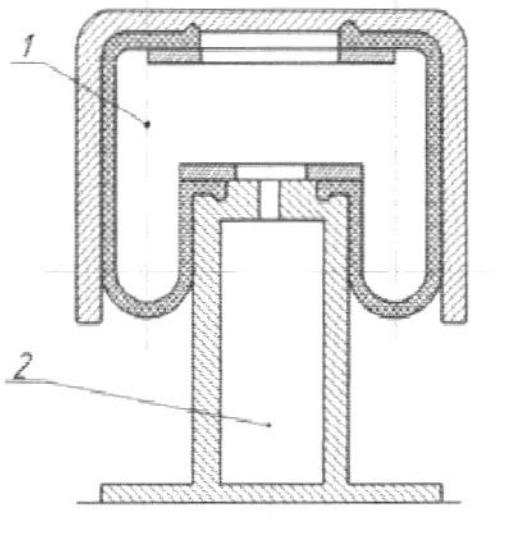

Figure 4. Pneumatic springs based on rubber-cord shell of diaphragm type with a throttling passageway between the main and supplementary chambers

5. Summary and conclusions

Thus the proposed analytical relationships allow making a preliminary selection of the parameters of the throttle passageway considering other given parameters of suspension on pneumatic spring to provide vibration isolation efficiency, both in the low-frequency and high-frequency oscillations.

References

[1] Tribelsky I A, Shalay V V, Zubarev A V and Tribelsky M I 2011 Computational-experimental methods of designing complex rubber-cord constructions 238

[2] Zubkov A I 1977 Characteristics approximation of pneumatic elastic element with rubber-cord shells. Calculation, design, production and operating . Collection of research papers (Moscow) pp 47–49.

[3] Elisev S V, Reznik U N and Chamenko A P 2011 Mechanotronic approach for dynamics mechanical oscillatory systems (Novosibirsk) 384

[4] Buryan Y A, Komarov Y P, Buryan Y A and S. N. Polyakov 2014 Vibration-isolated support with hydraulic inertial motion convertor based on rubber-cord shell *Dynamics of Systems, Mechanisms and Machines (Dynamics) IEEE International Conference*

[5] Klimentyev E V, Zvonov A O and Glazkova E U 2014 Rubber-cord cushions application in modern engeineeging *Dynamics of Systems, Mechanisms and Machines (Dynamics). IEEE International Conference* DOI: 10.1109/Dynamics.20144.7005706

[6] Loytsyansky A G, Lurye A. I. 1983 Course of theoretical mechanics 640

[7] Loytsyansky A. G and A. I. 1983 Lurye Course of theoretical mechanics (Moscow, Nauka Publishers) 640

[8] Pneumatic elements with rubber-cord shell http://www.progress-omsk.ru/constructor.php?act=group5. Date of the application 14.12.16.

IOP Conf. Series: Journal of Physics: Conf. Series **858** (2017) 012008 doi:10.1088/1742-6596/858/1/012008

Modeling of the processes of heating organic coolant in the system of heat losses recuperation of the mobile compressor unit on the basis of a low-speed single-stage

G I Chernov[1], V L Yusha[1], K V Sherban[1], A M Kalashnikov[1]

[1] Omsk State Technical University, 11, Mira ave., Omsk, 644050, Russia

E-mail: gi_chernov2002@mail.ru

Abstract. The article considers the modeling of the processes of heating organic coolant in the system of heat losses recuperation of the mobile compressor unit. The modeling of the liquid fuel heating, its boiling and vapour overheating was carried out in the ANSYS environment. The modeling results were compared with the available engineering methods, and proved the applicability of the considered modeling method.

1. Introduction

It is known that all the energy required for the mobile compressor unit (MCU) drive is released in the form of heat into the environment by cooling gas, lubricants and coolants, as well as communications [1, 2]. In mobile compressor units more than half of the energy supplied to the power unit (as a rule to the internal combustion engine) as a result of fuel combustion in its working chamber is discharged into the atmosphere together with the exhaust gases, as well as through the oil, coolant and charge air cooling system [3]. Slow-speed single-stage compressors with unique cooling conditions of the actuation gas in the cylinder are of special interest [4, 5].

The ways of solving the given problem for a mobile compressor unit (MCU) with the existing compressor units are proposed in papers [6, 7]. The present the results of the thermodynamic analysis of the various structural schemes of MCU heat losses recuperation. However, these studies have not addressed the coolant such as the fuel supplied to the power unit. Meanwhile, there are several ways to make internal combustion engines more efficient [3, 8], including an increase in the temperature of the fuel supplied to the engine, allowing to enhance the quality of the spray and the combustion efficiency [1, 9, 10, 11]. Obviously, the fuel may be one of the coolants in the MCU recovery heat losses recuperation system. At the same time, elements of the recuperation system may include similar in design cylindrical cooling cavities of fuel injectors, internal combustion engine cylinders and long-stroke slow-speed single-stage reciprocating compressors. In this regard, the task of developing a methodology for calculating organic fuel heating processes in the cooling systems of such units is urgent.

2. Statement of the problem

The given study investigates the processes of heating, boiling and overheating of organic fuel vapor (kerosene) in the channels of the elements of heat losses recuperation system in a mobile compressor unit on the basis of a slow-speed single-stage reciprocating compressor.

3. The object of the research

The object of the study is a cylindrical wall with inner cooling channel. Such an element is a base for such sources of heat loss, as cylinders, nozzles, and end nodes. Kerosene is taken as fuel.
The channel in the cylindrical wall, may generally be twisted with a variable radius loop along the length of the winding, as shown in Figure 1. In addition, the winding step may also vary.

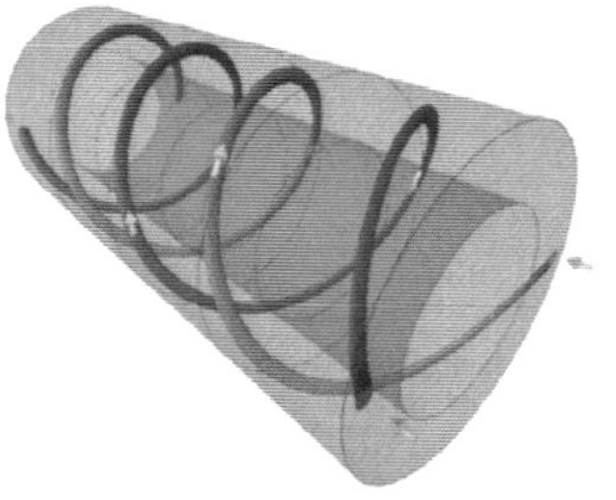

Figure 1. Fuel heat channel scheme

4. Research methodology

A well-known system of convective heat transfer equations is used as a method of calculating the kerosene heating in ANSYS environment. The main assumptions are the following:
1. The process of liquid flow and heat transfer are stationary. This assumption is determined by the fact that the calculation of the thermal-hydraulic processes is based on a complex systems of differential equations; while the process unsteadiness leads to a drastic complication of the equation system, making it difficult to obtain the solutions.
2. The temperature of the heating wall of the moving flow remains constant. This assumption follows from the first one.
3. In a liquid flow there are no internal heat sources.
The main computational equations used in the analysis of the flow are the following:
– The continuity equation expressing the law of conservation of mass.
– The Navier-Stokes equations, expressing the law of momentum change, on the basis of which the distribution of the velocity in the flow region is found out.
– The energy equation, which expresses the law of conservation of energy, which is the basis for the distribution of the temperature field in the liquid flow.
– The Newton–Richman equation, allowing to determine the heat flow, supplied to the flowing medium by the heating wall.
The above equations forms a system that describes the heating of a moving-phase medium, when liquid or gas moves in the channel. When the liquid is boiling, the system composed of two subsystems is built. One subsystem describes the motion of the liquid, the other describes the vapor equilibrium with the liquid. The relationship between these subsystems is described by the following equations concerning the interaction of the phases at their interface.

– The equation, describing mechanical equilibrium through the equality of the liquid and vapor shear stresses at the interface border [12]:

$$\mu' \cdot \left(\frac{\partial V'}{\partial n}\right)_{\text{int}} = \mu'' \cdot \left(\frac{\partial V''}{\partial n}\right)_{\text{int}}, \qquad (1)$$

IOP Conf. Series: Journal of Physics: Conf. Series **858** (2017) 012008 doi:10.1088/1742-6596/858/1/012008

where μ' is the coefficient of dynamic viscosity of the liquid phase;
V' is the line tangent to the interface component of the liquid phase velocity;
μ'' is the coefficient of dynamic viscosity of the vapour phase;
V'' is the line tangent to the interface component of the vapour phase velocity;
– The equation, describing mass transfer, while phase changing at the interface:

$$\left(\rho' \cdot V_n'\right)_{\text{int}} = \left(\rho'' \cdot V_n''\right)_{\text{int}}, \qquad (2)$$

where ρ' is the liquid phase density, kg/m3;
V_n' is the normal to the interface component liquid phase velocity;
ρ'' is the density of the vapour phase, kg/m3;
V_n'' is the normal to the interface component of the vapour phase velocity.
– The equation, describing heat balance during the heat transfer between the liquid and the vapour phases to the interface:

$$-\lambda' \cdot \left(\frac{\partial t}{\partial n}\right)_{\text{int}} = r \cdot \rho'' \cdot V_n'' \qquad (3)$$

where λ' is the heat transfer coefficient of the liquid phase ;
r is the evaporation specific heat, J/kg.

The left side of this equation determines the density of the heat flow supplied to (withdrawn from) the interface from the liquid phase. The right side of this equation describes the density of heat flow, withdrawn from (supplied to) the interface from the vapor phase.

The continuity, momentum change, energy, heat and Newton - Richman equations given above include thermal properties of the medium, which generally depend on the temperature.

The equations defining the dependence of the moving medium properties on the temperature when presented in an analytical form are quite complicated, so these dependencies are usually given in the form of tables, which are insert in the library of substance properties in special software systems such as ANSYS CFX.

The developed system of differential equations describe all possible cases of heat transfer during the flow moving in the channel. To select from an infinite number of possible solutions only the right set of equations, is should be supplemented by the single-valued conditions, which are divided into geometrical single-valued conditions, physical single-valued conditions and boundary conditions.

Geometrical single-valued conditions include the fact that the channel where the flow moves, should be considered with two types of cross-section (round and triangular one), the basis of triangular cross-section being an equilateral triangle. In both cases, the cross sectional area was assumed the same and equal to 176 mm^2. This corresponds to the diameter of 15 mm and side of the triangle 20.2 mm.

IOP Conf. Series: Journal of Physics: Conf. Series **858** (2017) 012008 doi:10.1088/1742-6596/858/1/012008

The physical conditions include establishing the dependencies of the moving medium properties (kerosene) on the temperature in the form of tables, which are insert in the library of substances properties in ANSYS CFX.

The boundary conditions for the calculation of heat transfer is the establishing of liquid kerosene temperature at the channel inlet at the level of 20 C, and the establishing of liquid kerosene velocity at the channel inlet as a set of values of 0.1; 0.5; 1.0 m/s. When calculating the liquid kerosene heating mode, its boiling and heating kerosene vapor, mass flow rate was assumed constant. The boundary conditions also included the establishing of the heating temperature of the channel side wall as a set of multiple values that changed when changing the flow heat mode. The pressure at the flow channel inlet was set in an arbitrary manner, as t not the absolute value of the pressure but its gradient at the heating area under consideration is of interest.

5. Analysis of the calculation results

The calculation of the given system of equations with the described single-value conditions was made in ANSYS CFX environment involving the CFX application. The building of the channel models geometry was made in the Geometry module, the construction of the calculated finite element mesh - in CFX-Mesh, the establishing of boundary conditions and calculation parameters - in CFX-Pre, the solution - in CFX-Solve, the visualization and analysis of the results - in CFX-Post. The units of all the values comply with SI.

The example of calculation of kerosene heating prior to the boiling for the special case at kerosene speed V=0.5 m/s and wall temperature t_{wall}=2000°C is shown in Figures 2, 3 and 4. The calculated dependencies of kerosene heating channel length before the boiling state on the velocity of kerosene and the temperature of the heating wall for different configurations of the channel cross-section are shown in Table 1.

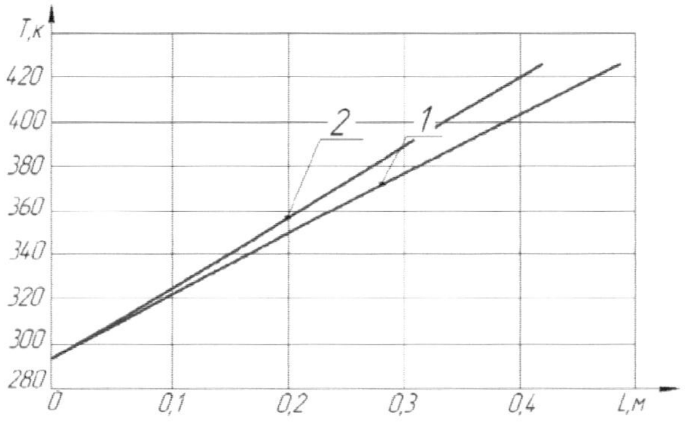

Figure 2. The flow temperature change along the channel length at kerosene velocity 0.5 m/s and wall temperature t_{wall}=2000°C: line 1 –channel with round cross-section; line 2 –channel with triangular cross-section

IOP Conf. Series: Journal of Physics: Conf. Series **858** (2017) 012008 doi:10.1088/1742-6596/858/1/012008

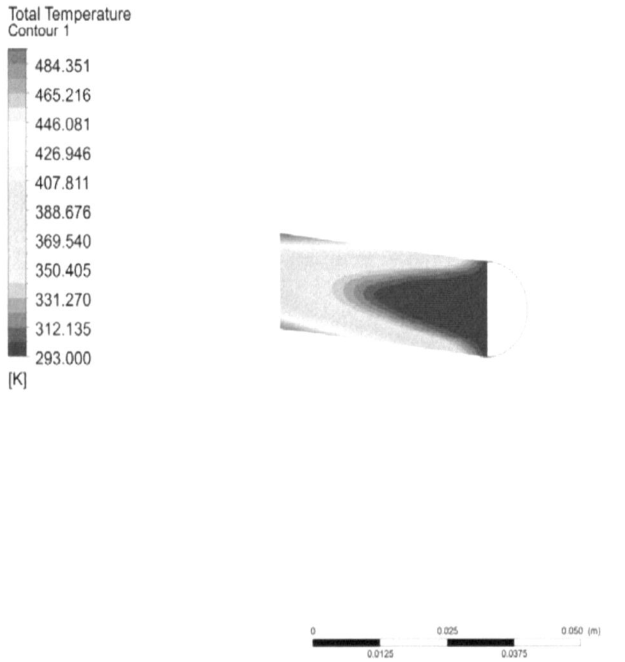

Figure 3. Temperature distribution in longitudinal plane, passing through the axis of round cross-section symmetry (V=0.5 m/s, t_{wall}=2000°C)

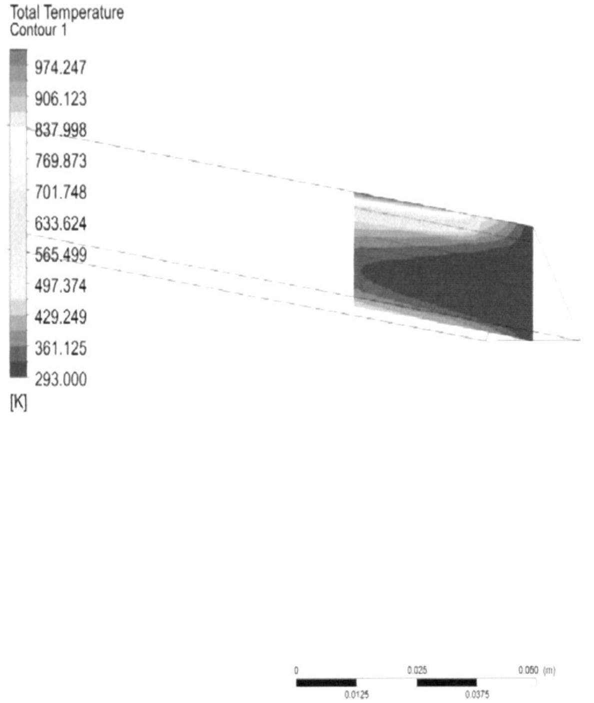

Figure 4. Temperature distribution in longitudinal plane, passing through the axis of triangular cross-section symmetry (V=0.5 m/s, t_{wall}=2000°C)

Table 1. Dependence of the kerosene boil-off section length on the flow rate and the heating wall temperature

Wall temperature, °C	Velocity V, m/sec	Channel length L, mm	
		Round cross-section	Triangular cross-section
800	0.1	855	496
	0.5	1860	1550
	1.0	2445	1890
1500	0.1	377	130
	0.5	938	660
	1.0	1165	830
2000	0.1	267	83
	0.5	680	442
	1.0	840	580

After the complete kerosene boiling-off at temperature range 280 – 300 °C with its further heating, the vapour heating occurs. The heating continues until the temperature reaches 430 – 450 °C, when kerosene vapour thermal decomposition into more light hydrocarbon fractions begins [13]. Both during liquid kerosene heating and boiling and kerosene vapors heating its movement in two types of channels cross sections was analyzed: the channel with a round cross section and the channel which cross-section is an equilateral triangle. The heating wall temperature took values 1000; 1500; and 2000 ° C. The liquid kerosene flow rate at the channel inlet was assumed to be equal to- 11; 55; 110 m/sec. This is determined by a decrease in the kerosene density during its transition to the vapor phase and by mass flow conservation corresponding to the velocities of kerosene movement in its liquid phase: 0.1; 0.5; 1 m/sec.

The example of calculation of kerosene vapor heating before the thermal cracking at heating wall temperature 2000 ° C and flow rate 55 m/sec are shown in Figures 11, 12 and 13. The calculated dependences of the kerosene vapor heating channel length before the beginning of the thermal cracking process on kerosene velocity and heating wall temperature for different configurations of the channel cross-section are presented in Table 3.

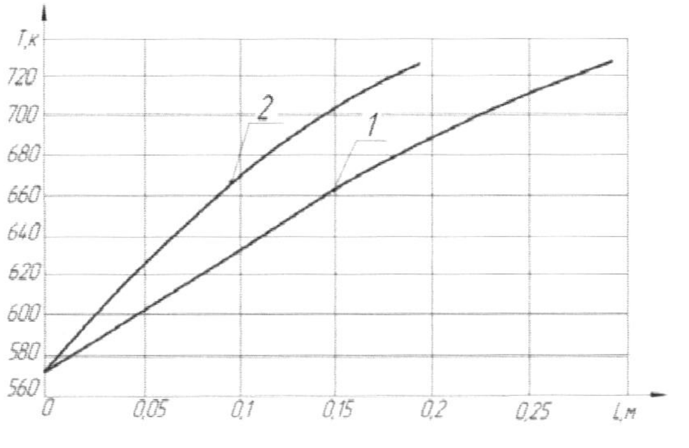

Figure 5. Temperature changing of the flow vapour phase along the channel length at kerosene velocity 55 m/s and wall temperature 2000°C: curve 1 – channel with round cross-section; curve 2 – channel with triangular cross-section

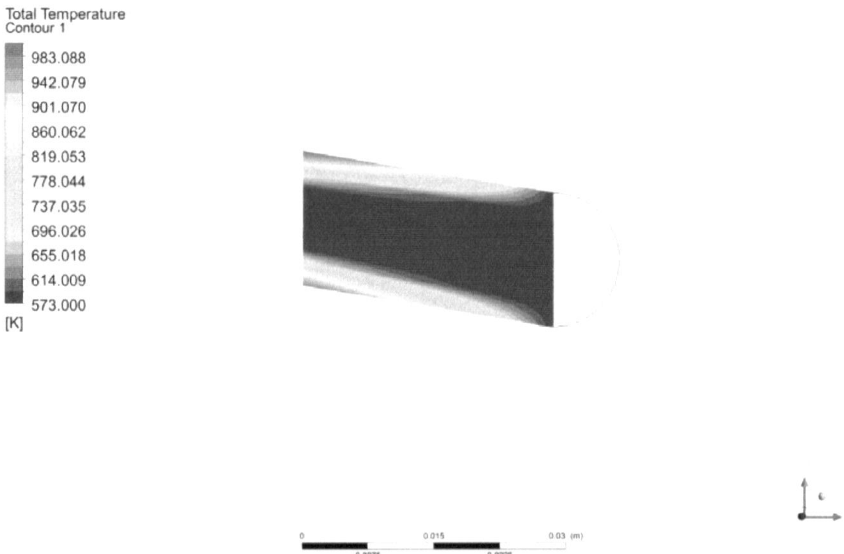

Figure 6. Temperature distribution of the flow vapour phase in longitudinal plane, passing through the axis of round cross-section symmetry (V=55 m/sec, t_{wall}=2000 °C)

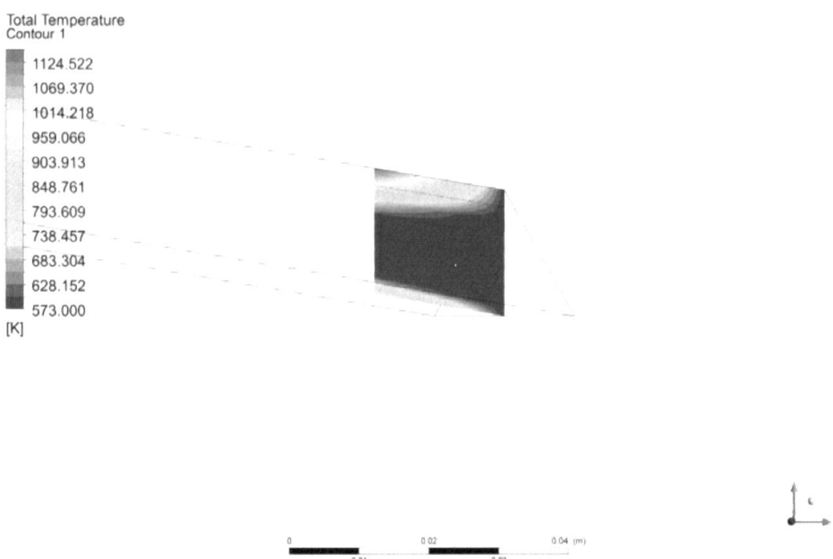

Figure 7. Temperature distribution of the flow vapour phase in longitudinal plane, passing through the axis of triangular cross-section symmetry (V=55 m/sec, t_{wall}=2000 °C)

Table 2. Dependence of the kerosene vapour heating section length before the beginning of its thermal decomposition on the flow rate and the heating wall temperature

Wall temperature T, °C	Velocity V, m/sec	Channel length L, mm	
		Round cross-section	Triangular cross-section
1000	0.1	492	373
	0.5	632	493
	1.0	699	550
1500	0.1	298	198
	0.5	387	266
	1.0	424	298
2000	0.1	208	135
	0.5	277	181
	1.0	308	203

6. Conclusion

The comparison of the results of integral calculation in the ANSYS environment with the results of calculation obtained by standard engineering techniques [14], has shown their satisfactory convergence up to 20%. The adequacy of kerosene boiling calculation is proved by satisfactory agreement between the interval of the boiling temperature limits (150 and 280^0 C) and the calculated limits interval of the boiling kerosene volume fraction change (1 and 0.4). Furthermore, the calculation analysis shows that the length of the heating area, which leads to the top of the kerosene boiling in a channel with a round cross-section is greater than in a channel with a triangular cross-section. This is determined by a big perimeter of triangular section. The flow rate growing leads to the increasing in the length of the heating section, that is a qualitative verification of the calculation results. All this shows that the technique presented in the article, which was designed on the basis of ANSYS package, is applicable for calculation of fuel heating in the channels of a complex configuration, which can be used in the system of heat losses recuperation in MCU with different types of compression equipment. The preliminary results suggest the possibility of the effective combination of MCU recuperation system with fuel preparation system, including the possibility of fuel thermal cracking.

References

[1] Guilherme B. Ribeiro 2016 Second law evaluation of a lightweight cooling unit *Case Studies in Thermal Engineering* **7** March pp 47–54

[2] Filkin N. Yu., Yusha, V. L. 2016 Theoretical analysis of changing gas dynamic characteristics of the dust filter with a short diffuser while in operation *Procedia Engineering* **152** pp 270–275.

[3] Lukanin, V.N., Buslaev, A.P., Novikov A.V. 2003 Traffic flows modelling and the evaluation of energy-ecological parameters *International Journal of Vehicle Design* **33(4)** pp 381–399.

[4] Yusha V L, Dengin V G, Busarov S S, Nedovenchanyi AV and Gromov AYu 2015 The estimation of thermal conditions of highly-cooled long-stroke stages in reciprocating compressors *International Conference on Oil and Gas Engineering, OGE-2015* pp 264–269.

[5] Caroline Willich, Christos N. Makides, Alexander J. White. 2017 An investigation of heat transfer losses in reciprocating devices *Applied Thermal Engineering* **111** pp 903–913

[6] Yusha V L and Chernov G 2013 Effectiveness analysis of using the Rankine cycle and cycle of refrigeration machine for recuperation of heat losses in mobile compressor unite *8th International Conference on Compressors and Coolants (Papiernička, Slovakia, 2013)* 45

IOP Conf. Series: Journal of Physics: Conf. Series **858** (2017) 012008 doi:10.1088/1742-6596/858/1/012008

[7] Yusha V L, Chernov G I and Sadvakasov D H 2016 The Efficiency Analysis for the Heat Losses Recuperation System of the Mobile Refrigeration Unit *Procedia Engineering* **152** pp 339–347.

[8] Kavtaradze R Z, Zelentsov A, Gladyshev S P, Kavtaradze Z and Onishchenko D. 2012 Heat insulating effect of soot deposit on local transient heat transfer in diesel engine combustion chamber *SAE Technical Papers* pp 68–79.

[9] Markov V A, Gladyshev S P, Devianin S and Mikhalsky L. 2010 Method of estimating the quality of the process of creating the fuel-air mixture in a high-speed diesel engine *SAE Technical Papers* pp 76–87

[10] Dent J C 1971 Basis for the comparison of various experimental methods for study spray penetration *SAE paper 710571* 18

[11] Hiroysu H and Arai M 1990 Structures of fuei spray in diesel engines penetration *SAE paper, 900475* 14

[12] Bessarabov A M, Avseev AV, Avseev V V and Kutepov A M 2004 Hierarchy of dynamic models of nonequilibrium chemically reactive plasma: The possibility of initiation of oscillating chemical reactions *Theoretical Foundations of Chemical Engineering* **38(2)** pp 214–218

Effect of composition on properties of In_2O_3–Ga_2O_3 thin films

I E Demin[1] and A G Kozlov[1]

[1]Omsk State Technical University, Mira ave., 11, 644050, Omsk, Russia
E-mail: IvanDemin3467@gmail.com

Abstract. The In_2O_3–Ga_2O_3 mixed oxide polycrystalline thin films with various ratios of components were obtained by pulsed laser deposition. The effect of films composition on surface morphology, electrophysical and gas sensing properties and energies of adsorption and desorption of combustible gases was studied. The films with50%In_2O_3–50%Ga_2O_3 composition showed maximum gas response (~25 times) combined with minimum optimal working temperature (~530 °C) as compared with the other films. The optical transmittance of the films in visible range was investigated. For 50%In_2O_3–50%Ga_2O_3 films, the transmittance is higher in comparison with the other films. The explanation of the dependency of films behaviors on their composition was presented.The In_2O_3–Ga_2O_3 films were assumed to have perspectives as gas sensing material for semiconducting gas sensors.

1. Introduction

Oxides of transition metals have various applications. Some examples aresemiconductor gas sensors and conductive transparent films.Semiconductor gas sensors, are actually one of the most investigated groups of gas sensors. They are low cost, have considerable long-term stability, provide a variety of detectable gases and able to operate in harsh environments [1]. All these advantages allow one to use semiconductor gas sensors in numerous fields [2].

An important trend in a field of semiconductor gas sensor research is the finding of for novel gas sensing materials [3]. One of the achievements on that way is the use of mixed metal oxides. The main idea of that method is based on fundamental principle of the resistance variation of the sensing layer. It involves two important functions: the recognition and the transducer function. The recognition includes all the interactions between gas and solid. The transducer function is responsible for transformation of the chemical stimulus into an electrical signal.

The most accepted explanation of mixed oxide gas sensing performance is the separation of those functions between two parts of the compound system [3]. One part of multicomponent system may be responsible for recognition having very high reactivity to target gas while the second part takes over the transducer duties. This effective combination allows optimizing both functions simultaneously, which lead to a possibility of sensors sensitivity increase [4].

The production of multicomponent films requires a technique with a good control and reproducibility of parameters of resulting films [5]. The pulsed laser deposition (PLD) method used in the present study satisfies these requirements very well. The uniform distribution of components in prepared films and low growth rates (5–10 nm/min) allow achieving very high reproducibility of a sensing element. Another important advantage of the PLD is the stoichiometric transfer of material from sputtering target to substrate, which enables very precise control over the composition of the film [6].

IOP Conf. Series: Journal of Physics: Conf. Series **858** (2017) 012009 doi:10.1088/1742-6596/858/1/012009

A huge variety of mixed oxides are employed as sensing materials in semiconductor gas sensors [1, 4, 7]. Some well-known examples are In_2O_3–SnO_2 (ITO), In_2O_3–Ga_2O_3–ZnO (IGZO), ZnO–SnO_2, SnO_2–WO_3 and many others. The In_2O_3-Ga_2O_3 films among them are poorly investigated although they have good gas sensing properties [7, 8]. Most of the available data show possibility of their application as conductive transparent films [9–11]. In [8] there is a study on gas sensing properties of In_2O_3-Ga_2O_3 thin films made by magnetron sputtering from In-Ga eutectic alloy followed by annealing in air. Also there are results [12] on agglomerative single-electrode gas sensors with In_2O_3-Ga_2O_3 sensing layer, coated on a platinum spiral from $In(OH)_3$, $Ga(OH)_3$ sols. However, while perspectives of In_2O_3-Ga_2O_3 thin films in a field of gas sensing are clearly marked, there is insufficient information to date on the electrical and gas sensing properties and methods of their production.

The aim of this paper is the study of electrophysical, optical and gas-sensing properties of In_2O_3-Ga_2O_3 thin films (made by PLD) with different ratios of components. Gas induced resistance response was investigated for the following combustible gases: alcohol, acetone, ammonia andliquefied petroleum gas (LPG).

2. Experimental

The In_2O_3-Ga_2O_3 thin films were obtained in two consecutive steps: fabrication of bulk targets followed by pulsed laser deposition of the films. The targets were formed by pressing the mixture of In_2O_3 and Ga_2O_3powders and $Ga(NO_3)_3{\cdot}8H_2O$. The obtained targets were annealed in natural air at 820 °C for 10 minutes to prepare them for subsequent pulsed laser deposition, which requires sufficient durability of targets [13].

The pulsed laser deposition was carried out with Nd:YAG pulsed laser «Melaz» at 1.06 µm in residual air with pressure 10^2 Pa. Pulse parameters: duration 16 ns, repetition rate 10 Hz, focal spot area 7.85 10^{-5} cm^2, laser pulse energy density 637 J/cm^2. Substrate parameters: temperature 400 °C, material CT 50-1 glass ceramics. The film thickness reaches 0.20–0.21 µm after 20 minutes of deposition.

Compositions of the targets were 100%In_2O_3, 75%In_2O_3–25%Ga_2O_3, 50%In_2O_3–50%Ga_2O_3, 25%In_2O_3–75%Ga_2O_3 and 100%Ga_2O_3. Surface morphology of the obtained films was investigated by atomic force microscopy (NT-MDT AFM «Solver Pro»). Optical transmittance spectra were obtained using spectrophotometer «SF-56» in the wavelength range 200–1100 nm.

Study of electrophysical and gas-sensing properties of the films was conducted in the temperature range 150–750 °C on a special experimental setup (two-electrode method of resistance measurement) placed in a sealed chamber with natural air. The sample gas was injected into chamber in small portions by medical injector to achieve required concentration of that gas in the volume of the chamber. The resistance response S was calculated as a ratio of R_0 to R_g, where R_0 and R_g are resistances of the film in the natural air (base resistance) and in natural air with sample gas, respectively.

$$S = R_0 / R_g, \qquad (1)$$

More details of the experiment may be found in [13, 14].

3. Results and discussion

3.1. Atomic force microscopy

Atomic-force microscopy (AFM) images of films surfaces allow calculating mean grain size, making assumptions of gas penetrability of films [15] and new phases formation. Results of AFM studies of films surfaces deposited from targets with 100%In_2O_3, 75%In_2O_3–25%Ga_2O_3, 50%In_2O_3–50%Ga_2O_3, 25%In_2O_3–75%Ga_2O_3, 100%Ga_2O_3 compositions are presented in figure 1 and in table 1. Mean grain size was calculated using watershed method of image processing. All films have polycrystalline structure with a developed surface consisting of grains with diameters 0.10–0.13 µm. Films with 100%In_2O_3 composition have least mean grain size and least standard deviation of surface

IOP Conf. Series: Journal of Physics: Conf. Series **858** (2017) 012009 doi:10.1088/1742-6596/858/1/012009

profile. Films made of pure Ga_2O_3 demonstrate another minimum of grains parameters. Maximum of mean grain size and standard deviation of surface profile is achieved for films with 75%In_2O_3–25%Ga_2O_3 and 50%In_2O_3–50%Ga_2O_3 composition.

Figure 1. Examples of AFM images of In_2O_3–Ga_2O_3 films.

Observed increase in grain size may be a result of new phase formation. Indium oxide and gallium oxides have nearly equal melting points: 1910 °C and 1900 °C, respectively. They may form $InGaO_3$ alloy during deposition procedure under certain conditions, because it have lower melting point, than pure oxides. Formation of this alloy is the only explanation of relatively large droplet-like grains appeared on the surface of films with composition close to 50%In_2O_3–50%Ga_2O_3, which is not observed for films with 100%In_2O_3 and 100%Ga_2O_3 composition (see figure 1). The $InGaO_3$ phase formation in obtained films may be the cause of other effects discussed below.

Table 1. Mean grain size and standard deviation of surface profile for In_2O_3–Ga_2O_3 films.

Composition of film	Mean grain size, nm	Standard deviation of surface profile, nm
100%In_2O_3	93	43
75%In_2O_3–25%Ga_2O_3	269	228
50%In_2O_3–50%Ga_2O_3	305	240
25%In_2O_3–75%Ga_2O_3	266	154
100%Ga_2O_3	231	138

3.2. Temperature dependencies of resistance

Temperature dependencies of resistance for all films are shown in figure 2. The resistance of films with 100%Ga_2O_3 composition in chosen temperature range (150–750 °C) exceeded 10 MΩ. That was the reason to exclude those films from further research. Among other films, the highest resistance in all temperature range is observed for films with 50%In_2O_3–50%Ga_2O_3 and 75%In_2O_3–25%Ga_2O_3 composition (see table 2). In a field of semiconductor gas sensors it is well known that high base resistance usually leads to high gas response [1]. Results of gas-inducedresistance response study (discussed below) confirm that trend.

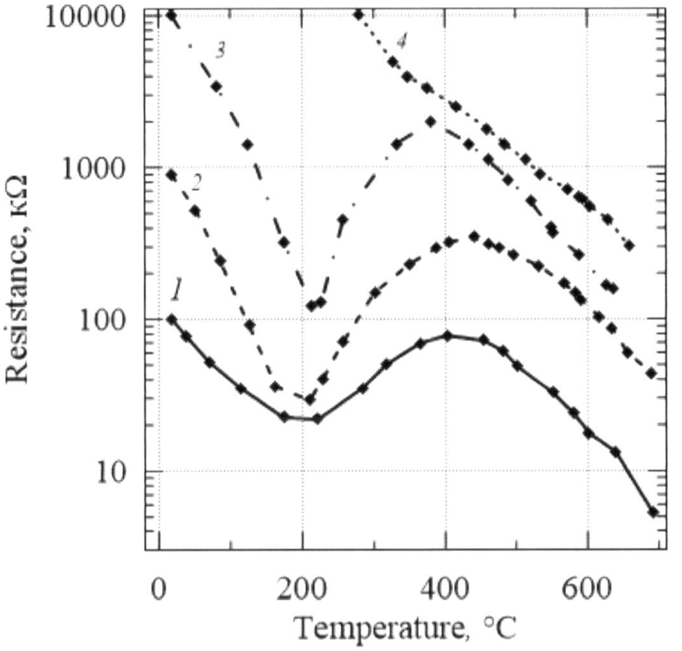

Figure 2. Temperature dependencies of resistance: 1 – 100%In_2O_3; 2 – 75%In_2O_3–25%Ga_2O_3; 3 – 50%In_2O_3–50%Ga_2O_3; 4 – 25%In_2O_3–75%Ga_2O_3.

One of important characteristics of films for semiconductor gas sensors is relative change of resistance per °C, because of usual problem with temperature-caused baseline instability [1, 16–18].Relative change of resistance per °C in temperature range 530–550 °C (range of maximum gas-induced resistance response)have values 1.0–1.5 % for films with different compositions.

There is a peculiarity on all the temperature dependencies of resistance in the temperature range 270–400 °C – a region of positive temperature coefficient of resistivity. This behavior of resistance is well-known for oxides of transition metals. It is explained by the change in the oxygen adsorption mechanism [13, 19, 20]. The transition from preferential adsorption of molecular oxygen O_2^- to preferential adsorption of atomic oxygen O^- and O^{--} occurs at these temperatures, which leads to increase in resistance.

Table 2. Maximum and minimum of resistance and corresponding temperature for In_2O_3–Ga_2O_3 films.

Composition of film	R_{min}, kΩ	R_{max}, kΩ	$T(R_{min})$, °C	$T(R_{max})$, °C
100%In_2O_3	21.9	77.8	221	403
75%In_2O_3–25%Ga_2O_3	28.9	345	210	441
50%In_2O_3–50%Ga_2O_3	122	1998	214	381
25%In_2O_3–75%Ga_2O_3	-	9998	-	280

3.3. Optical transmittance spectra

IOP Conf. Series: Journal of Physics: Conf. Series **858** (2017) 012009 doi:10.1088/1742-6596/858/1/012009

Optical transmittance spectra for all films are presented in figure3. Films with 100%In_2O_3 composition have the highest transmittance in visible range (60 %).In general, increase in Ga_2O_3 content reduces the film transmittance in visible range.The deviation from that trend is observed for films with 50%In_2O_3–50%Ga_2O_3which have 20–25 %transmittance in visible range composition. Optical transmittance of other films does not exceed 12 %.

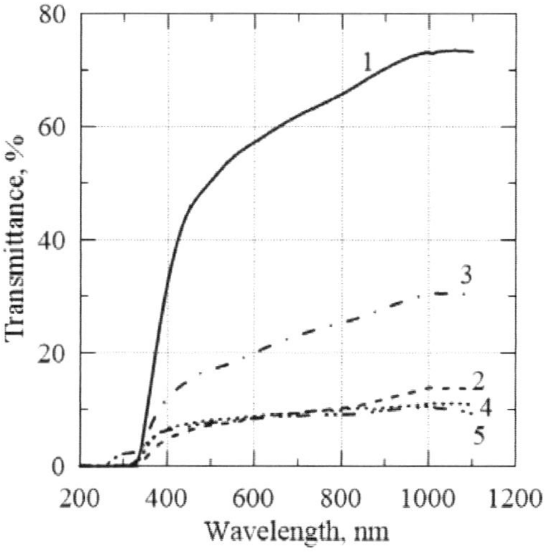

Figure 3. Optical transmittance spectra for In_2O_3–Ga_2O_3 films.
1 - 100%In_2O_3, 2 - 75%In_2O_3–25%Ga_2O_3, 3 - 50%In_2O_3–50%Ga_2O_3,
4 - 25%In_2O_3–75%Ga_2O_3, 5 - 100%Ga_2O_3.

All films were obtained with identical deposition parameters, so the only explanation of the optical transmittance increase is the formation of new phase in films with 50%In_2O_3–50%Ga_2O_3composition. It may be $InGaO_3$ phase, which is known for its good optical transmittance [10].

3.4. Temperature dependencies of gas-induced resistance response

Studies of temperature dependencies of gas-induced resistance response are essential for development of semiconductor gas sensors. Examples of that dependencies obtained for films with 100%In_2O_3, 75%In_2O_3–25%Ga_2O_3, 50%In_2O_3–50%Ga_2O_3 and 25%In_2O_3–75%Ga_2O_3 composition are shown in figure4 (a case of response on 25 ppm of acetone after 10 min.). The dependencies for other combustible gases have a similar bell-shaped form. Experimental error for temperature measurements did not exceed 5%, for resistance measurements – 3%.

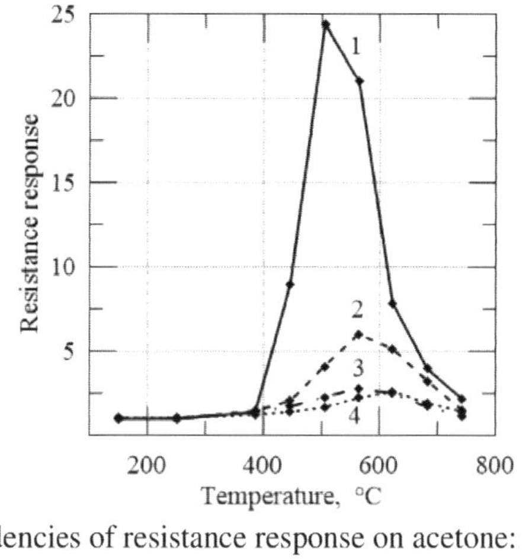

Figure 4. Temperature dependencies of resistance response on acetone:

1 – 50%In_2O_3–50%Ga_2O_3; 2 – 75%In_2O_3–25%Ga_2O_3; 3 – 100%In_2O_3;
4 – 25%In_2O_3–75%Ga_2O_3.

Temperature dependencies of gas response on the aforementioned combustible gases diluted in natural air have a maximum at a certain temperature, which is typical for semiconducting gas sensors. Amplitude of response on acetone and ethanol for all investigated films in temperature range 500–600 °C exceeds 3–4 times amplitude of their response on ammonia and LPG.
Table 3 represents data on amplitude of resistance response on acetone for all films and data on temperature of maximum response on that gas. Data for other gases are similar. The temperature of maximum response was determined by approximation of experimental results with Gauss function. The obtainedresults show that temperature of maximum response for all films differs no more than 7%. The temperature of maximum resistance response on ethanol and acetone for films with 50%In_2O_3–50%Ga_2O_3 composition is lower than that temperature for other films. Moreover, their amplitude of resistance response on those gases is significantly higher.

Table 3. Maximum of gas-induced resistance response and temperature of maximum response on acetone for In_2O_3–Ga_2O_3 films.

Composition of film	Temperature of maximum response, °C	Maximum resistance response
100%In_2O_3	566	2.8
75%In_2O_3–25%Ga_2O_3	579	6.0
50%In_2O_3–50%Ga_2O_3	528	21.0
25%In_2O_3–75%Ga_2O_3	609	2.6

3.5. Temperature dependencies of resistance response in coordinates ln(S)=f(1/kT)
The temperature dependencies of gas response on combustible gases in coordinates ln(S)=f(1/kT) have two linear regions (see figure 5 for cases of acetone and ethanol). Slope angles for those regions are proportional to activation energies of resistance response, ΔE. Table 4 represents obtained values of ΔE for all investigated films.

Table 4. Adsorption and desorption energies obtained from temperature dependencies of gas-induced resistance response on acetone for In_2O_3–Ga_2O_3 films.

Composition of film	ΔE_{LT}, eV	ΔE_{HT}, eV
100%In_2O_3	566	2.8
75%In_2O_3–25%Ga_2O_3	579	6.0
50%In_2O_3–50%Ga_2O_3	528	21.0
25%In_2O_3–75%Ga_2O_3	609	2.6

IOP Conf. Series: Journal of Physics: Conf. Series **858** (2017) 012009 doi:10.1088/1742-6596/858/1/012009

According to [1, 21] the main factor, influencing amplitude of resistance response in low temperature region (ΔE_{LT}), is difference between energy of combustible gas adsorption (E_R) and energy of oxygen adsorption (E_O):

$$\Delta E_{LT} = E_R - E_O. \qquad (2)$$

The activation energy of resistance response in low temperature region (ΔE_{HT}) is determined by the difference between energy of combustible gas desorption (q_R) and energy of its oxidation products desorption (q_{RO}) [1, 21].

$$\Delta E_{HT} = q_R - q_{RO}. \qquad (3)$$

Figure5 represents two of obtained dependencies. Maximum of amplitude of resistance response is observed for films with $50\%In_2O_3$–$50\%Ga_2O_3$ composition. Films with $75\%In_2O_3$–$25\%Ga_2O_3$ composition have 3.5 times lower amplitude of response. Amplitudes of response for other films are even lower.

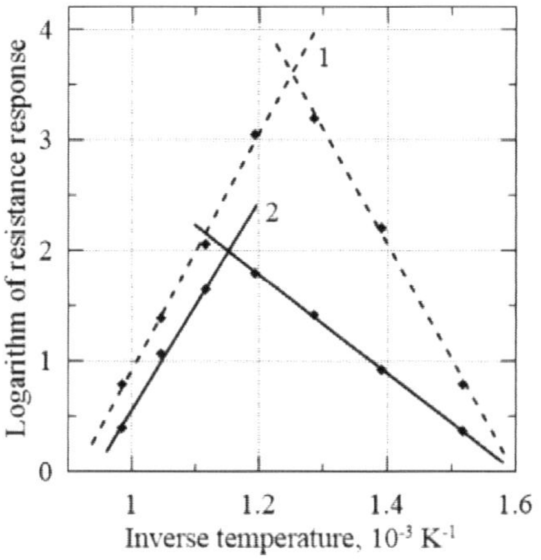

Figure 5. The dependencies of logarithm of resistance response on inverse temperature for films exposed to acetone diluted in air. 1 - $50\%In_2O_3$–$50\%Ga_2O_3$, 2 - $75\%In_2O_3$–$25\%Ga_2O_3$.

The slope angles of high temperature linear regions of the dependencies in figure 5 are approximately equal. According to (2) it indicates that $50\%In_2O_3$–$50\%Ga_2O_3$ and $75\%In_2O_3$–$25\%Ga_2O_3$ films have approximately equal difference between energy of combustible gas desorption (q_R) and energy of its oxidation products desorption (q_{RO}). The slope angles of low temperature linear regions of those dependencies are significantly different. According to (3) this is possible in two cases. In the first case energy of combustible gas adsorption (E_R) for film with $50\%In_2O_3$–$50\%Ga_2O_3$ composition is higher, than energy of that gas adsorption for film with $75\%In_2O_3$–$25\%Ga_2O_3$ composition. In the second case energy of oxygen adsorption (E_O) for film with $50\%In_2O_3$–$50\%Ga_2O_3$ composition is lower, than the same energy for film with $75\%In_2O_3$–$25\%Ga_2O_3$ composition.
The first case is not possible, because the energy of gas adsorption is connected with energy of its desorption. Those energies cannot be changed independently. However, energies of gas desorption for both films are approximately the same, based on previously made conclusion about slope angles of high temperature linear regions of the dependencies in figure5. Therefore, the difference between slope angles of low temperature linear regions of dependencies for films with $50\%In_2O_3$–$50\%Ga_2O_3$ and $75\%In_2O_3$–$25\%Ga_2O_3$ compositions is caused by the difference in their energies of oxygen adsorption.

3.6. Formation of $InGaO_3$ phase

Thus, films with $50\%In_2O_3$–$50\%Ga_2O_3$ composition have properties, which markedly differ from properties of films with other compositions. They have maximum of average grain size (table 1), considerable local maximum of optical transmittance in visible range (figure 3), highest among other films amplitude of resistance response on combustible gases (figure 4, table 3) and lowest oxygen adsorption energy (table 4).

A possible cause for all those properties may be a formation of some new objects in In_2O_3–Ga_2O_3 composite. There are several candidates for that role: $InGaO_3$ solid solution [8], rhombohedral In_2O_3 in the lattice of cubic In_2O_3 [12], area of heterojunction between the crystallites of In_2O_3 and Ga_2O_3 [3].

The most likely option is the formation of $InGaO_3$, because previously it was illustrated in [8] by the means of X-ray diffraction that In_2O_3–Ga_2O_3 films contain several phases:In_2O_3, $Ga_2O_3$and $InGaO_3$. It is obvious, that the highest concentration of $InGaO_3$ phase should be in films with composition close to 50%In_2O_3–50%Ga_2O_3. It was confirmed in [8].

4. Conclusion

Presented results demonstrate, that In_2O_3–Ga_2O_3 thin films may be considered as promising materials for sensitive layers of semiconductor gas sensors and conductive transparent layers. Production of these films by pulsed laser deposition allows precise control of their composition and thickness. Considered films are sensitive to a number of flammable gases. The highest sensitivity of the films is observed for the case of acetone and ethanol vapors. Study of effect of films composition on their properties showed, that highest gas-induced resistance response have films with 50%In_2O_3–50%Ga_2O_3 composition. In addition, these films have least temperature of maximum resistance response as compared with films with other compositions. Also, their optical transmittance in visible range have noticeable local maximum. The reason for all those phenomena may be the formation of $InGaO_3$ solid solution in films with composition close to 50%In_2O_3–50%Ga_2O_3.

Acknowledgements

This work was carried out within the project «Young scientist»;research effort № 17072B.Authors are gratefulof Omsk State Technical University for financial support.

References

[1] Korotcenkov G 2013 *Handbook of Gas Sensor Materials. Properties, Advantages and Shortcomings for Applications Volume 1: Conventional Approaches* (New York: Springer, 2013)
[2] Barsan N, Koziej D and Weimar U 2007 *Sens. Actuators, B***121**pp 18-35
[3] Gaskov A M and Rumyantseva M N 2001 *Russ. J. Appl. Chem.***74** pp 440–444
[4] Capone S, Forleo A, Francioso L, Rella R, Siciliano P, Spadavecchia J, Presicce D S and Taurino A M2003 *J. Optoelectronics Adv. Mater.***5** 1335–1348
[5] Korotcenkov G2007 *Mater. Sci. Eng. B***139**pp 1–23
[6] Lowndes D H, Geohegan D B, Puretzky A A, Norton D P and Rouleau C M1996 *Sci., New Series***273**pp 898–903
[7] Eranna G, Joshi B C, Runthala D P and Gupta R P 2004 *Crit. Rev. Solid State Mater. Sci.***29**pp 111–88
[8] Kozlov A G, Krivozubov O V, Kurdukova E A and Lila M N 2012 *Proc. 28th Int. Conf. Microelectronics (MIEL)* (Nis: University of Nis) pp 165-168
[9] Minami T, Takata S and Kakumu T 1996 *J. Vac. Sci. Technol., A***14** 1689–93
[10] Cava R J, Phillips J M, Kwo J, Thomas G A, van Dover R B, Carter S A, Krajewski J J, Peck W F, Jr., Marshall J H, and Rapkine D H1994 *Appl. Phys. Lett.***64** 2071–2
[11] Minami T, Takeda Y, Kakumu T, Takata S and Fukuda I1997 *J. Vac. Sci. Technol., A* **15** 958–62
[12] Korotcenkov G, Boris Iu, Brinzari V, Han S H, Cho B K, Lychkovsky Yu N 2015 *Ceram. Int.* **41** 7478–88
[13] Demin I E and Kozlov A G 2013 *Proc. 2013 Int. Siberian Conf. on Control and Communications (SIBCON)* (Hoboken: IEEE) 6693604

[14] Demin I E and Kozlov A G 2013 *Proc. 2015 Int. Siberian Conf. on Control and Communications (SIBCON)* (Hoboken: IEEE) 7147037
[15] Korotcenkov G2008 *Mater. Sci. Eng., R***61** 1–39
[16] G. Korotcenkov and B.K. Cho, 2017 *Sens. Actuators B***244** 182–210
[17] Hossein-Babaei F andGhafarinia V 2010 *Sens. Actuators B***143** 641–8
[18] Korotcenkov G and Cho B K 2011 *Sens. Actuators B***156** 527–38
[19] Carotta M C, Gherardi S, Guidi V, Malagù C, Martinelli G, Vendemiati B, Sacerdoti M, Ghiotti G, Morandi S 2011 *Thin Solid Films***517** 6176–83
[20] Batzill M and Diebold U 2005 *Prog. Surf. Sci.***79** 47–154
[21] Brynzari V, Korotchenkov G andDmitriev S 2000 *Electron Technology***33** 225–35

IOP Conf. Series: Journal of Physics: Conf. Series **858** (2017) 012010 doi:10.1088/1742-6596/858/1/012010

Universal size dependence of the physical properties of nanomaterials

E N Eremin[1], V M Yurov[2], S A Guchenko[2], V Ch Laurynas[2]

[1] Omsk State Technical University, 11, Mira ave., Omsk, 644050, Russia
[2] Karaganda State University named after E.A. Buketov, 28, Universitetskaya str., Karaganda, 100026, Kazakhstan

E-mail: weld_techn@mail.ru

Abstract. Dimensional analysis of the experimentally observed dependence of the physical properties of nanoparticles, nanofilms and nanomaterials showed that there is a universal equation that accurately describes the observed size effects. It is shown that the size factor is also a universal value and is determined only by the atomic structure of the nanomaterial. Discovered universal relationships enable us to calculate the physical properties (mechanical, electrical, magnetic, thermal, etc.) of small particles and thin films based on knowledge of the properties of bulk materials.

1. Introduction

A characteristic feature of the nanoparticles and nanomaterials is the size dependence of their physical properties. To distinguish two types of size effects [1]: own, or internal, associated with changes in properties such as melting temperature, heat capacity, and others, and the external, which is the size-dependent response to the external action of the forces that are independent of the internal effect (electric conductivity, magnetic susceptibility, and others).

2. Formulation of the problem

Currently there is no quantitative theory that allows to predict the desired properties of nanomaterials depending on the technological conditions. Therefore, even high quality models may help to avoid a large amount of routine, but expensive works.

In this paper we want to show that most of both internal and external dimensional effects described by one universal dependence. This makes it possible to calculate the physical properties (mechanical, electrical, magnetic, thermal, etc.) of small particles and thin films based on knowledge of the properties of bulk materials.

3. Theory

The physical nature of reducing the melting point of the particles with decreasing size is as follows: with a decrease in particle size increases the proportion of surface atoms, which is less than the binding energy of atoms than in volume, so that less energy is required for the thermal motion of the phase transition solid-liquid. However, to find a quantitative relation between the melting point of the nanoparticles and their size was not easy [2]. The melting point of the most accurate for the gold

IOP Conf. Series: Journal of Physics: Conf. Series **858** (2017) 012010 doi:10.1088/1742-6596/858/1/012010

nanoparticle size dependence was defined in the pioneering work [3]. A similar relationship was observed by many authors later for some other metal nanoparticles.
It was found that the experimental curves are well described by the equation [4]:

$$T_m(r) = T_\infty\left(1 - \frac{d}{r}\right), \tag{1}$$

where T_∞ - the melting point of the bulk sample.
The parameter d is determined by the formula [4]:

$$d = \frac{2\sigma\upsilon}{RT}. \tag{2}$$

Here σ - the surface tension of the bulk sample; u - molar volume, R - universal gas constant.
When r < d, instead of (1) need to use the equation:

$$T_m(r) = T_\infty\left(1 - \frac{d}{d + r}\right). \tag{3}$$

Equation (1) allows you to experimentally determine the surface tension of the solid-state σ [5].
Processing of the experimental curves by means of (1) has shown that for metals with great precision the relation:

$$\sigma = 0{,}7 \cdot 10^{-3} \cdot T_\infty. \tag{4}$$

In other words, between the value of the surface tension and pure metal melting point there is a simple universal dependence (4). Note that the relationship (4) was also obtained in [6] in another way.
When r>d melting temperature tends to the value characteristic of the bulk sample. Therefore, the parameter d is naturally called the thickness of the surface layer of metal.
From (2) and (4):

$$d = 2{,}41 \cdot 10^{-4} \cdot \upsilon \ (м^{-2} \cdot моль) \tag{5}$$

According to the (5), the thickness of the surface layer is defined by a single parameter - υ, i.e. the atomic structure of the metal. Equation (5) also allows you to estimate the thickness of the surface layer of metal on the known value of u. For example, for gold υ = 10,2 cm3/mol and the ratio of (5) gives d = 2,45 nm.
In [5] to the surface tension of the nanocrystal obtained by the following expression:

$$\sigma(r) = \sigma_\infty\left(1 - \frac{d}{r}\right), \tag{6}$$

where σ∞ = σ - the surface tension of the bulk sample, the parameter d has the same meaning as in the previous cases.
Formula (6) in the form coincides with the formula Gibbs - Tolman - Koenig - Buff:

$$\sigma(r) = \sigma_\infty\left(1 - \frac{2\delta}{r}\right), \tag{7}$$

where δ is called the Tolman constant or Tolman length [6].
It is believed that this option can not be determined experimentally. In this case, a constant Tolman δ = d/2, which can be determined experimentally. Formula (6) coincides with the mathematical formula (1).

4. Experimental results

Fig. 1 shows the size dependence of the luminescence intensity of the KCl-Tl phosphor, and Fig. 2 - size dependence of the magnetic susceptibility of magnetite.

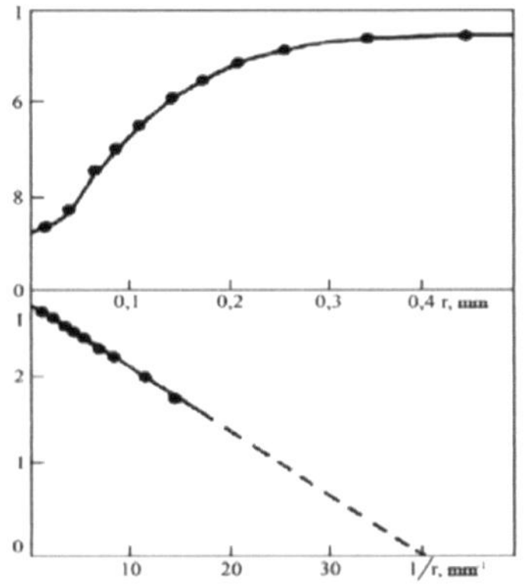

Figure 1. The dependence of KCl-Tl luminescence intensity of the phosphor grain size [7].

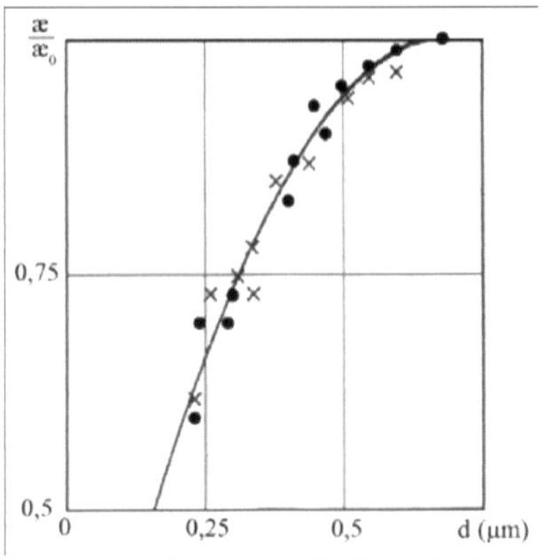

Figure 1. The dependence of the magnetic susceptibility of the diameter of the magnetite grains [8].

In both these cases, depending describe with great accuracy the formulas:

$$I = I_0\left(1 - \frac{d}{r}\right), \quad \chi = \chi_0\left(1 - \frac{d}{r}\right). \tag{8}$$

Here the parameter d is calculated by formula (2) and determines the corresponding surface of the material layer thickness. In the coordinates I/I0 ~ 1/r dependence in Fig. 1 and 2 give a straight tangent of the angle which is equal to d. This makes it possible to determine the surface tension of the dielectric and magnetic materials.

5. Discussion

For most metals, the melting temperature of nanoparticles of various sizes are presented in Table 1.

Table 1. The melting point of pure metal nanoparticles (M - metal)

M	T∞, K	Tm, K r = 1 nm	Tm, K r = 10 nm	Tm, K r = 50 nm
Li	452	188,3	396,5	440,1
Na	371	119,7	306,6	356,1
K	337	71,7	246,1	314,1
Rb	312	60,0	219,7	287,7
Cs	302	48,7	198,7	273.6
Be	1558	556,3	1320,4	1558
Mg	923	225,1	704,6	869,1

IOP Conf. Series: Journal of Physics: Conf. Series **858** (2017) 012010 doi:10.1088/1742-6596/858/1/012010

Ca	1118	139,8	657,6	980,7
Sr	1030	110,8	562,8	883,6
Ba	983	99,3	520,1	834,5
Al	933	291,6	764,8	933
Ga	302,8	159,3	277,6	297,3
In	429	165,0	369,8	415,7
Tl	576	169,4	464,5	549,6
Si	1686	285,8	1131,5	1535,9
Ge	1231	246,2	879,3	1140,0
Sn	505	168,3	420,8	485,6
Pb	600	166,7	600,0	600,0
Se	493	170,0	414,3	475,0
Te	725	161,1	537,0	676,7
Cu	1356	411,0	1102,4	1296,4
Ag	1234	301,0	942,0	1162,0
Au	1336	310,7	1004,5	1253,3
Zn	693	277,2	602,6	672,8
Cd	594	204,8	499,2	572,3
Hg	234	127,7	216,1	230,2
Cr	2173	452,7	1574,6	2019,5
Mo	2873	383,1	1741,2	2542,5
W	3673	390,7	1996,2	3144,7
Mn	1517	399,2	1185,2	1436,6
Tc	2473	405,4	1637,7	2244,1
Re	3423	422,6	2001.8	2997,4
Fe	1808	441,0	1380.2	1702,5
Co	1763	464,0	1377,3	1670,0
Ni	1726	466,5	1359,1	1637,6
Ce	1077	168,3	699.4	972,0
Pr	1208	172,6	755,0	10,78,6
Nd	1298	175,4	791,5	1150,7
Sm	1325	181,5	812,9	1176,7

From Table 1 it is seen that the melting temperature of nanoparticles with sizes of 1 nm is much lower than the melting temperature of the bulk sample. Part of the metal at such sizes are thermally unstable at room temperature.

Table 2 shows the values of d, calculated by the formula (5). From Table 2 it is seen that the thickness of the surface layer of pure metals is not more than 10 nm. This means that the surface layer is a pure metal nanostructure.

Consider the account of the size dependence of the physical properties of small particles on the example of the theory of homogeneous nucleation of a new phase.

Table 2. Surface layer thickness d pure metals

M	d, nm	M	d, nm	M	d, nm
Li	0,7	Sr	5,8	Sn	1,4
Na	1,5	Ba	6,2	Pb	1,8
K	2,6	Al	1,5	Se	1,3
Rb	2,9	Ga	0,6	Te	2,5
Cs	3,6	In	1,1	Cu	1,6
Be	1,3	Tl	1,9	Ag	2,2

Mg	2,2	Si	3,4	Au	2,4
Ca	4,9	Ge	2,8	Zn	1,1
Cd	1,3	Fe	2.2	Gd	5.3
Hg	0,6	Co	2,0	Tb	5,3
Cr	2,7	Ni	1,9	Dy	5,3
Mo	4,6	Ce	3,8	Ho	5,5
W	5,8	Pr	4,2	Er	5,5
Mn	2,0	Nd	4,5	Tm	5,2
Tc	3,6	Sm	4,4	Yb	4,6
Re	4,6	Eu	5,8	Lu	5,7

The main provisions of the theory of formation of new phase nuclei were advanced by Gibbs and then developed by Volmer, Becker and Doering, Stransky and Kaishev, Frenkel, Zeldovich et al. [9]. Despite various modifications and refinement of classical nucleation theory, and it is quite correct qualitative description of nucleation process, it is still far from matching the experimental and theoretical data.

According to the classical theory of nucleation of a new phase in the old metastable phase is considered as a fluctuation process. In the case of the crystallization of a supercooled melt critical radius of a nucleus can be expressed in terms of supercooling ΔT [9]:

$$r_{\kappa} = 2M\sigma T_0/\rho q\Delta T, \qquad (9)$$

where M - molecular weight; ρ - the density of crystal-nucleus; q - heat of fusion; T_0 - temperature equilibrium of the two phases of infinitely large radius, σ - interfacial tension of crystal-nucleus.

In most studies it is considered that homogeneous nucleation of metal crystals begins with the supercooling ΔT = 0.2Tm of the melting temperature. Assuming that T0 = Tm, we obtain:

$$r_{\kappa} = 10M\sigma/\rho q. \qquad (10)$$

All values in the expression (11) (except σ) are determined experimentally with great precision and are listed in directories. For pure metals surface tension value σ0 is defined in [4]. The interfacial surface tension at the crystal - melt is approximately equal to: σ ≈ 1/3σ0. Then the calculation of the classical formula (11) gives a value for the critical radius of the order of a micron or a fraction of that in the three orders of magnitude the experimentally observed values.

In [10] takes into account the size dependence of all the quantities in the equation (11) and calculated the critical radius of nucleation of solid phase of pure metals (Table 3).

Table 3. Critical radius of homogeneous formation of the pure metals [10]

M	rк, nm	M	rк, nm	M	rк, nm	M	rк, nm
Li	0,5	Sr	2,8	Sn	0,7	Cd	0,6
Na	0,7	Ba	2,9	Pb	0,8	Hg	0,3
K	1,2	Al	0,7	Se	0,6	Cr	1,3
Rb	1,4	Ga	0,3	Te	1,2	Mo	2,2
Cs	1,7	In	0,5	Cu	0,8	W	2,8
Be	0,6	Tl	0,8	Ag	1,0	Mn	0,9
Mg	1,0	Si	1,6	Au	1,1	Fe	1,0
Ca	2,3	Ge	1,3	Zn	0,5	Co	0,9

Table 4 shows the experimental values of the critical range from [11].

Table 4. Critical radius of homogeneous formation of the pure metals [11]

M	rк, nm	M	rк, nm

Ga	0,9	Ge	0,9
Sn	1,2	Ag	1,2
Cu	1,1	Pb	1,1
Bi	0,8	Hg	0,8

Compare values from Tables III and IV show good agreement, which speaks in favor of the developed approach.
From the above discussion, it follows that in the classical theory does not take into account the size dependence of the surface tension, the melting point and other physical properties of small particles. Therefore, the discrepancy between theory and experiment is so impressive, despite the fact that the approach is based on fluctuations of no objection to the theory.

6. Conclusion
From the above it follows universal dependence of physical properties of A (r) nanomaterial on its size:

$$A(r) = A_{\infty}\left(1 - \frac{d}{r}\right). \tag{11}$$

Discovered universal relationships enable us to calculate the physical properties (mechanical, electrical, magnetic, thermal, etc.) of small particles and thin films based on knowledge of the properties of bulk materials.

The work was supported by a grant from the RSF № 17-19-01224.

References
[1] Kreibig U, Vollmer M 1995 Optical Properties of Metal Clusters. Berlin, Heidelberg: Springer Verlag 532.
[2] Makarov G.N. 2010 *Advances of Physical Sciences* **180 (2)** 185–207
[3] Buffat Ph, Borel J P 1976 *Phys. Rev. A* **13 (6)** 2287–2298
[4] Yurov V M 2012 *Bulletin of the University. Physics* **3 (67)** 27–35
[5] Yurov V M, Ibraev N H, Guchenko S A 2011 *Russian Physics Journal* **54/3** 335–340
[6] Rekhviashvili S S, Kishtikova E V, Karmokova R Y et al. 2007 *Technical Physics Letters* **33 (2)** 1–7
[7] Aluker E D, Lucys D M, Chernov S A 1979 Electronic excitation and radioluminescence alkali halide crystals Riga: Zinatne 251.
[8] Yurov V M 2009 *Modern problems of science and education* **4** 152–156
[9] Danilov A I, Polukarov Y M 1987 *Russian Chemical Reviews* **56 (7)** 1082–1104
[10] Yurov V M 2015 *Physico-chemical aspects of the study of clusters, nanostructures and nanomaterials* **7** 548–554
[11] Skripov V P, Kaverda V P 1987 *Problems of modern crystallography* 232–246

Nanocrystalline coatings properties forecasting

E N Eremin[1], V M Yurov[2], S A Guchenko[2], V Ch Laurynas[2]

[1] Omsk State Technical University, 11, Mira ave., Omsk, 644050, Russia
[2] Karaganda State University named after E.A. Buketov, 28, Universitetskaya str., Karaganda, 100026, Kazakhstan

E-mail: weld_techn@mail.ru

Abstract. The paper considers various properties of nanocrystalline coatings. The methods of determining the surface tension of the deposited coating on the basis of the size dependence of their physical properties. It is shown that predict the mechanical properties of the coatings, their melting point, heat resistance, wear resistance, corrosion resistance, etc. It can be based on a theoretical evaluation of the surface tension.

1. Introduction

In the literature reviewed various methods of depositing nanostructured films [1]. However, most of the work is devoted to the traditional methods of film deposition: vapor deposition and plasma. The most promising for nanostructured coatings are vacuum ion-plasma methods: magnetron sputtering, ion and vacuum arc deposition. This is due to the fact that in addition to thermal factors there are other - high degree of ionization and the energy flux density of the particles.

Although methods for producing nanostructured materials and coatings are quite diverse, but they all are based on the mechanism of energy dissipation of intense, generalized in the three stages of formation. The first stage is a process of nucleation, which is due to lack of appropriate thermodynamic conditions, will not go into mass crystallization. The second step is the formation of nuclei around nanocrystalline amorphous clusters, which - in a third step - are combined in the intergranular phase to form a dissipative structure.

Each of these steps is a complex process. Suffice it to point out the process of nucleation of a new phase, a theory which has been developing for more than 100 years old, and wasps main points of which were laid by Gibbs and then developed by Volmer, Becker and Doering, and Stransky Kaishevym, Frenkel, Zeldovich.

Even more difficult is the third stage of the formation of the coating, when the growth of the film is determined by the condition, on the mobile interface. Such problems are called «problem Stefan» [2]. The above difficulties do not allow us to construct a rigorous theory of the formation of plasma coatings to date, including nanocrystalline coatings, where, along with the classic problems arise and the problems of the account size effects in the formation of new phase nuclei.

All this affects the prediction of plasma coating properties [3].

2. Formulation of the problem

In recent years, the concept has evolved high-entropy or multi-element alloys and coatings based on them [4].

IOP Conf. Series: Journal of Physics: Conf. Series **858** (2017) 012011 doi:10.1088/1742-6596/858/1/012011

The stability of the structure and composition, as well as high performance systems high-entropy create very attractive possibility of the formation based on these coatings in order to improve the surface characteristics or use them as protective films that prevent harmful impurities in the surface layers.
However, the quantitative theory of these effects yet. In the present work shows the possibility of predicting certain important properties of nanostructured coatings multielement.

3. Theory

In [5, 6], we have demonstrated the possibility of experimental determination of the surface tension of coatings deposited by the size dependence of their properties.
We used the method of measuring the surface tension by determining the dependence of the microhardness on the thickness of the deposited coating. The dependence of the microhardness of the deposited coating on its thickness is described by the formula [5]:

$$\mu = \mu_0 \cdot \left(1 - \frac{d}{h}\right), \tag{1}$$

where μ - micro hardness of deposited coatings; μ_0 - micro hardness «fat» sample; h - thickness of the deposited coating.

The parameter d is associated with a surface tension σ formula [5]:

$$d = \frac{2\sigma\upsilon}{RT}, \tag{2}$$

Here σ - the surface tension of the bulk sample; u - volume of one mole; R - gas constant; T - temperature.
The coordinates $\mu \sim 1/h$ ($1/h$ - reverse deposited coating thickness) obtained straight line the slope of which determines the d, and formula (2) is calculated as the surface tension of the deposited coating (σ).
One can use any size dependence of the properties of A(r) coatings (electrical conductivity, magnetic susceptibility, etc.), which are described by the equation:

$$A(r) = A_0 \cdot \left(1 - \frac{d}{h}\right). \tag{3}$$

In [7] defined by the surface tension of the pure metals. The results are shown in Table 1.

Table 1. Surface tension of pure metals (M) at a temperature of 300 K [7]

M	σ, J/m^2	M	σ, J/m^2	M	σ, J/m^2
Li	0,152	Sr	0,730	Sn	0,205
Na	0,071	Ba	0,683	Pb	0,300
K	0,037	Al	0,633	Se	0,193
Rb	0,012	Ga	0,003	Te	0,425
Cs	0,002	In	0,129	Cu	1,056
Be	1,258	Tl	0,276	Ag	0,934
Mg	0,623	Si	1,386	Au	1,036
Ca	0,818	Ge	0,931	Zn	0,399
Cd	0,294	Fe	1,508	Gd	1,285
Hg	0,07	Co	1,463	Tb	1,331
Cr	1,873	Ni	1,426	Dy	1,380
Mo	2,573	Ce	0,777	Ho	1,434

W	3,373	Pr	0,908	Er	1,470
Mn	1,217	Nd	0,998	Tm	1,518
Tc	2,173	Sm	1,025	Yb	0,797
Re	3,123	Eu	0,875	Lu	1,625

4. Experimental results

The surface tension of the coating Zn-Cu-Al was determined by the method described above. A quantitative analysis of the elemental composition of the composite coating was carried out on JEOL JSM-5910 electron microscope. Results stoichiometry calculation gave the following result: $Zn_{0,59}Cu_{0,01}Al_{0,40}$. The mean value of the surface tension is equal to: $\sigma_{Zn\text{-}Cu\text{-}Al}$ = 0, 351 J/m^2. The surface properties of coatings in most cases correspond to the principle of additivity. Then, the condition:

$$\sigma_{Zn_{0,59}Cu_{0,01}Al_{0,40}} = 0{,}59\sigma_{Zn} + 0{,}01\sigma_{Cu} + 0{,}40\sigma_{Al}. \quad (4)$$

Using the data in Table I for σ_{Zn}, σ_{Cu} and σ_A, obtain $\sigma_{Zn\text{-}Cu\text{-}Al}$ = 0, 412 J/m^2. Experimental and theoretical values for the coating Zn-Cu-Al were close to each other: ($\sigma_{Zn\text{-}Cu\text{-}Al}.\approx$ 0,4 J/m^2). Similar measurements were carried out for a number of multi-coatings. The equation of the type (4) turned out to be true for all investigated coatings.

This result leads to the following important conclusion: using the additive property of the surface tension, it is possible to select components of the coating so as to increase or decrease the surface tension of the coating compared to pure metals.

Experimental study of mechanical properties of nanomaterials and nanostructured coatings showed that the tensile strength, hardness, many metals (Pd, Cu, Ag, Ni et al.) is significantly higher than the corresponding bulk analogs [8]. Increased strength and hardness with decreasing grain size up to a certain critical size is typical for virtually all crystals. This follows from the known Hall-Petch equation that yield strength σ_T depends inversely on the average grain size d [8]:

$$\sigma_T = \sigma_M + kd^{-1/2}, \quad (5)$$

where σ_M- the limit strength of the single crystal, k - coefficient of some dimension.

Typically, Hall-Petch relationship (5) is performed for a large part of the nanomaterials studied only up to a certain grain size, while at lower values it is bucking the effects of hardness (strength) decreases with decreasing grain size.

Despite the large number of studies on the influence of the size factor on the mechanical properties of nanostructures, the physical mechanisms of this effect remain the subject of ongoing discussions. A comprehensive review of the problem is given in [8]. In [9], an equation for the yield point:

$$\sigma_T = \sigma_M + C\sigma d^{-1/2}. \quad (6)$$

Equation (6) coincides in shape with the Hall-Petch equation (5). However, the coefficients of proportionality differ in both formulas. In the case of the equation (6) the behavior of the yield strength of small particles is also determined by the size of their surface tension σ.

For small d Rusanov got asymptotic linear relationship [10]:

$$\sigma = Kd. \quad (7)$$

Here K - coefficient of proportionality.

Formula (7) is obtained on the basis of thermodynamic consideration and should be applicable to small objects of various nature.

In this case, the equation (6) takes the form:

$$\sigma_T = \sigma_M + CKd^{1/2}. \tag{8}$$

Equation (8) represents the opposite effect Hall-Petch.
From equation (6) that the Hall-Petch equation begins to be violated from the moment when it begins to appear the size dependence of the surface tension.

5. Discussion

In [11] it is shown that for pure metals with high accuracy, the relation:

$$\sigma = 0,7 \cdot 10^{-3} \cdot T_m, \tag{9}$$

where T_m - melting point metal.

Taking into account the additive coating the surface tension, the equation (9) can be rewritten as:

$$\grave{O}_{\text{ïë}} = 1,4 \cdot 10^3 \cdot \sigma \text{ (K)}. \tag{10}$$

As an example, Table 2 shows the melting temperature multielement coatings obtained while spraying us various cathodes.

Table 2. Melting point multielement coatings obtained in argon

Coating	T_m, K	Coating	T_m, K
12X18H10T+Zr	1358	12X18H10T+Zn-Al	1537
12X18H10T+Zn-Cu-Al	1530	12X18H10T+Al	1602
12X18H10T+Fe-Al	1809	12X18H10T+Cu	2023

Steel melting temperature depends on their chemical composition, but is in the range (1450 - 1520) K. As seen from Table 3 the coating 12Cr18Ni10Ti+Al, 12Cr18Ni10Ti+Fe-Al and 12Cr18Ni10Ti+Cu on the melting temperature significantly superior to many steel.

Currently refractory coating deposition is carried out mainly by arc or magnetron vacuum methods. Thus, various target compounds comprising metals such as chromium, titanium, zirconium in combination with non-ferrous metals. We investigated the coating deposited by ion-plasma method while spraying steel cathode 12Cr18Ni10Ti and composite cathodes. Method for determination of heat resistance is based on GOST 6130-71. The experimental results are presented in the table 3.

Table 3. Loss of coating weight, resulting in an argon atmosphere, after heat treatment at 600 ° C for 100 hours

Coating	Weight oxidized coating, mg
A sample of uncoated, steel 45	56,8
12Cr18Ni10Ti+Zr	24,4
12Cr18Ni10Ti+Zn-Cu-Al	14,4
12Cr18Ni10Ti+Fe-Al	5,6
12Cr18Ni10Ti+Zn-Al	14,2
12Cr18Ni10Ti+Al	4,8
12Cr18Ni10Ti+Cu	56,8

From a comparison of the results of Tables III and I, the conclusion: the greater the surface tension of the coating, the greater its heat resistance. If the heat resistance mark - ζ, then said mathematically can be expressed as a functional relationship:

$$\varsigma = f(\sigma) = C \cdot \sigma, \tag{11}$$

where C - a constant.
Anti-corrosion coatings obtained by magnetron method considered in [11-14]. In cases where the corrosion process proceeds as general corrosion, a change in the amount of metal can be used for evaluating corrosion rates in a process changing the amount of the reaction agent (oxidant), or one of the products of corrosion over time. Since the corrosion process is heterogeneous, the appropriate quantitative characteristics should be attributed to the surface of the unit.

Table 4 shows the corrosion rate at 600 °C studied coatings, which was determined by the formula:

$$v_{\kappa} = \Delta m / S \cdot t, \tag{12}$$

where Δm - decrease (increase) in weight; S - area of the sample; t - time.
From the comparison results tables IV and I, the conclusion: the greater the surface tension of the coating, the greater the corrosion resistance. If corrosion resistance mark - χ, then said mathematically be written as:

$$\chi = f(\sigma) = C_1 \cdot \sigma, \tag{13}$$

where C_1 - a constant.

Table 4. The corrosion rate at 600 ° C the coating obtained in argon

Coating	The corrosion rate, g/m^2 h
A sample of uncoated, steel 45	2,84
12Cr18Ni10Ti+Zr	1,22
12Cr18Ni10Ti+Zn-Cu-Al	0,72
12Cr18Ni10Ti+Fe-Al	0,28
12Cr18Ni10Ti+Zn-Al	0,71
12Cr18Ni10Ti+Al	0,24

The wear resistance of the coating is determined by the work of her destruction, which is equal to:

$$A = \sigma \cdot S, \tag{14}$$

where S - area of the sample surface.
As an example, in Table 5 shows the results of calculation of the fracture coating (S = 1 m^2) from the experimental value of the surface tension σ.

Table 5. Job destruction of some multi-element coatings

Coating	Energy destruction coating, J.
Zn–Cu–Al	2,43
Cr–Mn–Si–Cu–Fe–Al	7,11
Mn–Fe–Cu–Al	3,67

6. Conclusion

It follows from the above results that the majority of properties of the coatings is determined by their surface tension (surface energy). Using Table 1 and obtained in the ratio, it is possible to predict the performance properties of coatings based on their functionality.
Currently there is no quantitative theory that allows to predict the desired properties of the coating, depending on the technological conditions, the materials used and the target cathodes, etc. Therefore,

even high quality models proposed in this paper can help prevent a large volume of routine, but expensive works.

The work was supported by a grant from the RSF № 17-19-01224.

References

[1] Reshetnyak E N, Strel'nitskij V E 2008 *Problems of Atomic Science and Technology* **2** 119–130.
[2] Gupta S C 2003 The Classical Stefan Problem: Basic Concepts, Modelling and Analysis, Amsterdam: Elsevier 385.
[3] Verma N, Cadambi S, Jayarama V, Biswas S K 2012 *Acta Materialia* **60** 3063–3073.
[4] Otto F, Yang Y, Bei H, George E P 2013 *Acta Materialia* **61** 2628–2638.
[5] Yurov V M, Laurynas V Ch, Guchenko S A, Zavatsky O N 2014 *Reinforcing technology and coating* **1** 33–36.
[6] Eremin E N, Syzdykova A Sh, Guchenko S A et al. 2016 *IOP Conf. Series: Materials Science and Engineering* **110** art. 012017.
[7] Yurov V M 2011 *Eurasian Physical Technical journal* **8** **1(15)** 10–14.
[8] Malygin G A 2011 *Advances of Physical Sciences* **181** **(11)** 1129–1156.
[9] Yurov V M, Laurynas V Ch, Guchenko S A 2013 *Physico-chemical aspects of the study of clusters, nanostructures and nanomaterials* **5** 408–412.
[10] Rusanov A I 1967 Phase equilibria and surface phenomena, L.: Chemistry 346.
[11] Lorpez D, Alonso Falleiros N, Tschiptschin A P 2011 *Tribology International* **44** 610–616.
[12] Liu R, Li X, Hu X, Dong H 2013 *Surface and Coatings Technology* **232** 906–911.
[13] Jinlong Lv, Hongyun Luo 2013 *Surface and Coatings Technology* **235** 513–520.
[14] Lee Y J, Lee T H, Kim D Y et al. 2013 *Surface and Coatings Technology* **235** 819–826.

Geometrical specifications accuracy influence on the quality of electromechanical devices

V I Glukhov [1], M N Lakeenko [2] and S N Dolzhikov [3]

[1] Omsk State Technical University, 11, Mira ave., Omsk, 644050, Russia
[2] Department Experimental tool factory "Transport", 2A, Baranova st., Omsk, 644023, Russia
[3] Omsk State Transport University tracks messages, 35, Marksa ave., Omsk, 644046, Russia

E-mail: mips@omgtu.ru

Abstract. To improve the quality of electromechanical products is possible due to the geometrical specifications optimization of values and tolerances. Electromechanical products longevity designates the rolling-contact bearings of the armature shaft. Longevity of the rolling-contact bearings is less than designed one, since assembly and fitting alter gaps, sizes and geometric tolerances for the working parts of the basic rolling bearing details. Geometrical models of the rolling-contact bearing details for the armature shaft and the end shield are developed on the basis of an electric locomotive traction motor in the present work. The basic elements of the details conjugating with the adjacent details and materializing the generalized and auxiliary coordinate systems are determined. Function, informativeness and the number of geometrical specifications for the elements location are specified. The recommendations on amending the design documentation due to geometrical models to improve the accuracy and the quality of the products are developed: the replacement of the common axis of the shaft's technological datums by the common axis of the basic design datums; coaxiality tolerances for these design datums with respect to their common axis; the modifiers for these auxiliary datums and these datums location tolerances according to the principles of datums uniformity, inversion and the shortest dimension chains. The investigation demonstrated that the problem of enhancing the durability, longevity, and efficiency coefficient for electromechanical products can be solved with the systematic normalizations of geometrical specifications accuracy on the basis of the coordinate systems introduced in the standards on geometrical product specifications (GPS).

1. Introduction

The quality improvement of electromechanical products is a complex problem including the issue of increasing service time for electrical and mechanical parts. The issues of increasing service time for electrical parts are studied by a wide range of the scientists and not discussed in the present work. Whereas the issues of increasing the service time for mechanical parts are of design, technological and metrological nature. All investigations in this field are associated with the following scientists: Glukhov V I [1], Kovalev V P , Narodetsky M Z [2], Bell J C, Dyson A [3], Blok H [4], Schabel J S [5], and Tikvicki M [6].

Let us describe the way to construct graphical geometrical models [1] on the example of two details, i.e. an armature shaft and a traction motor end shield of a traction motor for an electric locomotive. The construction schematic of a traction motor is given in Figure 1. A traction motor is one of the most expensive parts of an electric locomotive. Therefore, to cut the number of unscheduled repairs and to increase the periods between them is a topical problem.

2. Problem statement

To investigate the parts convergence and optimization of their geometrical specifications accuracy, the following tasks were solved herein:

- the geometrical models of an armature shaft and an end shield of a traction motor were developed;
- function, informativeness and the number of geometrical specifications (initial errors) for dimensions, location and form of the elements in a generalized coordinate system of the detail are specified;
- the recommendations on making changes to geometrical specifications in the design documentation due to geometrical models of the details are developed.

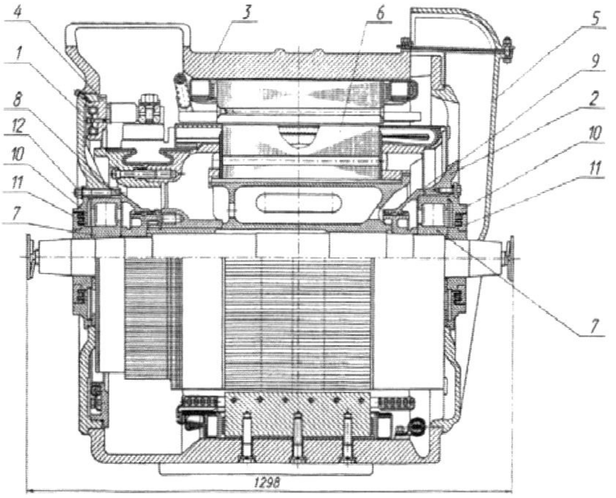

1 - end shield on the collector's side; 2 - second end shield; 3 - casing; 4 - traverse; 5 - jacket; 6 - armature; 7 - motor armature rolling bearing; 8, 9 - bearing cup, inner ; 10 - bearing cup, outer; 11 - rubber ring

Figure 1. Construction schematic of traction motor for electric locomotive

3. Geometrical model of the traction motor armature shaft

3.1. The shaft elements function is determined by its working drawing and assembly drawing of a traction motor. Shaft and armature 6 rotate on two rolling bearings 7 (Figure 1). The location of the shaft in a traction motor is determined by the axis of cylindrical elements *G2* and *I2*, each having informativeness 2. Two coaxiality datums *G2* and *I2* form together common basic axis *GI4* with informativeness 4 since they deprive the shaft of four degrees of freedom. Flat frontal datum *B1* with informativeness 1 is a dimensional supporting datum, and it deprives the shaft of one translation along the basic axis *GI4*. The plane of symmetry for a keyslot on the adjoining taper shank is a supporting datum depriving the shaft of one rotation around the common basic axis of shaft *GI4*. A set of basic datums deprives the shaft of six degrees of freedom: three linear translations and three angular rotations.

A set of basic datums forms a complete generalized coordinate system *OX2YZ4* of the shaft. Axis *Z4* is a common axis of basic cylindrical elements *G2* and *I2*. It comes through two centers of their midsections C_1 and C_2, and has informativeness 4. Origin point *O* is located on the intersection of axis *Z4* with flat supporting datum *B1*. Coordinate plane *ZOX* passes through axis *Z4* and the center of the symmetry plane of the keyslot and materializes axis *X2* with informativeness 2. Axis *Y* has informativeness being equal to 0 (zero) since it needs no supplementary datum. Axis *Y* starts from origin point *O* in perpendicular to *Z4* and *X2* of the generalized coordinate system *OX2YZ4.*

3.2. The axis of the auxiliary coordinate systems is found. A set of auxiliary design datums to press an armature socket is represented by the axis of basic cylindrical elements *J2* and *N2.* The datums form common basic axis *JN4* with informativeness 4 and deprive the armature socket of 4 degrees of freedom. The armature socket is deprived of another two movements along and around the axis *JN4* because of its tight fit on the shaft. A complete set of auxiliary datums is supplemented by a frontal datum collar *E2,* closest to origin point *O,* and basic design datum of the keyslot together with datum *JN4* materializing auxiliary coordinate system *O'X'2Y'Z'4.*

A set of auxiliary design datums for the attached cogwheel from the side of the collector is represented by the axis of conical surface *Q5* with informativeness 5 depriving the cogwheel of 5 movements and the supporting base, i.e. the plane of symmetry for the keyslot with informativeness 1 deprives the cogwheel of one rotation around axis *Q5*. The same prismatic datum of the keyslot functions both as basic and auxiliary design datums and mediates between two sets of datums and their coordinate systems - generalized *OX2YZ4* and auxiliary *O''X''2Y''Z''4.* A complete set of auxiliary datums materializes a complete auxiliary system *O''X''2Y''Z''4.* The similar set of datums materializes an auxiliary coordinate system *O'''X'''2Y'''Z'''4* for the second cogwheel.

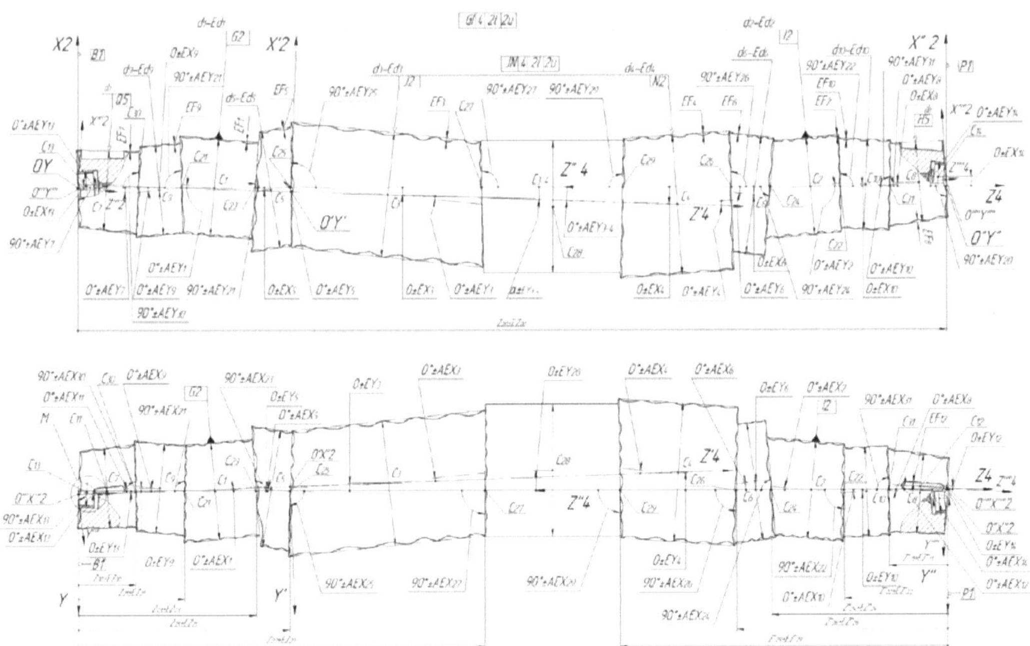

Figure 2. Geometrical model of the shaft: a) in plane *OX2Z4;* b) in plane *OYZ4*

3.3. Thus, as a generalized coordinate system for an armature shaft we choose coordinate system *OX2YZ4* from a set of basic design datums. An auxiliary coordinate system of the first order is represented by coordinate system *O'X'2Y'Z'4,* materialized by the basic shaft elements intended for pressing an armature socket. Auxiliary systems of the second order *O''X''2Y''Z''4* and *O'''X'''2Y'''Z'''4* are materialized by conic surfaces for cogwheels and their keyslots. Figure 2 demonstrates the graphical model of an armature shaft for a traction motor. The geometrical model specifications of the shaft are given in Table 1 representing a table form of the shaft geometrical model.

3.4. The number and the type of the elements location initial errors are determined by the number and the kind of degrees of freedom not used by the element under consideration for a generalized coordinate system of the detail. This rule is applied both to basic (BD) and auxiliary datums (AD).

Table 1. The geometrical model parameters for the shaft

Number and modifier, the detail element, and the coordinate system	Type, function, informativeness and the number of initial errors in element location	Dimensions and initial errors of location, dimensions and forms	Datums, dimensions, and location, dimension and form tolerances due to drawing	Suggestions, design parameters, and variants on the geometrical model basis
1 *G2* *OX2YZ4*	Cyl, BD, 2l 2l+2a+2l=2a	0°±AEX$_1$ 0°±AEY$_1$ d$_1$-Ed$_1$ EF$_1$	↗ 0,08 | L ⌭ 0,015 ○ 0,015	▶— G2 ↗ 0,02 | G/ ≡ 0,006 ○ 0,006
2 *I2* *OX2YZ4*	Cyl, BD, 2l 2l+2a+2l=2a	0°±AEX$_2$ 0°±AEY$_2$ d$_2$-Ed$_2$ EF$_2$	↗ 0,08 | L ⌭ 0,015 ○ 0,015	▶— I2 ↗ 0,02 | G/ ≡ 0,006 ○ 0,006
3 *J2* *OX2YZ4* C$_{3-4}$	Cyl, AD, 0d 2l+2a-0 = 2l+2a	0±EX$_3$ 0±EY$_3$ 0°±AEX$_3$ 0°±AEY$_3$ d$_3$-Ed$_3$ EF$_3$	↗ 0,08 | L ⌭ 0,02 ○ 0,02	◁— J2 ◎ 0,01 | G/ // 0,01 | G/ ≡ 0,006 — 0,006 ○ 0,006
4 *N2* *OX2YZ4* Z'4	Cyl, AD, 0d 2l+2a-0 = 2l+2a	0±EX$_4$ 0±EY$_4$ 0°±AEX$_4$ 0°±AEY$_4$ d$_4$-Ed$_4$ EF$_4$	↗ 0,08 | L ⌭ 0,02 ○ 0,02	◁— N2 ◎ 0,01 | G/ // 0,01 | G/ ≡ 0,006 — 0,006 ○ 0,006
7 C$_7$ *OX2YZ4*	Con, AD, 0d 3l+2a-0 = 3l+2a	0±EX$_7$ 0±EY$_7$ 0°±AEX$_7$ 0°±AEY$_7$ Z$_7$±EZ$_7$ d$_7$ EF$_7$	Not rated Not rated Not rated ▷ 1:10 82 Ø137.54$_{-0.08}$ Not rated	◎ 0,01 | G/ ▷ 1:10 AT8 82±0.11 (IT6) — 0,012 ○ 0,02
8 C$_8$ *OX2YZ4*	Con, AD, 0d 3l+2a-0 = 3l+2a	0±EX$_8$ 0±EY$_8$ 0°±AEX$_8$ 0°±AEY$_8$ Z$_7$±EZ$_8$ d$_8$ EF$_8$	Not rated Not rated Not rated ▷ 1:10 82 Ø137.54$_{-0.08}$ Not rated	◎ 0,01 | G/ ▷ 1:10 AT8 82±0.11 (IT6) — 0,012 ○ 0,02
11 *S1* Q	Pr, BD, 1l 2a+1l-1l=2a	0°±AEY$_{11}$ 0°±AEZ$_{11}$ W$_{11}$±EW$_{11}$	62^{+1} 10	▷— S1 // 0,006 | Q ⌯ 0,01 | Q

IOP Conf. Series: Journal of Physics: Conf. Series **858** (2017) 012012 doi:10.1088/1742-6596/858/1/012012

4. Geometrical model of the traction motor end shield

4.1. Functionality of the end shield elements is determined by working and assembly drawings of the traction motor.

End shields 1 and 2 (Figure 1) are mounted into the motor casing 3. A set of basic design datums determining the location of the end shield in the casing is represented by: flat supporting frontal datum *A3* with informativeness 3, since it deprives the end shield of 3 degrees of freedom: one linear and two angular; cylindrical datum *I 2* with informativeness 2, depriving the end shield of two translations; the axis of cylindrical surface of fixation hole *B1* with informativeness 1, depriving the end shield of one rotation around basic axis *I2*. Since the number of deprived degrees of freedom equals to six (3+2+1=6), the basic datums form a complete generalized coordinate system *OX4Y2Z* to the end shield: axis *X4* is located on flat datum element *A3*, corresponds to its plane, passes through origin point *O*, axis of base hole *B1* and has informativeness 4.

Origin point *O* is located on the intersection plane datum *A3* with the axis of cylindrical datum *I2*. Coordinate axis *Y2* with informativeness 2 is located on the datum surface *A3* in perpendicular to axis *X4*. Axis ***ZΘ*** has informativeness equal to 0 (zero) since it is a perpendicular starting from origin point *O* to coordinate plane *XOY*, containing axis *X4* and *Y2* with total informativeness 6.

4.2. Auxiliary design datum for outside ring of the rolling bearing is the axis of the cylindrical datum *G4* with informativeness 4 depriving the inside bearing ring of four movements. Since in traction motor assembly the inside ring abuts labyrinth ring from the inside, its abutting end is a flat supporting datum depriving the ring of one translation. When in operation, the inside ring should rotate slowly since the bearing is designed for circulation loading. So, a set of auxiliary datums deprives of 5 degrees of freedom. To construct complete auxiliary coordinate system *O'X'2Y'Z'4* one should use basic datum *B1* which brings the informativeness of an auxiliary coordinate system datum set to six degrees of freedom.

4.3. A coordinate system of a basic datum set is chosen as a generalized coordinate system *OX4Y2Z* to the end shield. Figure 3 shows the geometrical model of the end shield.

4.4. The number and the type of the geometrical specifications of the end shield elements location are determined by the number and the type of the movements not used by its elements for generalized coordinate system *OX4Y2Z*. The specifications of the geometrical model of the end shield are given in Table 2 representing a table form of the model.

5. Results and discussion

The systematic approach to normalization of the detail accuracy specifications on the basis of geometrical models constructed in coordinate systems of design datums sets allows the quantity, values and tolerances of geometrical specifications to be optimized. These specifications guarantee the quality and estimate indicators of the rolling-contact bearing longevity. The systematic approach is based on the engineering principles providing the quality of engineering products: uniformity principle of design, technological, measurement and construction datums, principle of the shortest dimension chains, and inversion principle. According to the inversion principle, operating conditions for each product detail in a design project should correspond to the operating conditions of the product both during the production and the verification. Operating conditions primarily include location, kinematics, acting forces and temperature.

Judging from the title of Technical committee for Standardization ISO/TC 213 - Dimensional and geometrical product specifications and verification, - standard ISO 5459:2011 - Datums and datum systems - developed by the committee includes only measurement datums for the verification process. Absence of engineering principles of quality control, refusal from systematic approach and materialization of datums sets coordinate systems lowers the standard quality and, subsequently, the quality of the products and puts the brake on the rate of technical progress.

Table 2. The geometrical model parameters for the end shield

Number and modifier, the detail element, and the coordinate system	Type, function, informativeness and the number of initial errors in element location	Dimensions and initial errors of location, dimensions and forms	Datums, dimensions, and location, dimension and form tolerances due to drawing	Suggestions, design parameters, and variants on the geometrical model basis
1 *A3* *OX4Y2Z*	F, BD, 2a+1l 1l+2a-1l-2a=0	EF_1	⊥ | 0,05 | G	▶— A3 ▱ | 0,02
2 *I2* *OX4Y2Z*	Cyl, BD, 2l 2l+2a+2l=2a	$90°\pm AEX_2$ $90°\pm AEY_2$ d_2-Ed_2 EF_2	↗ | 0,04 | G ○ | 0,1 ⌭ | 0,1	▶— I2 ⊥ | 0,06 | A
3 *G4* *OX4Y2Z*	Cyl, AD, 0d 2l+2a-0=2l+2a	$90°\pm AEX_3$ $90°\pm AEY_3$ D_3+ED_3 EF_3 $0\pm EX_3$ $0\pm EY_3$	▶— G ○ | 0,03 ⌭ | 0,03	◎ | 0,06 | I ⊥ | 0,04 | A ◁— G4
4 C_4 *OX4Y2Z*	F, ES, 0d 1l+2a-0=1l+2a	$0°\pm AEX_4$ $0°\pm AEY_4$ $Z_4\pm EZ_4$ EF_4	Not rated Not rated 36.5±1 Not rated	-
5 *B1*	Cyl, BD, 1l 2l-1l=1l	$X_5\pm EX_5$ D_5+ED_5 EF_5	Not rated Not rated Not rated	-

Based on the results of validating the working drawings of an armature shaft and an end shield for a traction motor with the help of geometrical models, the following recommendations for basic elements and geometrical specifications optimization were formulated:

- common axis of basic design datums of an armature shaft for inside bearing rings fitting (Figure 2) should be introduced as initial axis *OZ4* with informativeness 4 of a generalized coordinate system instead of the common axis of central shaft holes as technological datums are not used in product operation. This non-conformity results from contravention of the datums uniformity principle. Systematic approach allows enhancing the location accuracy for basic design datums of the shaft by 4 times (Table 1);
- improved location accuracy for basic design datums of the shaft will result in decreased misalignment of the axes for inside bearing rings at rolling-contact bearing assembly and will increase the bearing longevity;
- inside bearing ring should mate with bearing collar by its abut instead of intermediate detail abut (Figure 1), its abut run-out being much higher because of the contravention of the shortest dimension chains principle. The removal of this non-conformity will call for changing the design of the shaft, the rolling-contact bearing and the product itself;
- fitting plane for an end shield and a traction motor casing is a basic design datum with informativeness 3. All its informativeness is used for main coordinate system *X4OY2* with informativeness 3 in a generalized coordinate system (Figure 3). This datum can have no location tolerance, therefore, the perpendicularity tolerance on the working drawing should be assigned to an auxiliary datum, i.e. the axis of the hole for an inside bearing ring. This tolerance will limit the misalignment of the axis for the inside bearing ring in a rolling-contact bearing. The basic plane with informativeness 3 of the shield will have only the flatness tolerance (Table 2);

- similar recommendations are formulated for all details of the traction motor.

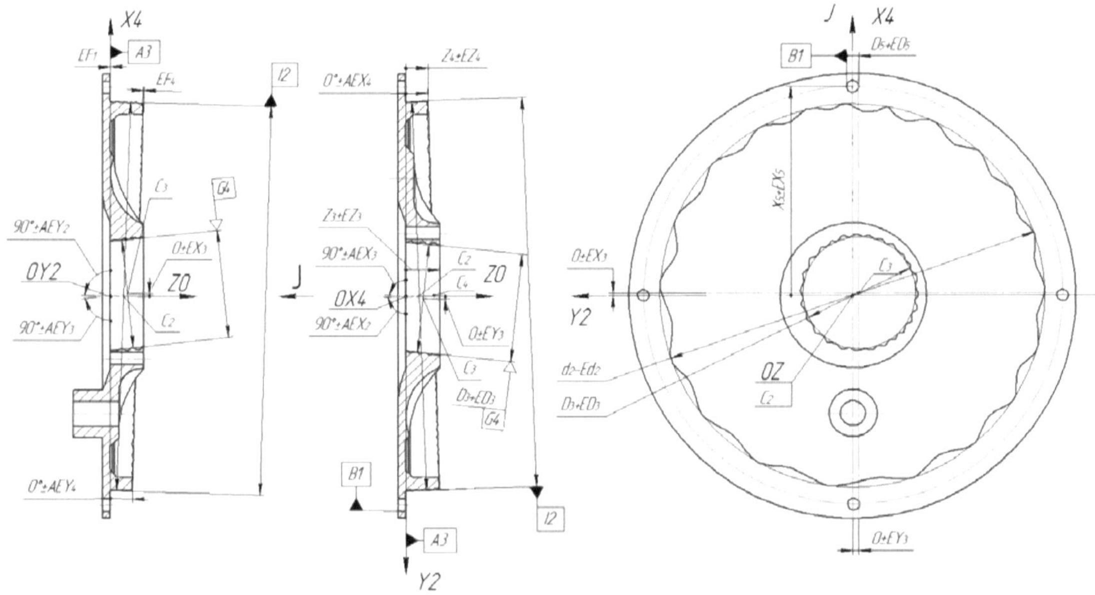

Figure 3. Geometrical model of the end shield a) in plane *OX4Z0*; b) in plane*OY2Z0*; c) in plane *OX4Y2*

6. Summary

6.1. Further improvement of the electromechanical products quality is possible due to systematic approach and geometrical modeling of the geometrical product specifications for the details in order to optimize their composition, values and tolerances.

6.2. An adequate geometrical model of the real detail shows relations of all geometrical specifications for working detail elements (coordinates, location deviations, the elements dimensions and their surfaces form deviations) in a generalized coordinate system formed by a set of basic construction datums. Functionality of every element (basic datum, auxiliary datum, executive element, binding element) and element informativeness (as a constraint sum of linear and angular degrees of freedom) are taken into account. The adequate geometrical model of the real detail has two forms - graphical and table ones, complementing each other.

6.3. The number of geometrical specifications affecting the quality of electromechanical products is 20-30% larger than the number of the electrical specifications.

7. Conclusion

The problem of enhancing the durability, longevity, and efficiency coefficient for electromechanical products can be solved with the systematic normalizations of geometrical specifications accuracy on the basis of coordinate systems introduced in the standards on geometrical product specifications (GPS).

References

[1] V I Glukhov 2014 Geometrical product specifications: Alternative standardization principles, coordinate systems, models, classification and verification *Dynamics of Systems, Mechanisms and Machines (Dynamics)* (IEEE Conf. Pub.) pp 1-9

[2] J C Bell and A Dyson 1972 Mixed friction in an elastohydrodynamic system

IOP Conf. Series: Journal of Physics: Conf. Series **858** (2017) 012012 doi:10.1088/1742-6596/858/1/012012

(Elastohydrodynamic lubrication Symp., London) pp 68-76

[3] M P Kovalev and M Z Narodetsky 1980 The calculation for high-speed ball-bearings (Moscow Machine-building) 373 p

[4] H Blok 1937 Les temperatures de surface dans des conditions de graissage sour pression extreme (Congr. mondial du petrole Paris v III) pp 471-86

[5] J S Schabel 1981 Theoretische und praktische Erkenntnisse der Walzlager-schmierung *Masch. Anlang. Verfahr.* **6**, pp 64-6, 70

[6] M Tikvicki 1980Erfahrungen iiber die Aussagegenauigkeit der SPM Megmetode fur elektrische Maschinen *Siemens energieteehn* **2** pp 55-8

Geometrical product specifications. Datums and coordinate systems

V I Glukhov [1], I A Ivleva [1] and O Y Zlatkina [1]

[1] Omsk State Technical University, 11, Mira ave., Omsk, 644050, Russia

E-mail: mips@omgtu.ru

Abstract. The work is devoted to the relevant topic such as the technical products quality improvement due to the geometrical specifications accuracy. The research purpose is to ensure the quality indicators on the basis of the systematic approach to the values normalization and geometrical specifications accuracy in the workpiece coordinate systems in the process of design. To achieve the goal two tasks are completed such as the datum features classification according to the number of linear and angular freedom degrees constraints, called the datums informativeness, and the rectangular coordinate systems identification, materialized by workpiece datums sets. The datum features informativeness characterizes the datums functional purpose to limit product workpiece linear and angular degrees of freedom. The datum features informativeness numerically coincides with the kinematic pairs classes and couplings in mechanics. The datum features informativeness identifies the coordinate system without the location redundancy. Each coordinate plane of a rectangular coordinate system has different informativeness $3 + 2 + 1$. Each coordinate axis also has different informativeness $4+2+\Theta$ (zero). It is possible to establish the associated workpiece position with three linear and three angular coordinates relative to two axes with the informativeness *4* and *2*. is higher, the more informativeness of the coordinate axis or a coordinate plane is, the higher is the linear and angular coordinates accuracy, the coordinate being plotted along the coordinate axis or plane. The systematic approach to the geometrical products specifications positioning in coordinate systems is the scientific basis for a natural transition to the functional dimensions of features position - coordinating dimensions and the size of the features form - feature dimensions of two measures: linear and angular ones. The products technical quality improving is possible due to the coordinate systems introduction materialized by datums sets in GPS standards and national standards. Standards on the datums and datums systems should cover all the geometrical specifications and all the processes of the products life cycle: designing, manufacturing, verification and exploitation.

1. Introduction

The regional standard "Locating and datums in mechanical engineering" [1] has been in operation in Russia since 1976. It has proved its working ability and effectiveness since then. According to this standard, datums are all real contacting surfaces of mating work pieces datum features. Contacting datums are divided into two types: main and auxiliary ones. The main datums specify the considered workpiece position in the product, the auxiliary datums specify the associated workpiece position to the given one. Auxiliary datums perform the support function for the main datums. A set of the main datums forms the workpiece generalized coordinate system. A set of auxiliary datums forms the auxiliary coordinate system for an associated workpiece. In the generalized coordinate system each

auxiliary coordinate system position is given by six coordinates: three linear coordinates of the coordinate point of origin and three angular coordinates of the auxiliary system coordinates axes.
To form a rectangular coordinate system, each set of datums should limit workpiece six degrees of freedom - three linear and three angular ones. Datums vary in their functional purpose and cover all processes of a product life cycle. A set of design datums specifies the workpiece position in the designing project. A set of technological datums specifies the workpiece position in the manufacturing process. A set of measurement datums specifies the workpiece position in a meter in the geometrical specifications verification process. A set of construction datums specifies the workpiece position in the construction and the product exploitation process.

2. Problem statement

According to the principle of datums unity [2] the design, technological, measuring and construction datums of a workpiece should coincide, in other words they should be unified for all life cycle processes. The principle of datums unity ensures the high-quality products production with the lowest cost in the shortest possible time. However, the standard [1] has found its application as the basis of the workpiece locating theory only in the standards system for technological documentation. Six datum coordinates *(3 + 2 + 1)* are the theoretical datums in mechanical engineering technology. The coordinates are located in three perpendicular planes forming the workpiece datum system. The standard is not included in the standards system for the construction documentation due to the influence of the previous versions on the international standard on datums and coordinate systems [3].
The renewal of the latest version of international standards on geometrical product specifications GPS [4] has not improved the standard quality [3] on datums and datum systems:
- the standard applies only to the geometrical tolerances of position and form normalization, it does not cover the feature dimensions regulation;
- the standard uses the datum features classification according to the number of degrees of freedom (invariance), which does not correspond to the datums functional purpose to limit normalized features degrees of freedom;
- the standard does not use the coordinate systems formed by datums - this has resulted in many cases in datum redundancy, or vice versa, to its deficiency.

The aim of this research is to provide the required quality of technical products in all life cycle processes based on the values normalization and geometrical specifications accuracy in a datum system, materialized by datums sets.
The following tasks are completed in this research:
- the classification of the products datum features on the degrees of freedom constraints number is performed;
- the rectangular coordinate systems identification rules of datums sets are developed;
- the dimensional geometrical specifications alternative classification, normalized in the product rectangular datum system in accordance with the informativeness priorities of coordinate axes and planes of the Cartesian coordinate system is proposed.

3. Classification of Workpiece Datum Features

The workpiece positioning accuracy in the product space in operation is achieved due to the constraint of all six degrees of freeom: three translational *(t)* and three rotational *(r)* along the movement *(t)* and round *(r)* three perpendicular directions. This positioning is done by associated workpiece main construction datums sets relative to an auxiliary construction datums set of a locating workpiece under the forces in the product operation.

IOP Conf. Series: Journal of Physics: Conf. Series **858** (2017) 012013 doi:10.1088/1742-6596/858/1/012013

Table 1. Classification of datum features on information content

Pair class	Datums locating			Information content	Invariance
θ	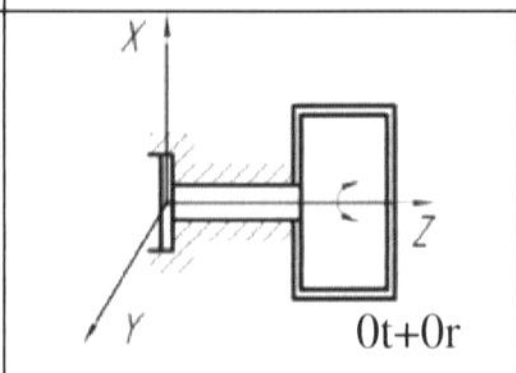 0t+0r	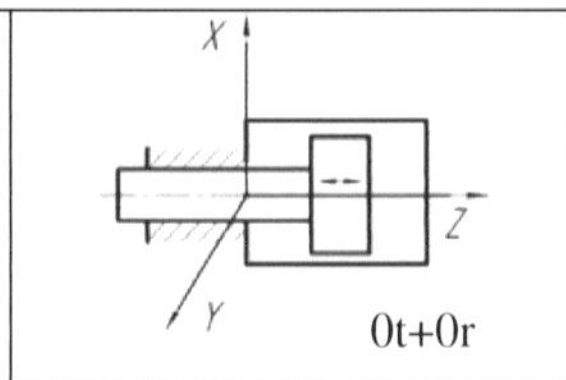 0t+0r	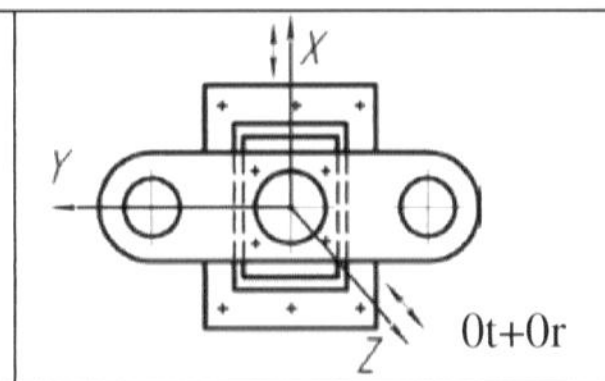 0t+0r	θ	θ
I	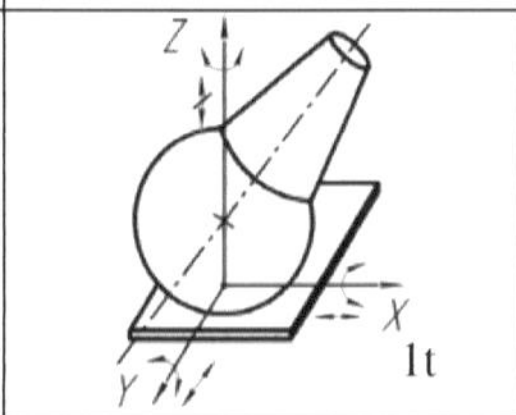 1t	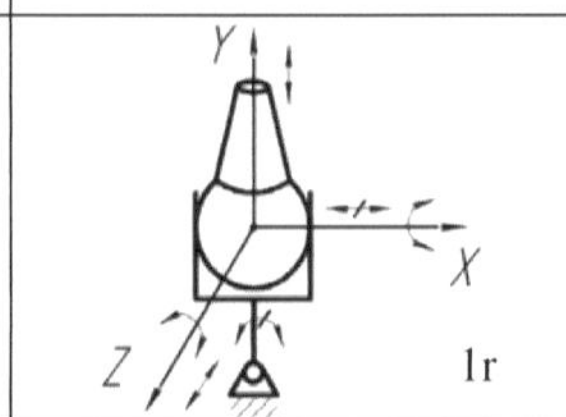 1r	–	1	5
II	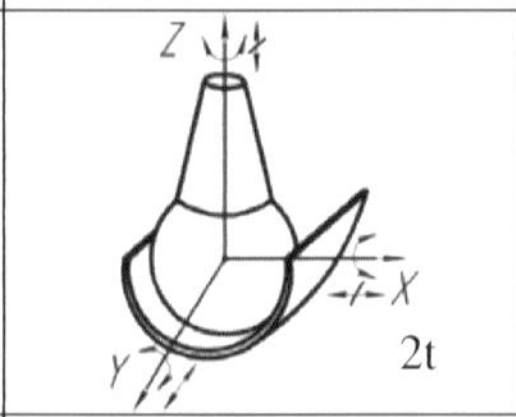 2t	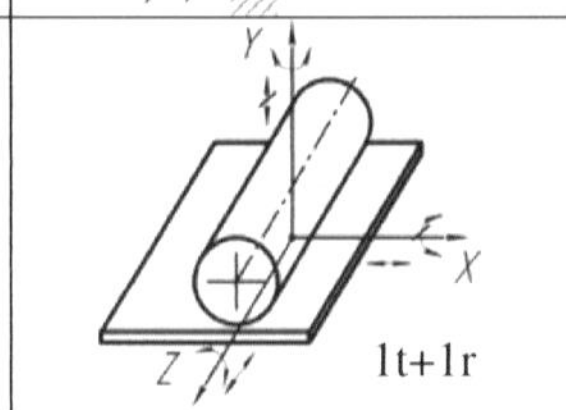 1t+1r	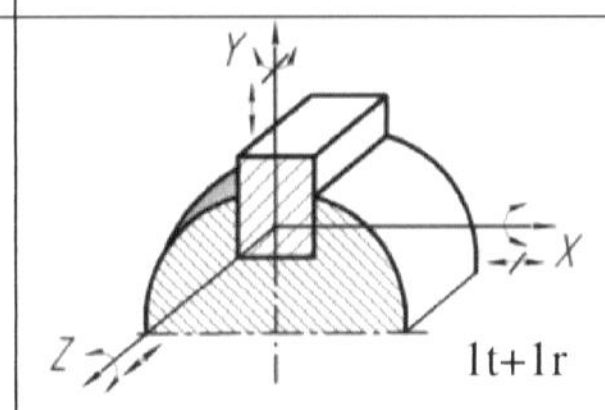 1t+1r	2	4
III	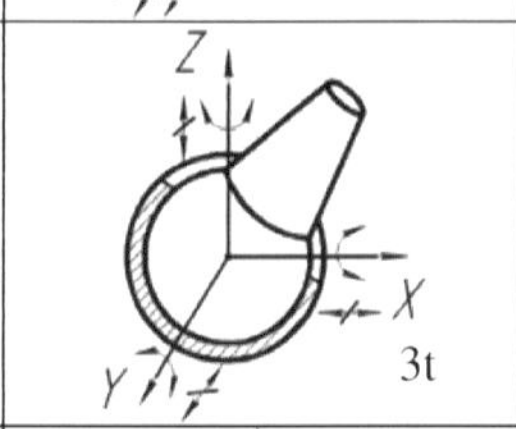 3t	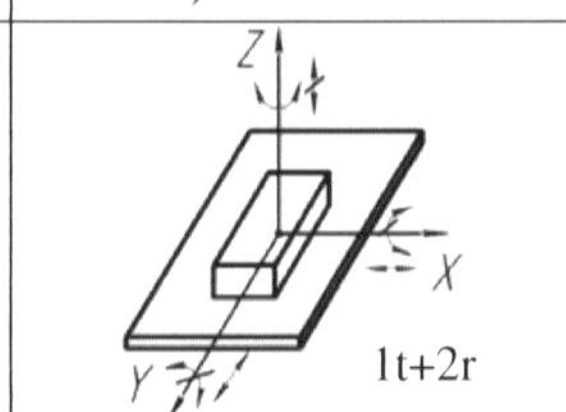 1t+2r	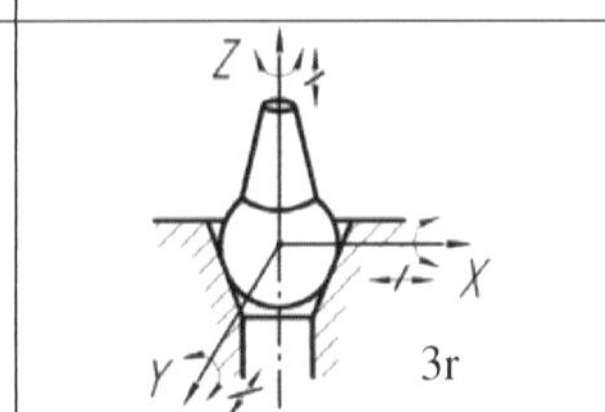 3r	3	3
IV	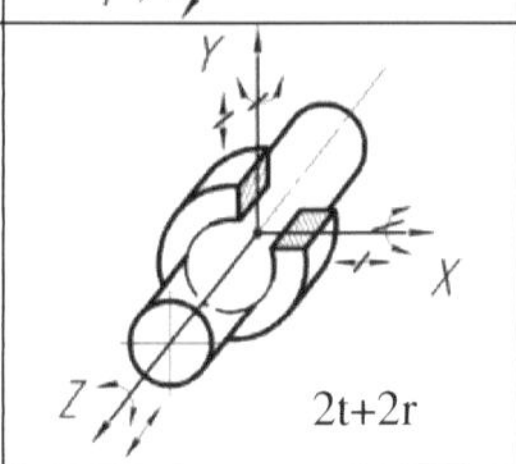 2t+2r	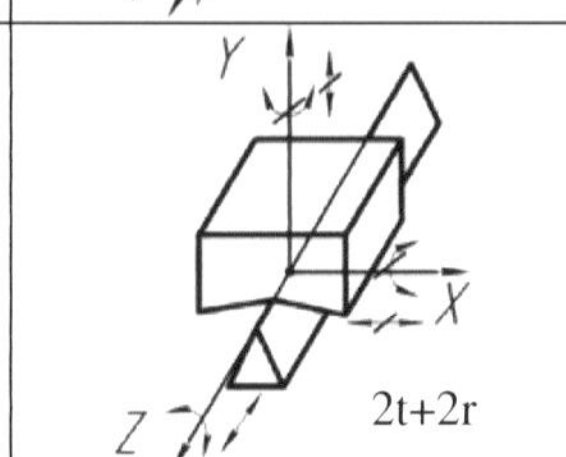 2t+2r	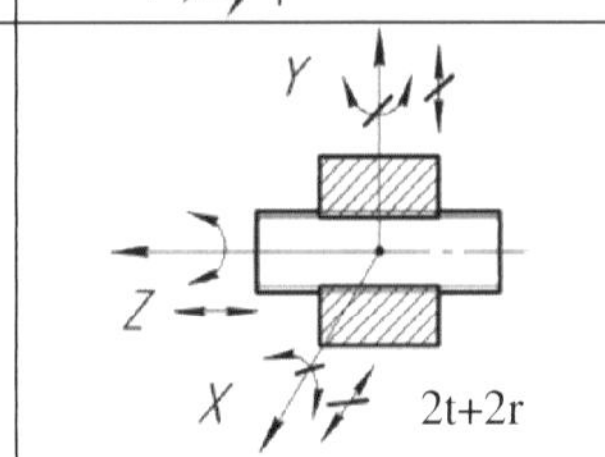 2t+2r	4	2
V	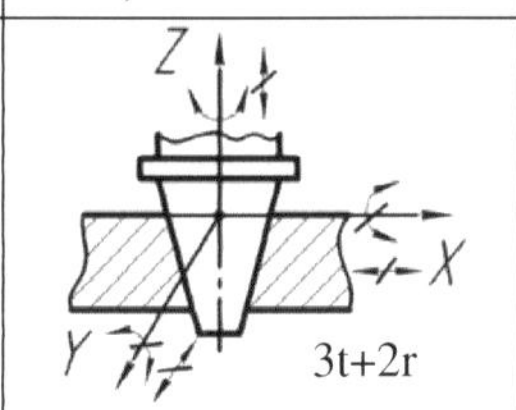 3t+2r	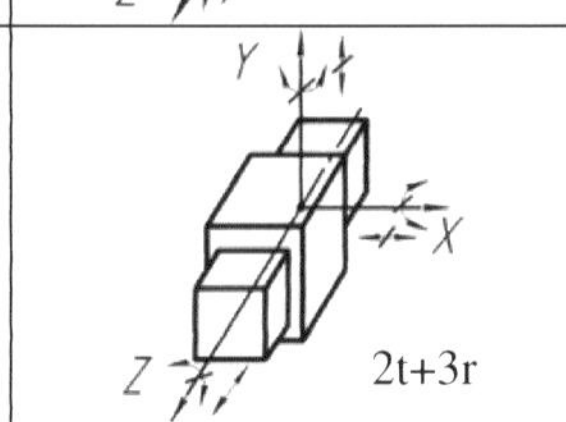 2t+3r	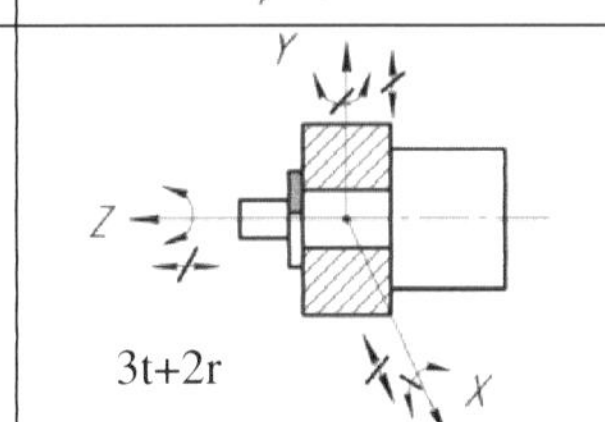 3t+2r	5	1
VI	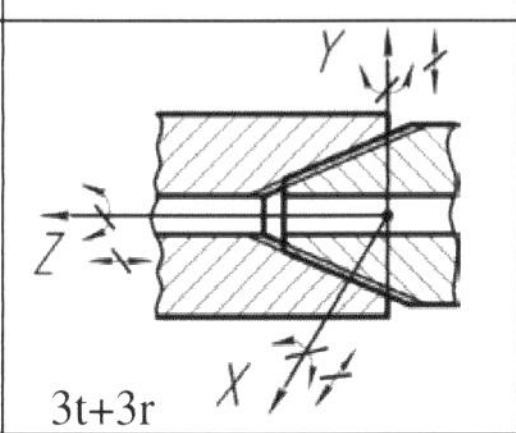 3t+3r	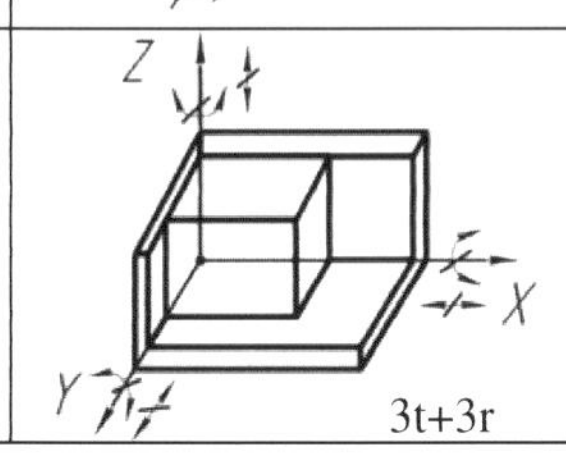 3t+3r	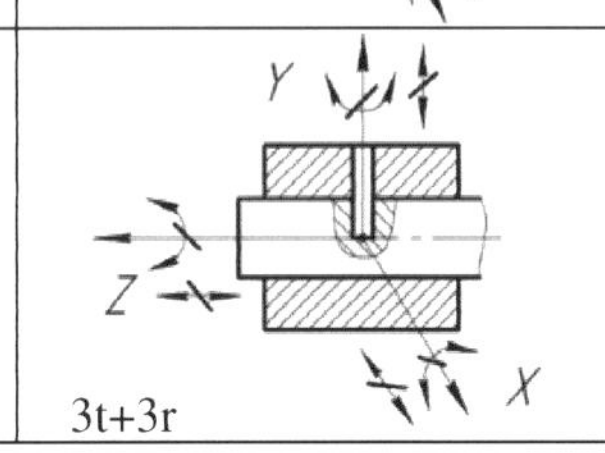 3t+3r	6	θ

IOP Conf. Series: Journal of Physics: Conf. Series **858** (2017) 012013 doi:10.1088/1742-6596/858/1/012013

In Theoretical mechanics contacting main and auxiliary datum features form the couplings - moveable kinematic pairs or fixed locating joints (see Table 1). Depending on the coupling datum features surface type (plane, sphere, cylinder, cone, torus, screw, geometrically complex surface), the dimensions and the datum features number, the couplings are divided into six classes according to the number of datums limited degrees of freedom (from one to six) which contain linear *(t)* and angular *(r)* constraints *(c)*.

Limited degrees of freedom of datum features have more informativeness than invariance, they define couplings classes, therefore, it is proposed to call the degrees of freedom sum limited by a datum feature as "the informativeness". The informativeness numerical value is specified by the degrees of freedom number limited by a datum feature (Table 1), which coincides with the coupling pair class number. The informativeness verification and datum invariance accuracy is simple: their sum must be equal to six.

The informativeness of datum features coincides with the kinematic pairs first-fifth classes number formed by main and auxiliary datums moveable joints of associated and datum features of a moveable workpiece relative to the fixed one. The kinematic pair class number is specified by the linear and angular freedom degrees sum limited by the associated features being the main and auxiliary datums.

It is proposed to supplement the classification of the.

The Cartesian coordinate system formed by three perpendicular planes also has the excessive informativeness $9c = 3\ (1t + 2r)$ instead of six $6c = 3t + 3r$. This requires three angular tolerances normalization of the coordinate planes perpendicularity: $3\ (1t + 2r) - (3t + 3r) = 3r$.

The rectangular coordinate system formed by planes, therefore, requires the position tolerances normalization two times less than that formed by axes.

In (Figure 1b) an alternative rectangular coordinate system is presented which takes account of the different coordinate planes informativeness $3 + 2 + 1$ and coordinate axes $4+2+\Theta$ (zero) and different directions of positive angles reading from the coordinate axis with the maximum informativeness *4* (two angles A and B) and one angle C from the coordinate axis with informativeness *2*. Precisely this workpiece coordinate system should be used for the workpiece geometrical specifications normalization: position and form tolerances, linear and angular coordinates, linear and angular workpiece dimensions.

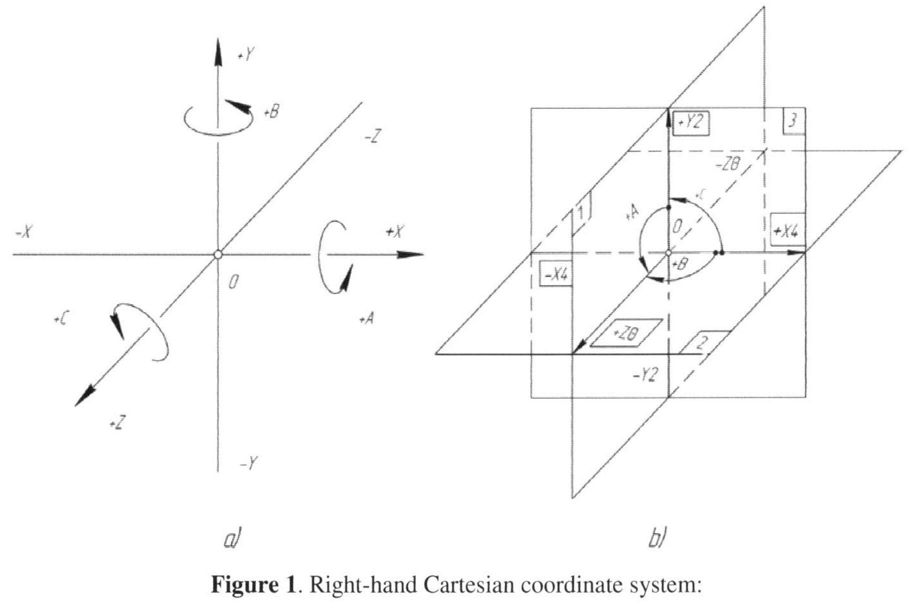

Figure 1. Right-hand Cartesian coordinate system:
a) for machine tools with CNC in accordance with ISO 841: 2002;
b) for geometrical products specifications (alternative).

A set of datums can form a coordinate system, if it limits a workpiece by six degrees of freedom: three linear and three angular ones. Set datums should not duplicate the constraints of the same degrees of freedom. The real coordinate systems models are shown in Figure 2.

A set of three flat datums G3, K2, N1 (see Figure 2a) limits the workpiece of six degrees of freedom set on a datums indicator *4 (1t + 2r) + (1t + 1r) + 1t = 3t + 3r*. If the datums dimensions of K and N are not limited, the datums set informativeness is *(1t + 2r) + (1t + 2r) + (1t + 2r) = 3t + 6r*.

Such informativeness leads to the locating redundancy and the need to increase normalization of three additional perpendicularity tolerances between the datums materializing a rectangular coordinate system.

Similarly, a set of two cylindrical datums *J4* and *S2* (Figure 2b) forms a rectangular coordinate system without redundant locating due to reducing the datum length *S2: (2t + 2r) + (1t + 1r) = 3t + 3r*. Without the datum length reduction the datums informativeness will be *(2t + 2r) + (2t + 2r) = 4t + 4r*. This results in the datum redundancy: *(4t + 4r) - (3t + 3r) = 1t + 1r* and the need to normalize two geometrical tolerances: the datum axes intersection and the perpendicularity between the axes.

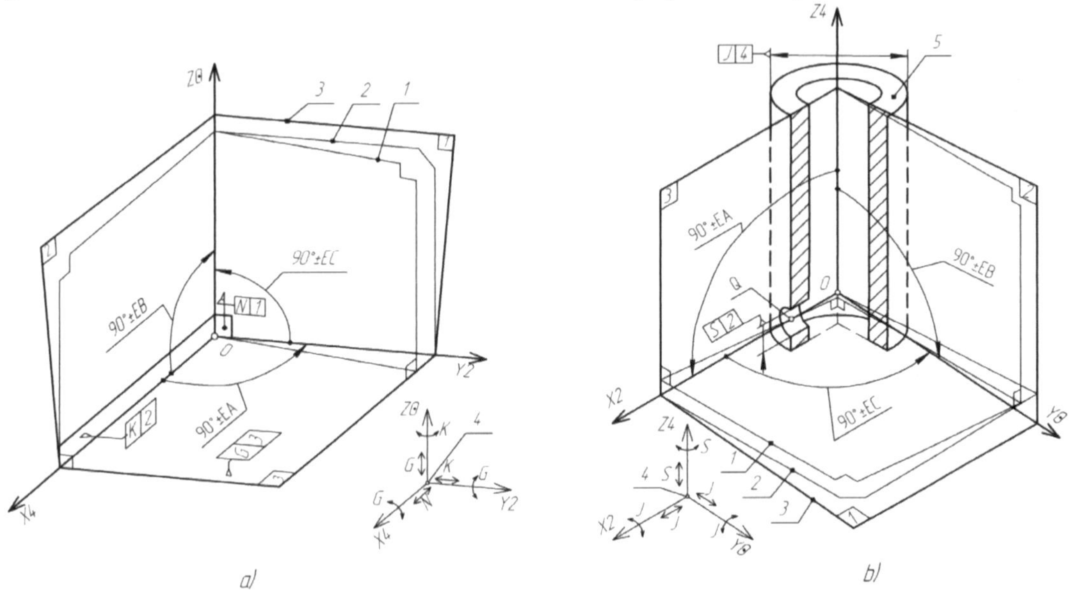

1 - coordinate planes nominal position; 2, 3 - coordinate planes real position;
4 - datum set informativeness indicator; 5 - product (workpiece).
Figure.2. Real coordinate systems models formed by datums sets:
a) by a set of three flat datums; b) by a set of two cylindrical datums.

In the geometrical product specifications verification process it is impossible to materialize theoretically exact nominal right angles between the coordinate planes and the coordinate axes. The accuracy of the rectangular coordinate system should be rated by three angular coordinates 90 ° ± EA, 90 ° ± EB, 90 ° ± EC, shown in Figure2.

A set of datums can form a coordinate system, if it limits a workpiece of six freedom degrees: three linear and three angular ones. Set datums should not duplicate the same freedom degrees constraint. Real coordinate systems models are shown in Figure 2.

Such information content leads to the locating redundancy and the need to increase three additional perpendicularity tolerances normalization between the datums materializing a rectangular coordinate system.

A set of three flat datums G3, K2, N1 (see Figure 2a) limits the six freedom degrees workpiece dimensioned on a datums indicator $4\ (1t + 2r) + (1t + 1r) + 1t = 3t + 3r$. If the datums dimensions of K and N are not limited, the datums set informativeness is $(1t + 2r) + (1t + 2r) + (1t + 2r) = 3t + 6r$. Such information content leads to the locating redundancy and the need to increase three additional perpendicularity tolerances normalization between the datums materializing a rectangular coordinate system.

Similarly, a set of two cylindrical datums *J4* and *S2* (Figure 2b) forms a rectangular coordinate system without redundant locating due to reducing the datum length *S2:* $(2t + 2r) + (1t + 1r) = 3t + 3r$. Without the datum length reducing the datums informativeness will be $(2t + 2r) + (2t + 2r) = 4t + 4r$. These results in the locating redundancy: $(4t + 4r) - (3t + 3r) = 1t + 1r$ and the need to normalize two geometrical tolerances: the datum axes intersection and the perpendicularity between the axes.

Similarly, a set of two cylindrical datums *J4* and *S2* (Figure 2b) forms a rectangular coordinate system without redundant locating due to reducing the datum length *S2:* $(2t + 2r) + (1t + 1r) = 3t + 3r$. Without the datum length reducing the datums informativeness will be $(2t + 2r) + (2t + 2r) = 4t + 4r$. These results in the locating redundancy: $(4t + 4r) - (3t + 3r) = 1t + 1r$ and the need to normalize two geometrical tolerances: the datum axes intersection and the perpendicularity between the axes.

Products geometrical specifications verification process does not allow theoretically exact nominal right angles between the coordinate planes and the coordinate axes to be materialized. The accuracy of the rectangular coordinate system should be rated by three angular coordinates 90 ° ± EA, 90 ° ± EB, 90 ° ± EC shown in Figure 2.

4. Conclusions

- The datum features informativeness characterizes the datums functional purpose to limit linear and angular freedom degrees of a product workpiece. The basic elements informativeness numerically coincides with the classes of kinematic pairs and couplings in mechanics. The datum features informativeness identifies the coordinate system without the datum redundancy.
- Three planes intersecting at right angles are producers of the solid rectangular coordinate system in which the coordinate axes are the plane intersection lines and the origin is the planes intersection point.
- Each coordinate plane of the Cartesian coordinate system has different informativeness 3 + *2* + *1*. It is sufficient to establish the associated workpiece position by three linear and three angular coordinates, depriving a workpiece of all six degrees of freedom. Coordinate planes informativeness specifies their priority.
- Each coordinate axis has different informativeness $4+2+\Theta$ **(zero)**. This means that relative to two coordinate axes with informativeness *4* and *2* located in a coordinate plane with informativeness *3* it is possible to specify the associated workpiece position by three linear and three angular coordinates. The coordinate axes informativeness specifies their priority.
- Linear coordinates are measured on the coordinate axes from the origin point and the angular coordinates are determined in the coordinate planes. The coordinate axes with informativeness *4* and *2* are the origin of angular coordinates. The vertexes of coordinate angles lie at the coordinate system origin.
- The more informativeness the coordinate axis or a coordinate plane has, the higher is the linear and angular coordinates accuracy, the coordinate being rated relative to the coordinate axis or a coordinate plane.
- The systematic approach to the products geometrical specifications positioning in the coordinate systems is the scientific basis for a natural transition to the functional dimensions of features positioning - coordinating dimensions and features form dimensions - two measures features dimensions: linear and angular ones.

IOP Conf. Series: Journal of Physics: Conf. Series **858** (2017) 012013 doi:10.1088/1742-6596/858/1/012013

5. Summary

- The products technical quality improvement is possible due to the coordinate systems introduction materialized by datums sets in GPS standards and national standards.
- The standards on the datums and datums systems should cover all the geometrical specifications and all the processes of products life cycle such as designing, manufacturing, verification and exploitation.

References

[1] GOST 21495-76 *Locating and datums in Mechanical Engineering. Terms and definitions*

[2] V I Glukhov 2014 Geometrical product specifications: Alternative standardization principles, coordinate systems, models, classification and verification *Dynamics of Systems, Mechanisms and Machines (Dynamics) (IEEE Conf. Pub.)*

[3] ISO 5459:2011. *GPS Geometrical product specifications (GPS)* – Geometrical Tolerances – Datums and datum systems

[4] V A Srinivasan 1999 Geometrical product specification language based on a classification of symmetry groups *Computer – aided design* **31** **11** pp 659-68

[5] ISO 841:2001. *Industrial automation systems and integration* – Numerical control of machines – Coordinate system and motion nomenclature

[6] ISO 13584-101:2003. *Industrial automation systems and integration* Part 101: Geometric view exchange protocol by parametric program

IOP Conf. Series: Journal of Physics: Conf. Series **858** (2017) 012014 doi:10.1088/1742-6596/858/1/012014

Geometric modeling in the problem of ball bearing accuracy

V I Glukhov [1], V V Pushkarev [1] and V G Khomchenko[1]

[1] Omsk State Technical University, 11, Mira ave., Omsk, 644050, Russia

E-mail: mips@omgtu.ru

Abstract. The manufacturing quality of ball bearings is an urgent problem for machine-building industry. The aim of the research is to improve the geometric specifications accuracy of bearings based on evidence-based systematic approach and the method of adequate size, location and form deviations modeling of the rings and assembled ball bearings. The present work addressed the problem of bearing geometric specifications identification and the study of these specifications. The deviation from symmetric planar of rings and bearings assembly and mounting width are among these specifications. A systematic approach to geometric specifications values and ball bearings tolerances normalization in coordinate systems will improve the quality of bearings by optimizing and minimizing the number of specifications. The introduction of systematic approach to the international standards on rolling bearings is a guarantee of a significant increase in accuracy of bearings and the quality of products where they are applied.

1. Introduction

Rolling bearings are characterized by high manufacturing accuracy. Manufacture of bearings is one of the fundamental criteria for the national industry development. Ball bearings have been in use for over a century. There is lot of experience accumulated in the ball bearings design and application. The issue of ball bearings quality increase is currently a priority for producers and consumers. The efforts of producers are focused on the improvement of production technology, search for and application of new materials, methods of calculation, defects diagnostics.

When analysing the requirements of international standards [1, 2] on tolerances of geometric specifications of rings and bearing assemblies some discrepancies associated with normalization of optimum accuracy of geometric specifications and complexity of utilization of presented definitions emerge. For example, to describe the tolerance diameter of the outer ring there are 11 various specifications. Even experts on rolling bearings have no clear confidence in the correctness of interpretations and applications of tolerances [3].

In the latest version of the international standard [1], it was decided to use the standard notation GPS "Geometrical product specification" to indicate the accuracy of the tolerances on the specifications of the rolling bearings (Fig. 1). However, these innovations also do not fully meet the requirements of high-quality standard normalization. Such important specifications as symmetry raceway profile, width of the ball bearing assembly are absent in the standard, as well as in its earlier versions [1, 2].

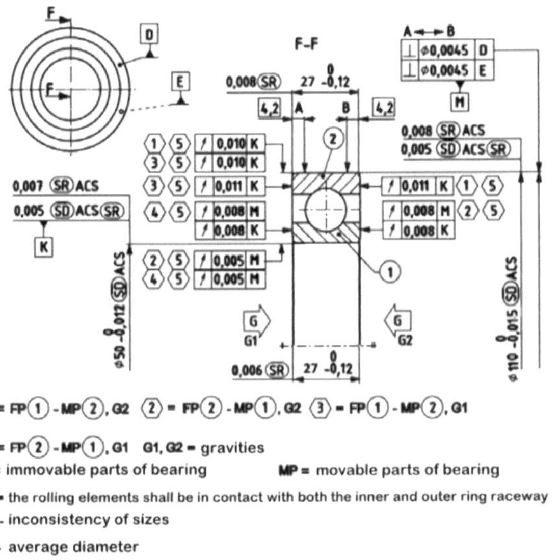

Figure 1. Single row radial ball bearing 310, normal accuracy class - dimensioning and geometric tolerances according to standard ISO492:2014

2. Problem statement

Despite a large number of studies carried out by researchers from different countries, some inconsistencies in the field of geometric specifications of ball bearings accuracy normalization still remain. Standard requirements [1, 2] do not fully solve the problem of high accuracy of bearing specifications due to the lack of systematic approach to their normalization.

The purpose of the present work is to provide high quality of ball bearings based on geometric modeling of real specifications of rings and bearing assembly in the coordinate systems formed by the datum features.

The objectives of the research are as follows:

1) Identification of geometrical specifications of bearing and its rings based on system geometric modeling;

2) Experimental study of non-standardized geometric specifications of bearings and rings;

3) Optimizing and minimizing the number of geometric specifications of bearings and rings for alternative standardization.

3. Geometric Modeling of Study Subject Accuracy Specifications

The study subject is a single-row radial ball bearing. The total output of such bearings is more than 50% of all manufactured types of bearings. A tight ring determines a bearing position in the connection. The inner ring is a tight ring in 95% of the products. The position and orientation of the bearing ring in the product space occurs by means of the balls which are disposed in raceways. The accuracy of the location of the raceways profile with respect to the datum ends directly affects the accuracy of the bearing rings, bearing mounting width and durability of bearing compounds. In practice the location of the ball bearing rings raceways symmetry plane is provided by the coordinating size relative to the datum end. However, the coordinating size control requires identification of the ring datum end and racetrack symmetry plane materialization, which is irrational.

Ring datum ends belong to the same geometric plane-parallel prismatic feature. The ends flat datums of ball rings are functionally symmetrical which makes the prismatic feature symmetrical with a plane of symmetry available, which becomes a datum feature of the ring. The raceway of the ring is a geometric feature - a torus which also has a plane of symmetry. Two symmetrical geometric features of the bearing rings in combining symmetry planes make them symmetrical parts. This property of the rings defined the symmetry of the ball bearings assembly.

Geometric model of the rings and bearings assembly constructed according to method [4] is shown in (Figure 2 ... 4). The model of geometric specifications of the internal tight ring (Figure 2a) is constructed in a generalized *OX4Y2Z0* Cartesian coordinate system formed by a set of key datum features: the plane of symmetry *E3* of the plane-parallel prismatic feature ends, the axis of symmetry of the cylindrical feature *F2*, and indicator *G1* of the eccentricity vector of the raceway center. The set of the main datums positions the ring on the shaft. The digits after the letter designation of datums reveal their informativeness and mean the sum of the constraints of linear degrees of freedom (the first figure in the box after the datum case) and angle (the second figure) on the projection of the model. Specific datums limits are shown on the indicator of the main datums set (Figure 2b). The auxiliary set of datum features determines the position of the inner ring of the outer ring via the balls through the center and the plane of symmetry *H5* of the raceway and indicator *G1* of the eccentricity vector, together forming the auxiliary coordinate system *O'X'4Y'2Z'0*. The indicator of the outer ring tolerance by the set of auxiliary datums is shown in (Figure 2c).

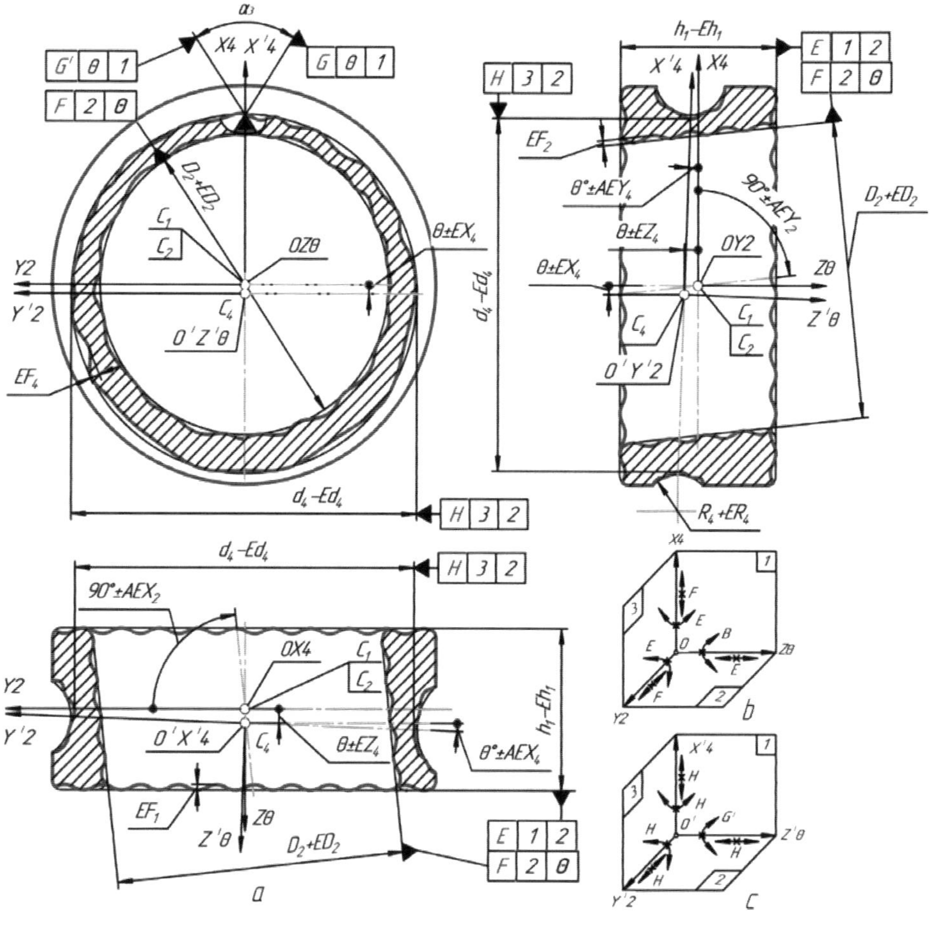

Figure 2. Ball bearing inner ring - the model of geometric specifications
a) geometric model in three projections; *b*) indicator of the main datums set;
c) indicator of auxiliary datums set.

The outer free ring is the part attachable to the basing detail – inner ring – through the balls. The model of geometric specifications of the outer ring (Figure 3a) is built in generalized coordinate system *OX4Y2Z0* formed by the set of main datum features of the ring: *E5* – the center and the raceway plane of symmetry and *F1* – eccentricity vector indicator of the actuating cylindrical surface center *G4*, fulfilling the main functionality of the cylindrical shaft datum. The actuating surface *G4* together with the datum features *H1* – the plane of symmetry of the prismatic feature ends and *F1* –

the indicator of the eccentricity vector of the actuating surface *G4* center form an auxiliary coordinate system *O'X'2Y'θZ'4*. The indicators of tolerance of linear and angular degrees of freedom for the generalized system of coordinates are shown in (Figure 3b) and for the auxiliary system of coordinates in (Figure 3c).

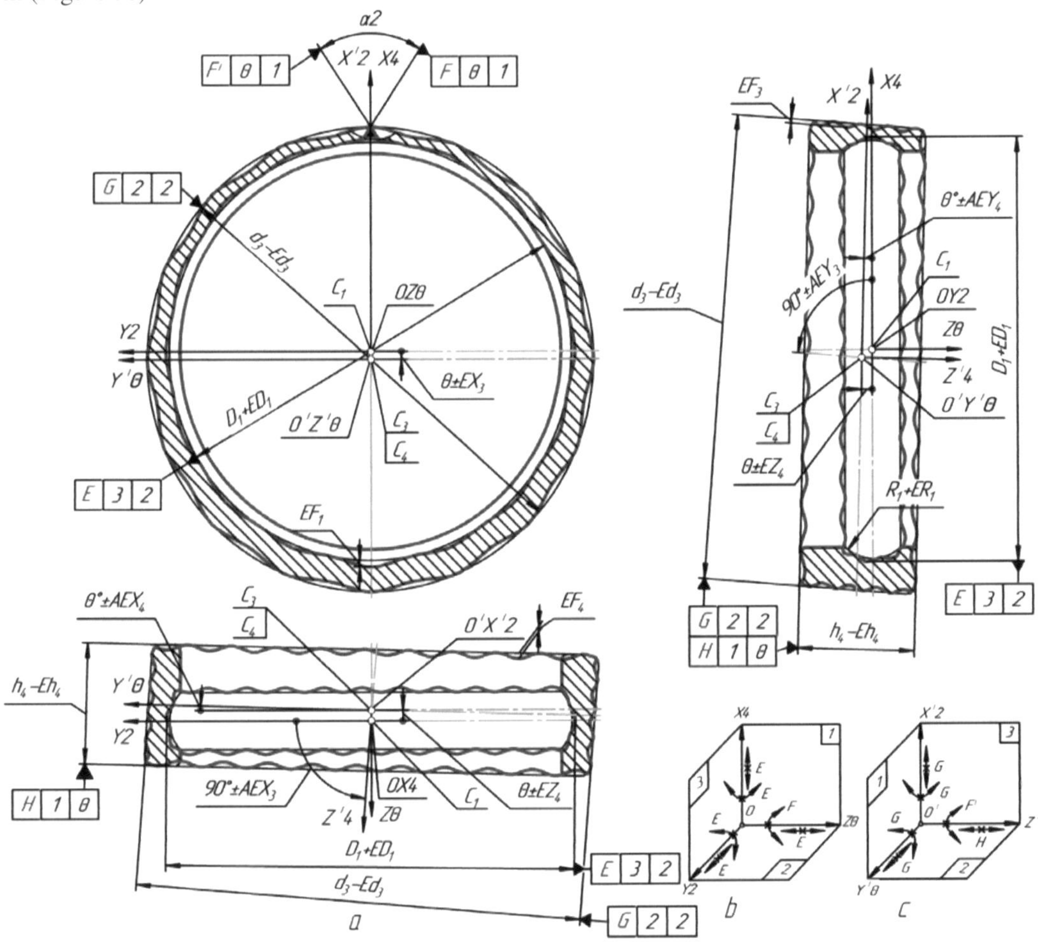

Figure 3. Ball bearing outer ring-model of geometric specifications
a) geometric model in three projections; *b*) indicator of main datums set;
c) indicator of auxiliary datums set.

The geometric model of the bearing assembly (Figure 4) is built in the generalized coordinate system *OX4Y2Zθ* of the tight inner ring basing the bearing on one of the two main datums of the cylindrical shaft.

The inner ring forms an interference fit with the rotary shaft restricting the ring with its main datums of all six degrees of freedom: three - by one of two datum ends of the *E3*; two – by the axis of cylindrical hole *F2*, and one – by the eccentricity index *G1*, displaceable in phase by 180° with respect to the eccentricity vector of the main cylindrical shaft datum under the assembly.

The free outer ring has only one-rotation motion around axis *Z*. The location of the outer ring relative to the inner ring should be normalized by five coordinates: three linear and two angular. Standard [5] declares only one deviation – the radial clearance of bearing which restricts the ring of two translational displacements along axes *X* and *Y*, expressed in linear coordinates with zero nominal values. Translational displacement along axis *Z* is limited by the linear error $\theta \pm EZ_5$, but in standards

IOP Conf. Series: Journal of Physics: Conf. Series **858** (2017) 012014 doi:10.1088/1742-6596/858/1/012014

[1, 2] it is not regulated by the tolerance despite the deviation of the axial gap, which is in the empirical dependence on the radial gap.
Standards [1; 2] also do not regulate two angular coordinates of the outer ring relative to the inner bearing assembly $90^{\circ}\pm AEX_4$, $90^{\circ}\pm AEY_4$.

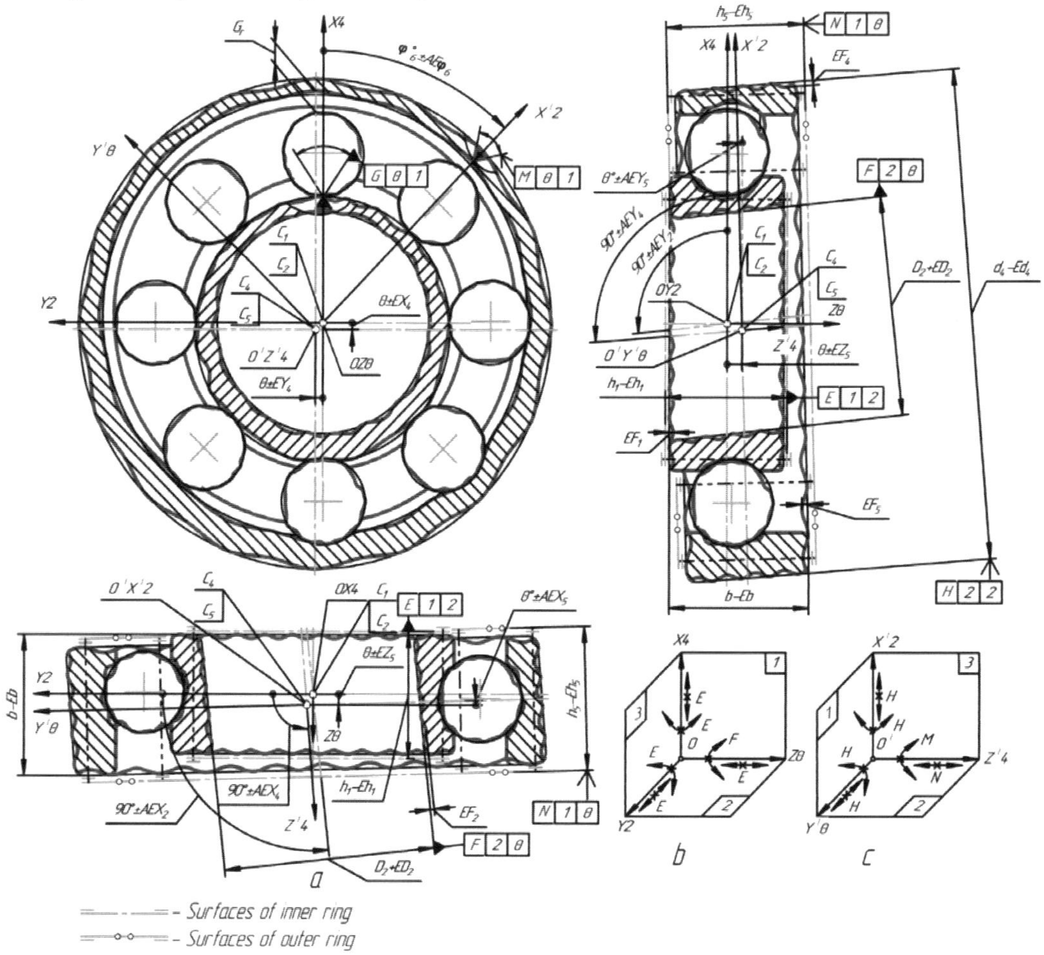

Figure 4. The single row radial ball bearing, model of geometric characteristics
a) geometric model in three projections; *b*) indicator of main datums set;
c) indicator of auxiliary datums set.

Tolerances perpendicularity axes and cylindrical features are normalized with respect to the ends of the outer ring, not to the datum end of the inner ring.
Due to geometrical deviations from the symmetry of the bearing rings and gaps existing between the rolling features and the rings the mounting width of ball bearing assembly *b-Eb* will differ from the width of the inner ring and the width of the outer ring, but in standards [1; 2] the mounting width of the ball bearing assembly is not specified.

4. Research Results

To evaluate deviation of the rings raceway symmetry from the symmetry plane with respect to the plane of symmetry of the prismatic feature formed by the two flat ends of the ball bearing the experimental research of the distance difference from the raceway plane of symmetry to the datum ends of the ball bearing ring were carried out.

The measurements of the ring bearing were carried out on the control unit with the transmitter scale division of 1 μm. The difference between the measured values of the distances to the raceway plane of symmetry in 24 points on the angle of rotation of the ring characterizes its deviation from the symmetry of the track relative to the track ends and is represented on a graph (Figure 5).

In order to establish the mounting width of the ball bearing assembly the measurements have been performed with successive installation of rings on the basic surface in a horizontal plane. In the course of measurement one of the rings was hang out in the free state and under the influence of gravity it shifted vertically. The results of measurement of the bearing assembly mounting width in the six end points of the free ring without rotation is described in graph (Figure 6).

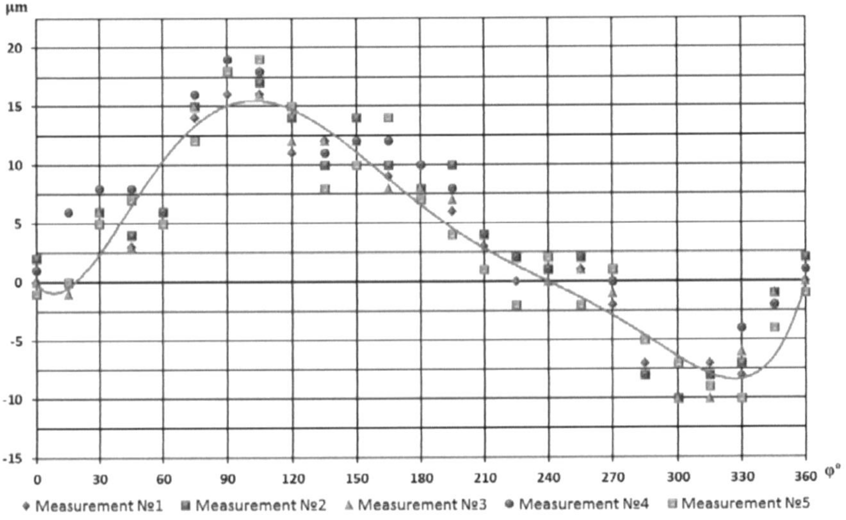

Figure 5. Protocol of measurement the racetrack plane symmetry deviations relative to the plane of symmetry of the outer ring ends.

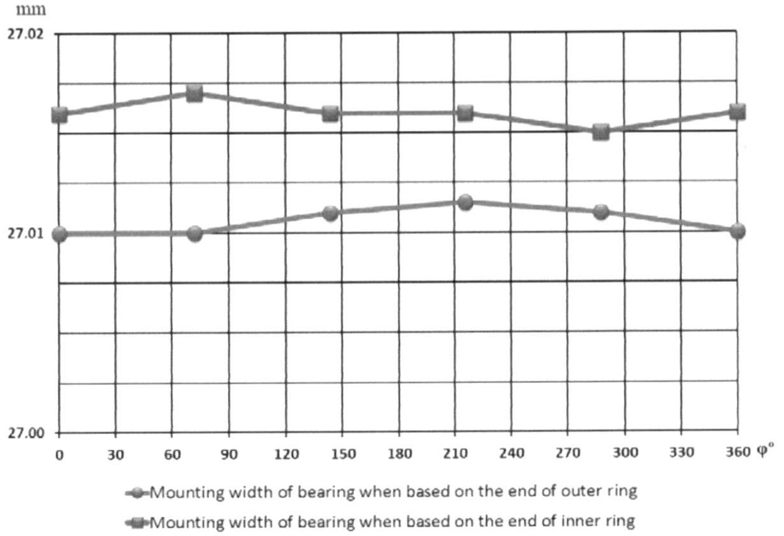

Figure 6. Protocol of measurements of ball bearing assembly mounting width.

5. Results Discussion

On the basis of adequate geometric modeling and experimental studies the optimal geometric specifications and dimensional tolerances and deviations of the bearing rings and arrangement assembly installed in rectangular coordinate systems have been established. A systematic approach has minimized the number of standardized geometric specifications. Simultaneously, new geometric specifications affecting the accuracy of the bearing have been determined. These include symmetry tolerances of the raceways, symmetry planes tolerance of the rings symmetry, and the mounting width of the ball bearing.

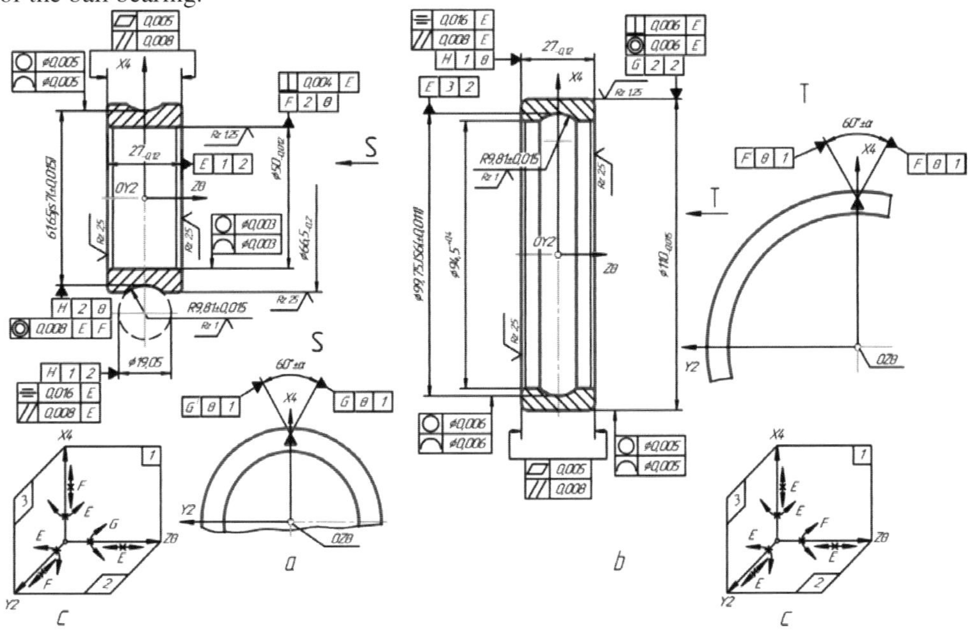

Figure 7. Rings of ball bearing 310, normal accuracy class - system setting of geometric specifications values and tolerances
a) inner ring; *b)* outer ring; *c)* informativeness indicators of main datums set of ring.

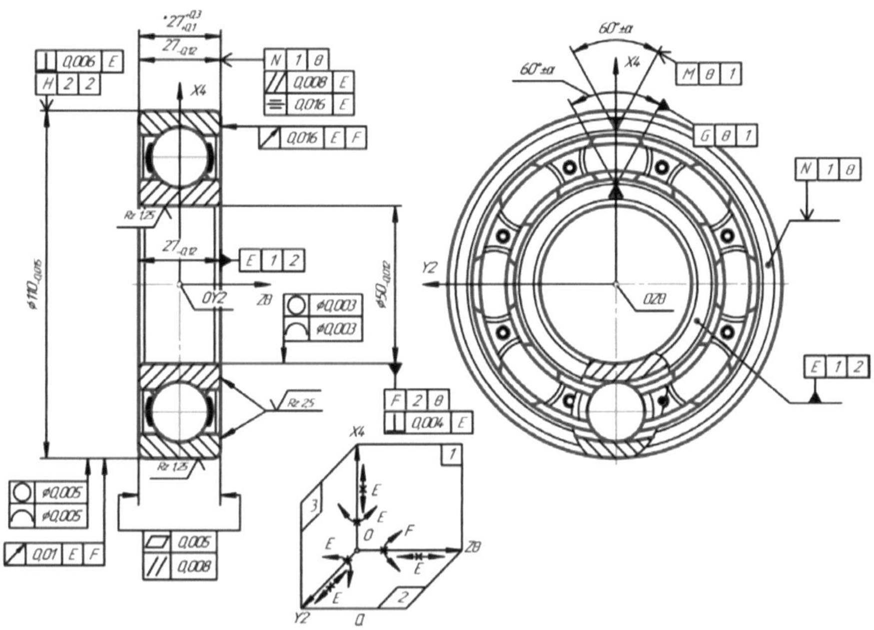

Figure 8. Single row radial ball bearing 310, normal accuracy class - system setting of geometric specifications values and tolerances
a) Indicator of main datums set; * – mounting width.

System setting of the rings values and geometric specifications is shown in (Figure 7), and the ball bearing assembly is presented in (Figure 8). The tolerances of geometric sizes limit all primary errors in the coordinate systems of main datums in full compliance with the geometric models of rings and ball bearing assembly (Figure 2...4).

The eccentricity vectors of the rings raceways identify one of the axes with informativeness 4 or 2 of the rings rectangular system of coordinates and the bearing assembly. The experimental studies of the geometric specifications of the rings and the bearing assembly have confirmed the existence of deviations from the plane symmetry and the ball bearing mounting width.

6. Summary

- Adequate geometric models allow the optimal composition of geometric specifications in the rings and bearing assembly coordinate systems to be identified.
- Coordinate systems of rings and bearing assembly are materialized by the datum features sets performing as primary design datums, defining the position of the bearing in operation.
- Eccentricity vectors of the rings raceways identify one of the axes of the Cartesian coordinate system and should be marked on the face of the ring with the indicator.
- The coordinate system allows one to optimize and minimize the number of geometric specifications that determine quality performance of bearings.
- The flat ends of the bearings have the same function and form a symmetrical plane-parallel prismatic feature, plane of its symmetry being the datum feature.
- Experimental studies have confirmed the presence of deviation of rings raceways and the ball bearing mounting width symmetry exceeding the width of the rings.

7. Conclusion

A systematic approach to the normalization of values and tolerances of ball bearings geometric specifications in the coordinate systems can improve the quality of the bearings by optimizing and minimizing the number of specifications. The introduction of a systematic approach to the international standards on rolling bearings is a guarantee of significant increase in accuracy of the bearings and the quality of products in which they are applied.

References

[1] ISO 492:2014 *Rolling bearings – Radial bearings – Geometrical product specifications (GPS) and tolerance values*
[2] ISO 492:2002 *Rolling bearings – Radial bearings —Tolerances*
[3] H Wiesner 2015 ISO 199:2014 and ISO 492:2014 standards – complexity versus unambiguity *journal SKF Evolution* **1** pp 27-30
[4] V I Glukhov 2014 Geometrical product specifications: Alternative standardization principles, coordinate systems, models, classification and verification *Dynamics of Systems, Mechanisms and Machines (Dynamics) (IEEE Conf. Pub.)*
[5] ISO 1101:2012 *Geometrical product specifications (GPS) – Geometrical tolerancing - Tolerances of form, orientation, location and run-out*
[6] ISO 5459:2011 *Geometrical product specifications (GPS) – Geometrical tolerancing – Datums and datum systems*

IOP Conf. Series: Journal of Physics: Conf. Series **858** (2017) 012015 doi:10.1088/1742-6596/858/1/012015

Geometric modeling of controlled third-class hinged mechanisms with a stand in one extreme position for cyclic automatic machines

V G Khomchenko [1], L G Varepo[1], V I Glukhov[1], E A Krivokhatko[1]

[1] Omsk State Technical University, 11, Mira ave., Omsk, 644050, Russia

E-mail: v_khomchenko@mail.ru

Abstract. The geometric model for the synthesis of third-class lever mechanisms is proposed, which allows, by changing the length of the auxiliary link and the position of its fixed hinge, to rearrange the movement of the working organ onto the cyclograms with different predetermined dwell times. It is noted that with the help of the proposed model, at the expense of the corresponding geometric constructions, the best uniform Chebyshev approximation can be achieved at the interval of the standstill.

1. Introduction

Cycle machines are widely used in a number of industries, in particular, machine-building, printing, food, textile and others [1, 2]. As a rule, such automatic machines have high productivity, provided by the speed of their executive mechanisms. The automatic machine composition usually includes a sufficiently large number of executive mechanisms, which requires strict coordination of the movements of their working organs in time and space.

2. Formulation of the problem

One of the most promising executive mechanisms for the implementation of fast-flowing technological processes by cyclic automatic machines are lever mechanisms that have a high load capacity and the possibility of a purposeful conversion of motion and transfer it to sufficient distances from the leading link to the working body within the machine. Lever mechanisms in such industries as printing and food can be considered as an alternative to cam mechanisms, which until now are quite common actuators of automatic machines with an unchanging cyclogram.

One of the most significant problems of cyclic automatic machines is their relative "rigidity" in the transition to a slightly different process from the previous process, associated with changes in the size of the manufactured product, the duration of individual operations, etc., and, consequently, with the corresponding change cycle of operation of individual actuators.

In the works [3-5], the methods of synthesis of hinged lever mechanisms were considered that ensure a given duration of the working element's standing in one and at its two extreme positions. There are also articles that deal with the issues of adjusting the length of stay and the amount of movement of the working body under certain conditions [6, 7].

IOP Conf. Series: Journal of Physics: Conf. Series **858** (2017) 012015 doi:10.1088/1742-6596/858/1/012015

It should be noted that geometric modeling is widely used in solving general problems of analysis and synthesis in the theory of mechanisms and machines [8-12], especially for lever (rod) mechanisms [9, 11].

In this paper, we propose a constructive method for geometric modeling of third-class hinged link mechanisms according to the Assur-Artobolevsky classification, oriented to obtaining such a displacement of the working part of a cyclic automatic machine, which has an approximate stopping of the output link in one of the extreme positions. In this case, such a stop (delay) occurs at the time points specified by the cycle and has a predetermined duration. In addition, the proposed method makes it possible to synthesize such mechanisms for the realization of a family of cyclograms formed by a number of fixed preset values for the duration of an organ working in one of its extreme positions.

3. Methods of solution

The geometric design of the lever mechanism of the third class for cyclograms with several fixed dwell times should be carried out in two stages.

The initial basic stage should be considered as the synthesis of the initial mechanism corresponding to some average duration of the standstill (to some intermediate specified value of the crank angle for the interval of the standstill). During this stage, all the necessary kinematic parameters of the lever mechanism are determined, as well as the initial length of the auxiliary link and the position of its fixed hinge on the rack of the machine.

The second stage consists in graphically constructing some additional points of the mechanism with other (smaller and larger) fixed values of the crank angle for the intervals of stability and allowing to determine the corresponding length and position of the fixed hinge of the auxiliary link.

The geometrical construction of the mechanism will be carried out in relative units of length, assuming the crank length equal to one. This will allow us to obtain more general solutions of the problem of synthesis of the mechanism and by scaling to integrate the obtained kinematic scheme into the design of the machine.

First phase. From the center A of the crank AB rotate the circle with the radius $a = 1$ (figure 1), draw the crank in the initial position AB_0 and in the positions AB_1 and AB_2 respectively at the angles φ_1 and φ_B, laid counterclockwise (here φ_1 and φ_B respectively the crank rotation angles for the interval working stroke and for the interval of standstill). Recall that in the first step, one of the prescribed intermediate elevation angles is adopted as the angle φ_B. We arbitrarily assign the length b of the connecting rod BE and, making the notches of radius b from the points B_1 and B_2, we find the position of the point E_1 at the beginning and at the end of the interval of finite stability. The position of the point E_0 with instantaneous growth is defined as the extension of the crank in the initial position AB_0. As a result, we obtained the positions of the links AB and BE at the beginning of the working stroke and at the beginning and end of the interval of stability. In the interval of extinction, the point E will approach the point A, reaching a minimum distance from it equal to $b - 1$. To determine the position E_p of the point E at the minimum distance from the point A, draw from this point a circle of radius $b - 1$ and a straight line AE_p under with some small angle ρ. The intersection of the named circle and the straight line will be taken as the position where the point E_p will have momentary exponentials at the interval of finite stability.

The foregoing construction corresponds to the definition of the characteristic positions of the so-called two-link drive [4].

We determine the position of the C_1 point of the output link of the CD mechanism at the beginning and end of the interval of the gap φ_B. It is determined from the condition of ensuring a uniform Chebyshev approximation, in accordance with which the point C_1 must occupy the same position both at the beginning and the end of the interval of extinction, and at the moment of instantaneous stability of the point E_p. Therefore, the point E_p can be graphically found as the point of intersection of circles drawn from the points E_1 and E_p by the preset length e of the EU side of the triangle link ECF.

We find on the triangular link ECF the point F in position F_1 along the previously assigned length *f* of the side CF of this link and the angle μ.

Let us now turn to the construction of the positions of the links of the mechanism at the instant of instantaneous standing of the working organ. From the point E_0 under the preselected angle γ we draw the side CF of the triangular link ECF and finish this link in the position $E_0C_0F_0$. To complete the construction of the kinematic scheme of the mechanism, it is now sufficient to find the positions of the rotation centers D and H of the fixed hinges, respectively, of the output link CD and the auxiliary link FH. The position of the point D and, therefore, the length of the CD are at a given angle of the span ψ^m with the use of the chord C_1C_0. To determine the point H and the length FH, it is necessary to draw perpendiculars from the middle of the segments E_1E_p and F_1F_0. The intersection of these perpendiculars is the center of the fixed hinge H, and the distance FH is the length of this link.

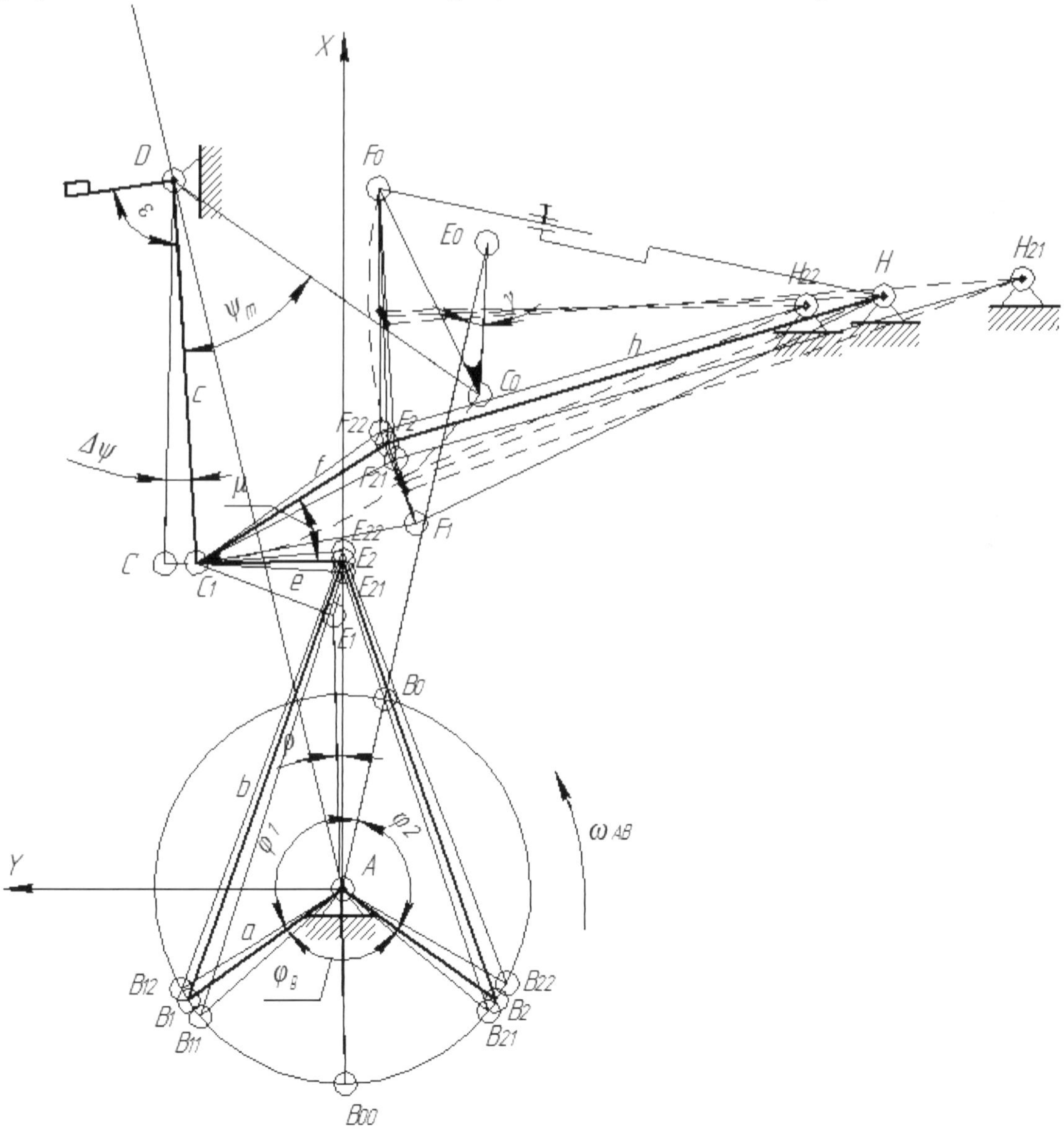

Figure 1. Geometric model of the kinematic scheme of the hinged mechanism of the third class with a stand in one extreme position.

As a result of the geometric constructions carried out at the first stage, a kinematic scheme of the hinged mechanism is realized, realizing the cyclogram with the adopted intermediate prescribed angle of the gap φ_B. The sequence of the presented constructions is the content of the first stage of synthesis. *Second phase.* In the second stage of the synthesis, it is necessary to find the position of the point H for all other (smaller and larger) values of the corners of the stability. The points B_1 and B_2 for other angles of stability will be laid symmetrically to the bisector of the angle B_1AB_2 and from these points we will determine the positions of the points E_1 for each of the corners of the dwell, and then the position of the point F_1. In accordance with the new positions of the point F_1, as before, the positions of the center H of the fixed hinge of the auxiliary link FH are determined on the perpendiculars drawn from the middle of the segments E_1E_p and F_1F_0.

As an example, in the figure 1, the constructions for angles of stability equal to 100, 110 and 120 angular degrees were made (in the first stage, 110° was assumed as the intermediate value of the standoff angle). The positions of the characteristic points B_1, B_2, E_2, F_2 and H of the mechanism obtained in the second stage of the synthesis for the angles of stability equal to 100 and 120 angular degrees are marked with an additional lower index 2.

To determine the error of the working element's standing, it is sufficient to make a notch on the circle of radius CD with the center at point D equal to the sum of the lengths of the links FH and CF. The angle C_1DC_P is the deviation of the working element from the standing position. With a suitable selection of the parameters of the mechanism, it is possible to ensure a sufficiently small value of this angle [1].

To switch the mechanism from one cycle to another, it is sufficient to release the fastener on the telescopic link FH and to re-attach the hinge H to the position corresponding to the required stop angle (in the figure 1 of position H_{21} H_{22}).

It is important to note that when switching from one cycle to another, the angular working stroke of the output link ψm remains unchanged and is equal to the originally set value. The required position of the operating element in the automatic machine relative to the output link of the mechanism can be set by changing the angle ε.

4. Results

A geometric model for the synthesis of lever mechanisms of the third class is obtained, allowing in two stages to design the hinge mechanisms for a number of fixed, preset values of the corners of the stability.

5. The discussion of the results

The proposed geometric model contains all the necessary data for the transition, if necessary, to the analytical method of geometric modeling in order to obtain the corresponding mathematical dependencies.

6. Conclusion

The proposed method of geometric design can be used when developing the appropriate nomograms or tables for setting up the considered mechanisms for reproducing the cyclogram with the required length of the standby due to the change in the length of the auxiliary lead and changing the position of its fixed hinge on the body of the machine with the remaining lengths of the remaining links. The method has the property of visibility, which makes it suitable for use in engineering practice.

References

[1] Wilhelm S R, Sullivan T and Van de Ven J D 2017 Solution rectification of slider-crank mechanisms with transmission angle control *Mechanism and Machine Theory* **107** pp 37–45

[2] Nadezhdin I V 2010 Designing of level mechanisms of cyclic automatic machines (Moscow: *Mechanical Engineering)* 232

[3] Chaudhary K and Chaudhary H 2015 Optimal dynamic balancing and shape synthesis of links in planar mechanisms *Mechanism and Machine Theory* **93** pp 127–146
[4] Gogate G R 2016 Inverse kinematic and dynamic analysis of planar path generating adjustable mechanism *Mechanism and Machine Theory* **102** pp 103–122
[5] Chanekar P V and Ghosal A 2013 Optimal synthesis of adjustable planar four-bar crank-rocker type mechanisms for approximate multi-path generation *Mechanism and Machine Theory* **69** pp 263–277
[6] Khomchenko V G, Bakcheev V A, Khorunzhin V S and Skabkin N G 2002 Synthesis of level mechanisms of the third class with an adjustable stand from the condition for minimizing the deviation of the asymmetry angle *Mechanics of processes and machines: collection of scientific papers*, ed V V Evstifeeva (Omsk: OmSTU) pp 8–17
[7] Qu X and Kim D 2012 Kinematic design and analysis of a four-bar linkage-type continuously variable valve actuation mechanism *Mechanism and Machine Theory* **57** pp 111–125
[8] Suna J, Chena L and Chu J 2016 Motion generation of spherical four-bar mechanism using harmonic characteristic parameters *Mechanism and Machine Theory* **95** pp 76–92
[9] Tiana Y, Shirinzadeha B, Zhangb D, Liuc X and Chetwynd D 2009 Design and forward kinematics of the compliant micro-manipulator with lever mechanisms *Precision Engineering* **33(4)** pp 466–475
[10] Volkert van der Wijk 2017 Design and analysis of closed-chain principal vector linkages for dynamic balance with a new method for mass equivalent modeling *Mechanism and Machine Theory* **107** pp 283–304
[11] Balchanowski J 2016 General method of structural synthesis of parallel mechanisms *Archives of Civil and Mechanical Engineering* **16(3)** pp 256–268
[12] Bronsvoort W, Gravesen J and Keyser J 2011 Theory and practice of geometric and physical modeling *Computer-Aided Design* **43(7)** pp 739–740
[13] Zhaoa H, Hana D, Zhangb L and Bi S 2017 Design of a stiffness-adjustable compliant linear-motion mechanism *Precision Engineering* **48** pp 305–314
[14] Imbach R, Schreck P and Mathis P 2014 Leading a continuation method by geometry for solving geometric constraints *Computer-Aided Design* **46** pp 138–147

IOP Conf. Series: Journal of Physics: Conf. Series **858** (2017) 012016 doi:10.1088/1742-6596/858/1/012016

Fractal parametrization in erosion process and surface investigation received by electrospark modification

D N Korotaev [1], **E V Ivanova**[2], **V A Kim**[3]

[1] Omsk State Technical University, 11, Mira ave., Omsk, 644050, Russia
[2] Omsk Tank-Automotive Engineering Institute, 119, 14th military town, Omsk, 644098, Russia
[3] Komsomolsk on Amur State Technical University, 27, Lenina ave., Komsomolsk on Amur, 681013 , Russia

E-mail:korotaevd99@mail.ru

Abstract. High-energy metal surfaces processing by electrospark modification (ESM) leads to formation of the difficult structural state of work materials in non-equilibrium thermodynamic conditions by the self-organizing laws. The article deals with the fractal geometry approaches for a quantitative assessment of the erosion processes by using ESM and the surface structure states after treatment. The authors determine the synergies between fractal dimensions of the surface structures, received by ESM of steel, and their physical and mechanical properties. They point out the surface formation with the maximal gauge, microhardness and durability in the case of definite energy of a spark impulse. In this condition, the fractal dimension of structures is maximum too.

1. Introduction

Maintainability engineering products and increasing their resource become possible due to hardening their detail surface layers. The authors introduce several technologies of the surface modification with high concentrated energy flows. In the industry, the method of electrospark modification (ESM) is widely known. This method is based on the material transfer phenomenon of an alloying electrode (anode) on a processed surface (cathode) at the spark discharge. The new surface forms because of such external actions. Processes, which lead to formation of a new structural state of the material, are activated in the surface layer [1, 2].

The processes, occurring at the ESM, carry in the extremely non-equilibrium thermodynamic conditions. The material erosion of an alloying electrode and surface layers with the modifying structure are the basic processes, which influence physical and mechanical properties of materials. These processes regard to dynamic self-organizing processes.

Research and quantitative assessment of the self-organizing processes in materials is based on the fractal parametrization. It is a fresh approach on the analysis, modelling and forecasting of material properties. The fractal structures are used for a geometrical assessment of self-similar and different-scale objects. They can be multipurpose tool for analysis of the self-organizing process while material investigation [3, 4].

The fractal parametrization allows getting several fractal correlation in any objects and the processes having complex multi-stage structure. In this regard, the fractal parametrization, except information

IOP Conf. Series: Journal of Physics: Conf. Series **858** (2017) 012016 doi:10.1088/1742-6596/858/1/012016

loading, reveals physics of the processes. To sum up, using the fractal formal description to the process analysis with the ESM materials is topical enough. Erosion of the particle stream and the case with complex grain and block structures are characterized by the fractal properties of the ESM.

The purpose of the research is evaluation of optimum technological parameters of the ESM. They control structural state and material properties by the surface treatment in terms of the contact between these parameters with the fractal dimension of the structure.

2. Results and discussion

The central technological parameters and conditions of the ESM are electrode material, pulse energy and machine time. They determine the structure and properties of the surface layer. These parameters define quality of the surface and their characteristics such as adhesive strength with a foundation, a structure and coating thickness, and durability.

The erosion mass stream composes with a burst form scattering substance of an electrode material under the ESM. Development of the erosion particle stream with the high internal energy proceeds in the non-equilibrium conditions. It comes amid further size reduction to the state wherein the total surface energy of disintegration products of the particles will completely counterbalance their internal energy. Energy conditions of such particle organization form are expressed by the equation

$$\gamma \cdot \sum F_i = \mu \cdot V , \tag{1}$$

where γ is the specific surface energy; μ is the specific internal energy; ΣF_i is the total area of all erosion stream particles; V is the total volume of the particles.

Make a transformation of the equation (1) to the form

$$\frac{\mu}{\gamma} = \frac{\sum F_i}{V} . \tag{2}$$

From (Equation 2) follows that with increase of the internal energy, erosion substance flow will have "the developed" surface, i.e. small erosion particles.

The mass erosion stream is non-homogeneous in its dimensional range and characterized by the fractal proportion

$$N = \frac{C}{\upsilon^D} , \tag{3}$$

where N is the total number of erosion particles; D is the fractal dimension of the erosion product distribution by volume; υ is the volume of a separate erosion particle; C is the constant characterizing by self-similarity of the received particles.

If the system is non-equilibrium, the frequency of the particle distribution on the volume n, we can present it in the form of differential exponential function. As a result, we receive following expression

$$n = -\frac{dN}{d\upsilon} = \frac{DC}{\upsilon^{D+1}} . \tag{4}$$

The erosion particles have the sphere form. It is possible to use the correlation from geometry, such as each particle with the volume υ has the surface area proportional to $\upsilon^{2/3}$.

Radius r expresses in terms of the sphere volume formula, we can substitute it in the formula of semisphere surface area and we receive

$$S = 4\pi \left(\frac{3 \cdot \upsilon}{4 \cdot \pi} \right)^{\frac{2}{3}} .$$

The total area of all erosion particles can be defined under the formula

$$\sum F_i = \sum \upsilon_i^{2/3} \cdot n_i = CD \int_0^{\upsilon_{\max}} \frac{d\upsilon}{\upsilon^{D+\frac{1}{3}}} = \frac{CD}{\frac{2}{3} - D} \cdot \upsilon_{\max}^{\frac{2}{3}-D}. \tag{5}$$

The total volume of the erosion particles can be defined on the expression

$$V = \sum \upsilon_i n_i = CD \int_0^{\upsilon_{\max}} \frac{d\upsilon}{\upsilon^D} = \frac{CD}{1-D} \cdot \upsilon_{\max}^{1-D}. \tag{6}$$

In this case, the proportion of the total area of all erosion particles to their total volume is

$$\frac{\sum F_i}{V} = \left(\frac{D-1}{D-\frac{2}{3}} \right) \cdot \frac{1}{\sqrt[3]{\upsilon}} = \frac{C_*}{\sqrt[3]{\upsilon}}, \tag{7}$$

where C_* is the constant, characterised by the fractal dimension of particles.
In figure 1, percentile curve of microwells diameters are presented. They are formed after interaction of the erosion particles with a processed surface and represented ordinary scaling dependence. The dependence data are resulted from measurement of hundred microwells.
The fractal dimension are counted off the diameters erosion microwells using the program Image Pro Plus 5.1. [4]. It is defined under the formula

$$D = \frac{\Delta\left(\ln \sum P\right)}{\Delta\left(\ln d\right)},$$

where $\Delta(\ln \Sigma P)$ is the logarithm increment of cumulateive frequency, $\Delta(\ln d)$ is the logarithm increment of the microwell diameters.

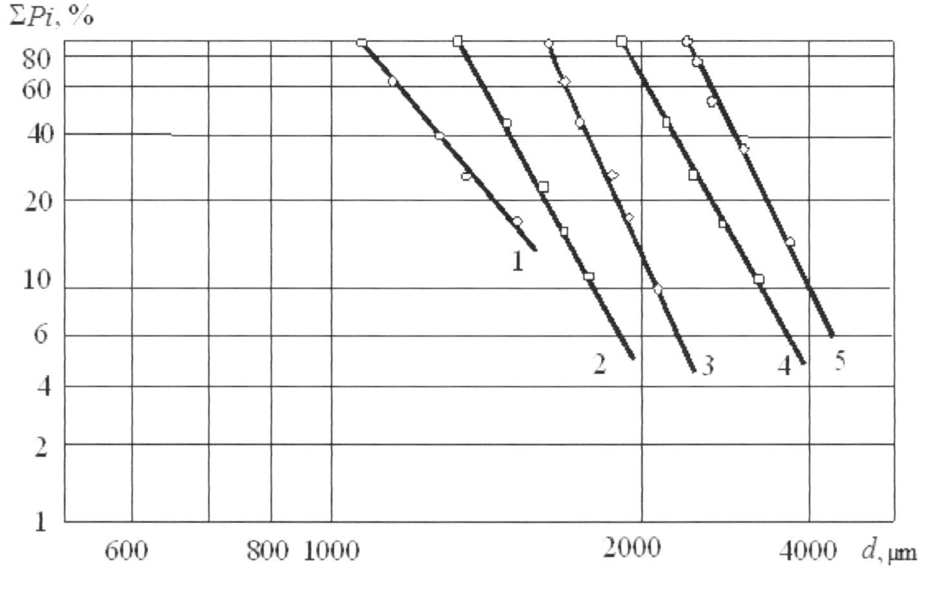

Figure 1. Integral distribution of the microwell diameters under the steel ESM with various pulse energies: $E_1 = 0{,}022$ J; $E_2 = 0{,}09$ J; $E_3 = 0{,}25$ J; $E_4 = 0{,}73$ J; $E_5 = 0{,}86$ J.

IOP Conf. Series: Journal of Physics: Conf. Series **858** (2017) 012016 doi:10.1088/1742-6596/858/1/012016

There are the fractal dimensions of self-similarity (D_o), coating thickness (δ), microhardness of a surface layer (*HV*) and ratio distortion of interplanar distance of a crystal lattice ($\Delta d/d$) depending on energy of the singular spark discharge (*E*) in the table 1.

Table 1. Research results of the steel surface after the ESM treatment by the electrode of BK6M mark.

Pulse energy, *E*, J	Specifics of the surface-modified layer			
	D_o	*HV, MPa*	δ, μm	$(\Delta d/d)\cdot 10^{-3}$
0,022	2,130	8450	20	1,5
0,09	2,571	11450	25	1,5
0,25	2,453	13820	35	2,5
0,73	2,605	16400	55	3,5
1,15	2,444	14610	40	2,7

The tendency of the nonmonotone fractal dimension change D_o becomes apparent with energy increment of the singular spark discharge. Value of the fractal dimensions $D_o > 2$ verifies by high maturity (activation) of the surface. Microhardness, coating thickness and ratio distortion of a crystal lattice depend on the pulse energy and they are nonmonotone by experimental results. The range of the maximum values arrangement for investigated properties reside in the same area of the electrospark impulse energy. The microhardness and the crystal lattice degree of distortion grow with increase in energy of the electric discharge. The specified physical and mechanical properties depend on defect concentration of a crystal structure [5, 6].

When we increase the defect concentration of the crystal lattice adhesion and diffusion processes become more active and the surface adhesive strength with a base enhances. Stability, bigger gauge and wear resistance increase of the surface layer characterize the changed surface.

Figure 2 presents the connection of steel plate wearing value, which is strengthened of ESM by steel 50 turning.

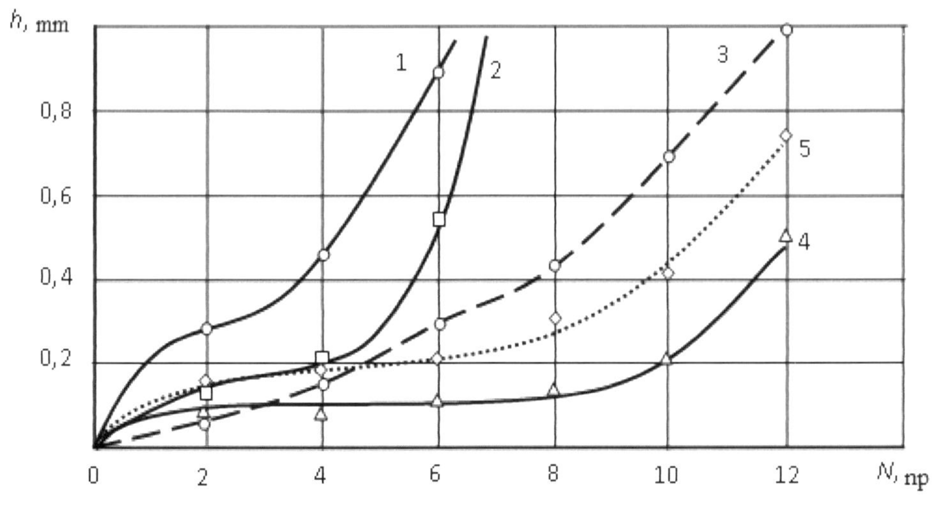

Figure 2. Dependences of steel plate wearing by steel 50 turning: *1* – after quenching; *2* – $E = 0{,}022$ J; *3* – $E = 0{,}25$ J; *4* – $E = 0{,}73$ J; *5* – $E = 1{,}15$ J.

The initial steel surface after heat treatment and strengthened with the spark discharge energy $E = 0{,}022$ J are characterised by the short period of the stationary wear and abrupt junction in the area of

"pest" degradation. There are wearing distinctive features of the modify sample with higher values in the pulse energy. These features are small wearing value at the stage of running-in (h = 0,15 … 0,175 mm) and smooth transition in the area of the catastrophic wearing. Steel plate hardness enhances with the pulse energy increasing, has a maximum at E = 0,73 J and then starts to decrease.
We have analyzed the received linear connections of wearing and the sequels of the table. As a result, the biggest coating thickness and the high density of the crystal structure defects support the maximum hardness of the steel surface reinforcement at E = of 0,73 J. Besides, the fractal dimension of the received structure D_o is maximum at this value of the pulse energy. It verifies workability of the fractal dimension as a criterion for an optimum choice of technological modes and conditions of the ESM.

3. Conclusion

1. Scaling invariance of the fractal structure parameters means the correlation with it physical and mechanical properties, including durability. The connection between the fractal dimension of the hardened surface and its durability is appeared in identical change of these characteristics with increase of the singular discharge energy. Extreme points are situated in the same range of energies.
2. The fractal dimension can be considered as the topological characterization of the area. Internal state energy enhances with the fractal dimension increase. Connection between the fractal dimension and structural energetic properties are verified with the fractal size rising with increase of the crystal structure defect concentration. The comparative analysis of the change Do and Δd/d confirms the conformity for purposes of the surfaces strengthened by the electrospark high-energy influence.
3. The analysis shows the presence of the fractal correlation of the surface characteristics with the spark discharge energy. It is possible to use the fractal dimension of the material structural state as a criterion for a choice of the optimum technological modes of the ESM and formation of the case.

References

[1] Mashkov Y K, Malij O V, Korotaev D N, Alimbaeva B S and Baybaratskaya M Y 2016 The effect of electro-spark machining on the structure and properties of modified friction surfaces *J. of Friction and Wear* **37(1)** pp 66–70
[2] Demminger C, Klose C and Maier H J 2016 Microstructure and Magnetic Properties of Cobalt and Zinc Containing Magnesium AlloysOriginal Research Article *3rd Int. Conference on System-Integrated Intelligence: New Challenges for Product and Production Engineering* 1 **26** pp 35–42
[3] Blednova Zh M and Stepanenko M A 2013 Structure development at the steel surface modifying by the intelligent material based on TiNi *Fundamental problems of the modern materials technology* **10(1)** pp 60–67
[4] Kim V A, Katuntseva N V and Kochetkov M S 2016 The multifractal analysis of the armko iron structural transformations at laser treatment *Scientific notes of Komsomolsk on Amur State Technical University* **1(25)** pp 97–104
[5] Mashkov Yu K, Korotaev D N, Baybaratskaya M Yu and Alimbaeva B Sh 2015 Research nanostructured surfaces synthesized by the method of electro-spark machining *Periodical of applied physics* **85(10)** pp 75–79
[6] Ghosha C, Aranas Jr C and Jonas J J 2016 Dynamic transformation of deformed austenite at temperatures above the Ae3 *Progress in Materials Science* **82** pp 151–233

IOP Conf. Series: Journal of Physics: Conf. Series **858** (2017) 012017 doi:10.1088/1742-6596/858/1/012017

Development of 3D modeling technology for manufacturing finned ribbons from heat-resistant steels

A A Lyashkov[1], E V Vasil'ev [1]and A Y Popov[1]

[1]Omsk State Technical University, 11, Mira ave., Omsk, 644050, Russia

E-mail: 3dogimod@mail.ru

Abstract. The process of shaping a workpiece by a tool using the rolling method is, from the geometric point of view, a process of interaction of two conjugate surfaces. The technology of rolling finned stainless steel ribbons is close to the technology of shaping details by cutting. However, the problems of its practical implementation in the well-known papers analyzing this issue are practically not considered. As a result of the analysis of conjugate surfaces profiling methods in relation to the problem, it was concluded that it seems urgent to develop a methodology for the formation of corrugated ribbon based on 3D modeling use. The implementation of this methodology includes the creation of solid models of the product and the tool, as well as computer simulation of their shaping processes using rolling method. So, at the first stage, a 3D model of finned ribbon was developed, which was then used to produce a profile of a rolling tool. The modeling of this profile was carried out on the basis of the proposed software package in the CAD environment. The created theoretical model of the tool profile was replaced from the technological point of view by a rectilinear profile. To carry out the analysis of the obtained results, the inverse shaping problem was solved - according to the corrected profile of the tool, real profile of the corrugated ribbon is obtained. Computer modeling of extruded volumes in the process of shaping was performed. The analysis of qualitative and quantitative parameters of the extruded volumes made it possible to give recommendations on setting the increment of the tool motion parameter. Based on the results of the studies, profile parameters of the roller are assigned for its practical implementation. The proposed methodology, based on 3D-modeling, allowed to develop a technology for manufacturing finned ribbons from heat-resistant steels by rolling with high productivity, accuracy and stability of the sizes obtained.

1.Introduction

There is a group of industrial heat exchangers where the heating of the medium is carried out by cylindrical elements equipped with corrugated stainless ribbon and forced feeding of the heated gas mixture is provided. Figure 1 shows a sketch of a corrugated ribbon made of heat-resistant steel of thickness B, height H and pitch T. The finned corrugations are inclined at angle α. To reduce hogging of finned ribbons, cuts are made along them. Making corrugated ribbons tens of meters long is a technologically complicated process which requires the profile precision and ribbon straightness.

The technology of rolling finned stainless steel ribbons is close to the technology of shaping parts by cutting using the rolling method. If a lot of theoretical approaches are known in the field of the cutting tool profiling, though the problems of practical implementation of this method with regard to shaping corrugated ribbons are practically not considered in the existing studies. The problems are caused by operations of corrugations shaping. In this regard, the task is urgent. As a result of the analysis of

IOP Conf. Series: Journal of Physics: Conf. Series **858** (2017) 012017 doi:10.1088/1742-6596/858/1/012017

conjugate surfaces profiling methods, it was concluded that it would be advisable to develop a methodology based on the use of 3D modeling with respect to corrugated ribbon shaping.

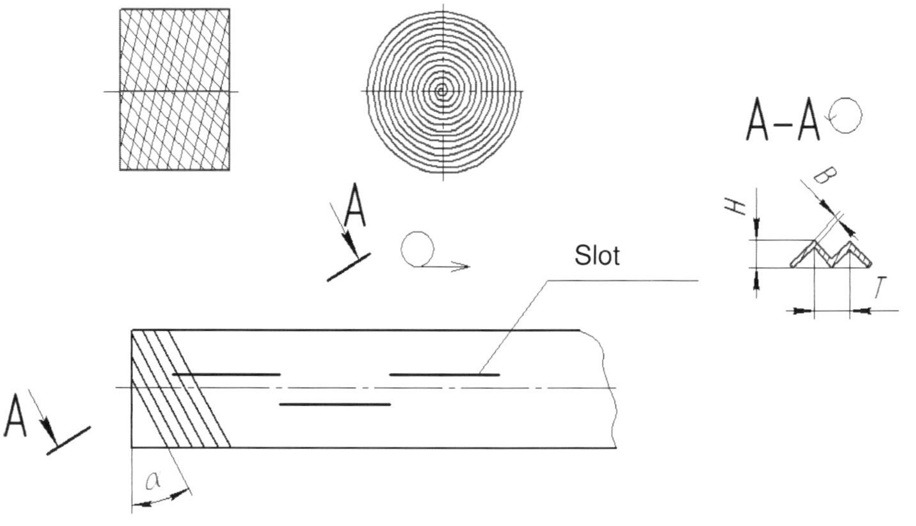

Figure 1. Corrugated ribbon.

2. Statement of the problem

The purpose of this work is to develop a technology for processing corrugated ribbons with high efficiency, accuracy and stability of the resulting sizes based on the creation of geometric and solid models of a finned ribbon and a tool for rolling it using 3D modeling.

The research tasks are the following: creation of 3D-model of corrugated ribbon; creation of a software package providing, in an automated mode, modeling the shaping of a tool for processing a ribbon; replacement, from technological considerations, the obtained profile by the rectilinear one; solution of the inverse problem of shaping - modeling the ribbon profile by a tool obtained in the previous stage; analysis of the obtained data and developing a resulting technology for corrugated ribbons processing.

3. Computer modeling of ribbon shaping by rolling method

3.1. Statement of the shaping problem

The process of item shaping with respect to surface treatment by rolling and rolling is the interaction of two conjugate surfaces - the workpiece and the tool. These processes are quite similar by their kinematic schemes. In both cases, the shaping is carried out under certain relative movements of a tool and a workpiece. In these schemes, the combinations of rectilinear-forward and rotational motion are used. The rolling method is based on the principle of gearing, for which the kinematics is studied and the developed apparatus can be used to determine the parameters of the tool [1, 2].

Let us consider the rolling process from the point of view of the centroid envelope method. This method is characterized by the fact that during the process of shaping the workpiece, the centroids of the tool and the processed wheel roll over each other without sliding. The profile of the workpiece is obtained as the envelope curve of different positions of the tool profiles. The centroids in these cases are the initial line and the initial circle.

A lot of papers have tried to solve the problems of manufacturing workpieces with a cutting tool. During their solving by analytical methods for calculation the computational dependences with reference to various initial data are required. Such dependencies have often a form of transcendental equations. Their solution involves the use of numerical methods, which complicates the process of profiling.

Modern CAD-systems make it possible to perform the modeling of shaping and to study the influence of various tool parameters on the workpiece profile shape and vice versa [3 - 8].

To solve this problem, algorithms and programs are proposed that perform in the automated mode solid modeling of the shaping of a rolling tool for manufacturing ribbon-like items which have the geometry of a rack-type tool. The inverse problem is also solved that is modeling the shaping of products of the ribbon type according to the obtained tool - the roller. In addition, in the process of shaping, solid modeling of the extruded layers is provided, which makes it possible to obtain some of their quantitative and qualitative characteristics. These characteristics are used for the optimal shape sizing. The proposed solutions are mainly based on the results of [8, 9]. A generalized algorithm for solving the problems based on the modeling of shaping workpieces by the method of centroid enveloping by means of computer graphics, proposed in [9], was also used to solve the described problem.

3.2. Creating a solid-state ribbon model

The initial data for modeling is the ribbon profile in its normal section, shown in Figure 5. However, for computer modeling of the ribbon profile, additional parameters α or *m* are required (Figure 3).

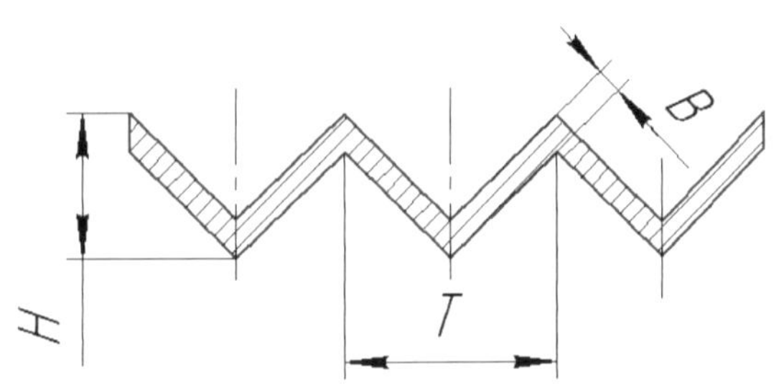

Figure 2. Initial ribbon profile and its parameters.

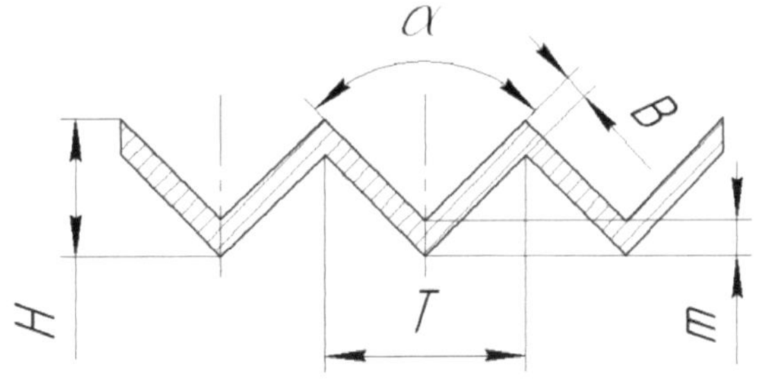

Figure 3. Ribbon profile and its parameters for modeling.

It follows from Figure 3 that

$$m = B / \sin(\alpha / 2),$$

$$\sin(\alpha/2) = \left\{B * H \pm \sqrt{B^2 * H^2 - \left[H^2 + (T/2)^2\right] * \left[B^2 - (T/2)^2\right]}\right\} / (H^2 + T^2)\ .$$

Let us take the following values of the initial data for further modeling: B = 0.4 mm; H = 0.1 mm and T = 3 mm. Then we get: m = 0.428088 mm; $\alpha = 69.1294^\circ$ or $\alpha = 69^\circ 7' 46''$ (Figure 4).

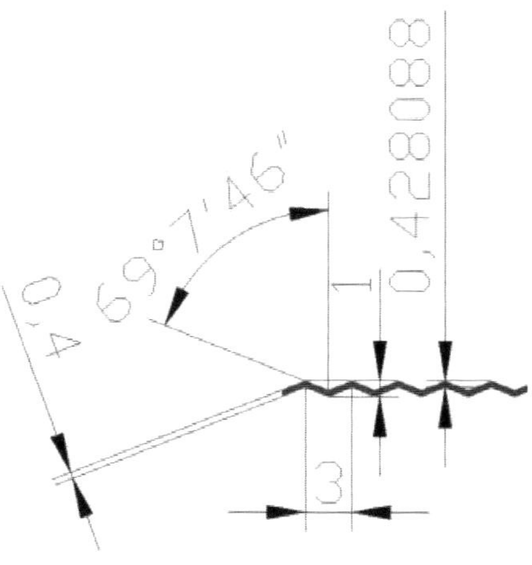

Figure 4. Ribbon profile parameters in the normal section.

The resulting profile is the initial one for the production of a solid-state ribbon model. At the first stage, a solid model of a workpiece is created for the ribbon, on the basis of which the ribbon model part is formed. The result of modeling the ribbon compartment is shown in Figure 5. It presents the projection of the model and its section: oblique – A-A and normal - B-B. Section A-A is the starting point for modeling the shaping of the tool - the roller. This section and its parameters is shown in Figure 6. Before modeling, let us determine the diameter of the initial cylinder D_{in} of the roller, knowing its outer diameter D and pitch H_k of the ribbon profile in its oblique section. The value of H_k is determined by the model of the ribbon profile ribbon (Figure 6) and is 3.916222 mm. D_{in}, H_k, and the number of teeth Z are related by the dependence $\pi * D_{in} = Z * H_k$, so we get: Z = 64 and D_{in} = 79.7806 mm.

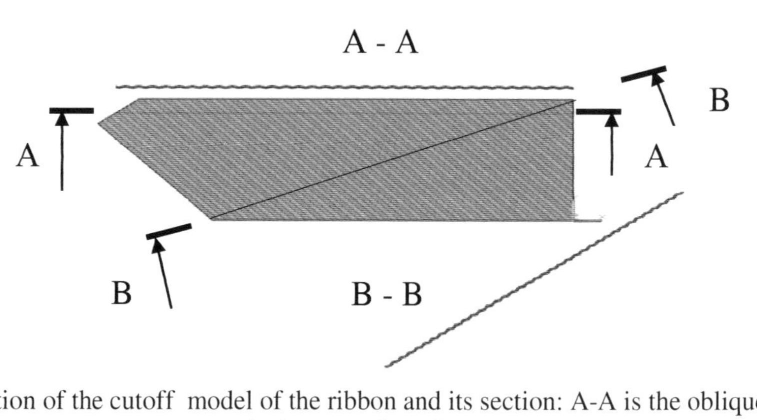

Figure 5. Projection of the cutoff model of the ribbon and its section: A-A is the oblique section; B-B is the normal section.

IOP Conf. Series: Journal of Physics: Conf. Series **858** (2017) 012017 doi:10.1088/1742-6596/858/1/012017

The resulting cross section model and its installation parameters are used later for modeling a tool for manufacturing the ribbon.

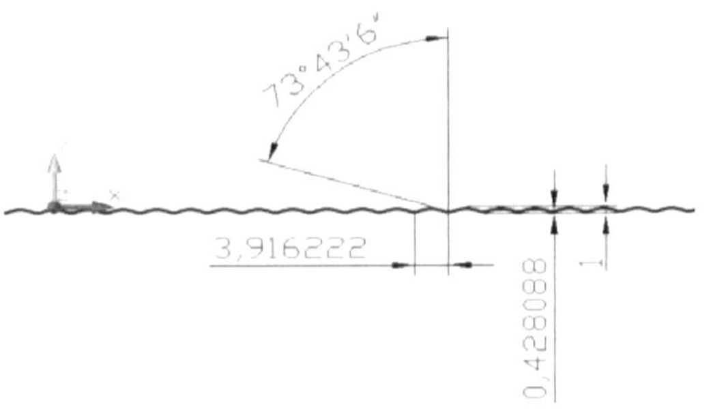

Figure 6. Computer model of the ribbon oblique section and its parameters.

3.3. Computer and solid modeling of direct and inverse problem shaping

3.3.1. Solid modeling of the tool for shaping the ribbon

In the developed software package, it is proposed to carry out the tool surface shaping on the basis of solid modeling to solve the described problem. Then, in accordance with the algorithm of shaping modeling [9], at first, solid models of the tool and the detail workpiece are created, each in its coordinate system. The location of these coordinate systems is determined by the installation parameters of the workpiece relatively to the tool.

The relative movement of the ribbon model, which determines the kinematic pattern of the shaping, is similar to the kinematic scheme of shaping the workpiece with a tool rack.

The sequence of solving the task on the basis of the developed algorithms and programs is presented below. At the first stage of shaping, the models of the initial data are set: the model of the ribbon and the workpiece for the roller that is its end section (Figure. 7).

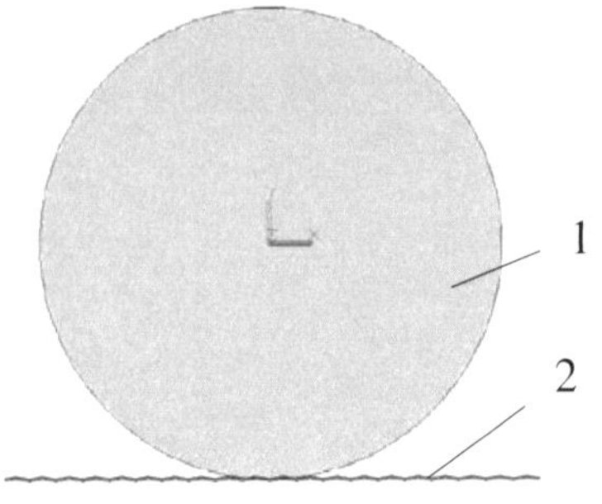

Figure 7. Models of a ribbon profile - 1 and a workpiece - 2 for a roller.

At the second stage, to obtain preliminary modeling results, a family of ribbon models is built by rolling its initial straight line along the initial circumference of the roller (Figure 8). The scaled-up fragment of the obtained family is shown in Figure 9. As it is known, the profile of the ribbon is outlined by the involute curve, which is confirmed by Figure 9. No transitional curves are observed. This means that the profile of the roller does not have sections outlined by special curves, which allows us to proceed to the next stage - solid modeling.

IOP Conf. Series: Journal of Physics: Conf. Series **858** (2017) 012017 doi:10.1088/1742-6596/858/1/012017

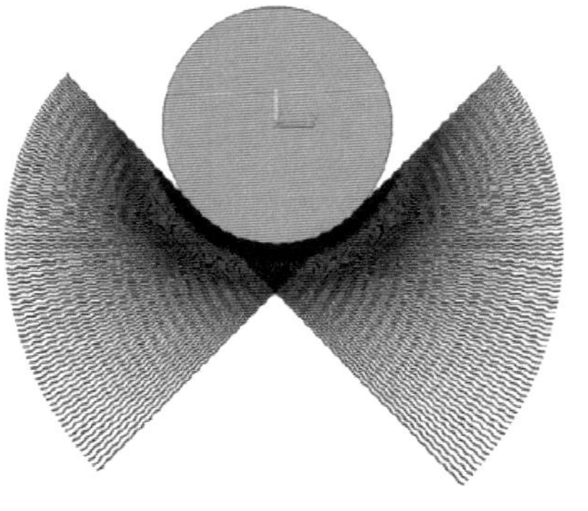

Figure 8. A family of ribbon profiles in its relative motion.

Figure 10 shows the shaping of roller profiles by a ribbon model. At this stage, the designer can vary the value of the family parameter to obtain the required accuracy of the tool profile.

The final result of the modeling is shown in Figure 11a. A scaled-up fragment of the ribbon and the roller profiles is shown in Figure 11b. According to the figure, practically these profiles contact not in a point touch (theoretically), but along the part of their profiles.

Due to the fact that it is technologically easier to produce a roller tooth having a rectilinear profile than the involute outlined (theoretically), the corresponding approximation is performed, as shown in Figure 12. In this figure, the profile side is replaced by a segment passing through the points of the vertex and the tooth hollows.

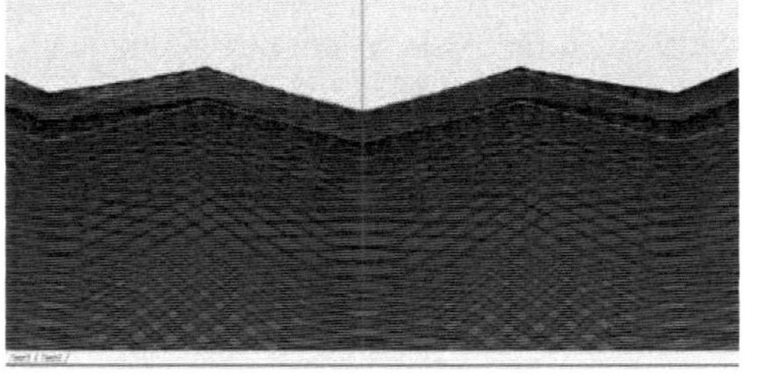

Figure 9. Scaled-up fragment of ribbon profiles family on an enlarged scale.

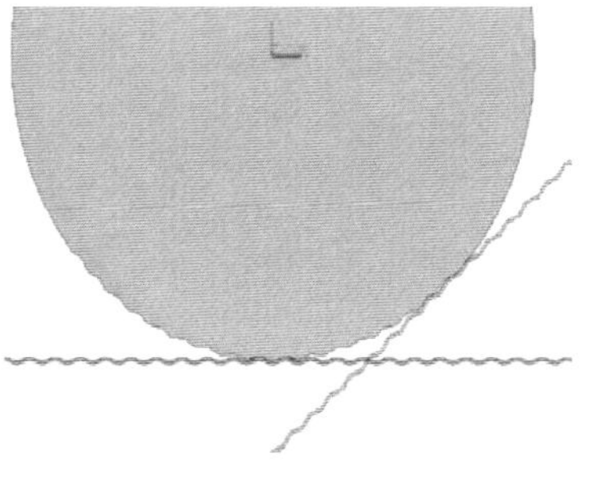

Figure 10. Shaping a roller by a ribbon model.

IOP Conf. Series: Journal of Physics: Conf. Series **858** (2017) 012017 doi:10.1088/1742-6596/858/1/012017

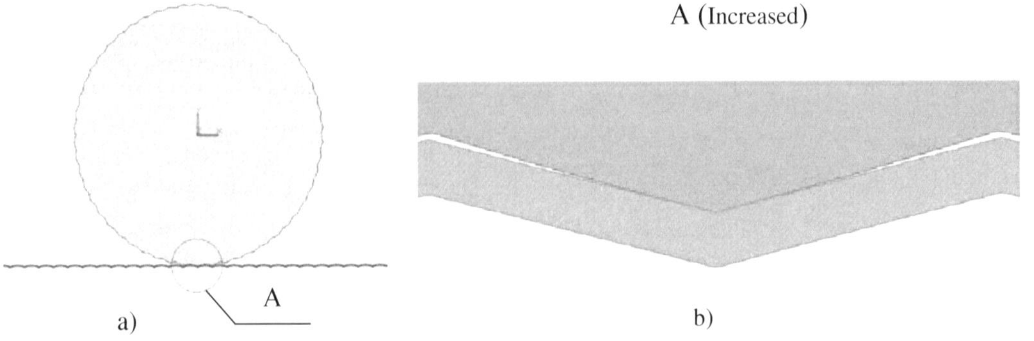

Figure 11. Result of roller profiles modeling: a) general view; b) scaled-up fragment of models.

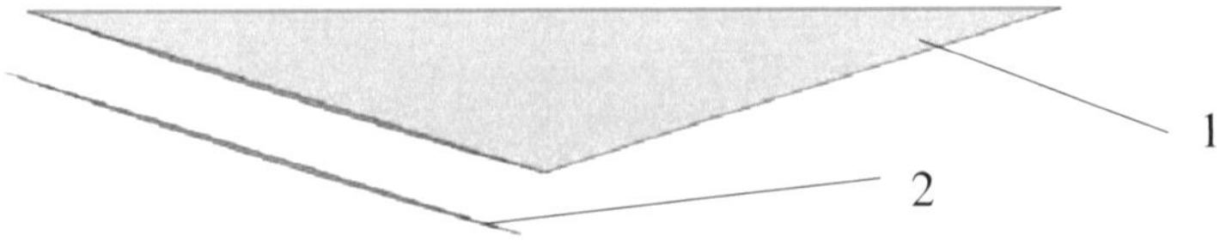

Figure 12. Profile of the roller tooth and the part to be removed: 1 - tooth profile; 2 - extruded part.

In this case, the area of the removed part is 0.0097 mm^2, which is 0.6% of the total area of the tooth. As a result of the replacement, the real profile of the roller has the parameters shown in Figure 13. The obtained profile is further used to solve the inverse problem of shaping - obtaining the ribbon profile to the designed and corrected roller model.

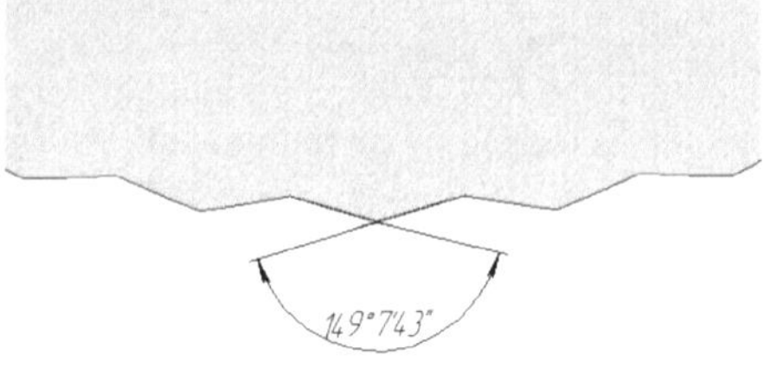

Figure 13. Profile of the roller tooth after revision.

3.3.2.Solid-state ribbon modeling with the obtained tool

After correcting the tooth profile and obtaining the end section of the roller, a number of stages of ribbon profile modeling are performed being the solution of the inverse problem of shaping. At the first stage, like in the previous problem, a family of end sections of the roller is built. The initial data is the roller model and the workpiece model for the ribbon having rectangular profile. After analyzing the resulting family, solid-state modeling of the ribbon by the obtained tool is carried out (Figure 14). Scaled-up fragment of this modeling is shown in Figure 15. The resulting ribbon profile is used for its comparison with the original one. If the resulting profile meets the necessary requirements, the modeling is completed.

In some cases, there is a need to model the extruded volumes, which is provided for by the developed programs. The results of this modeling allow us to correct the technology of ribbons manufacturing. Figure 16 shows the fragments of roller and ribbon models, as well as models of deleted layers. The increment of the family parameter for the case under consideration is 3°. The analysis of the obtained

IOP Conf. Series: Journal of Physics: Conf. Series **858** (2017) 012017 doi:10.1088/1742-6596/858/1/012017

models allows us to make a series of qualitative and quantitative conclusions. So, it follows from Figure 16 that at a given family parameter at each instant of time, the shaping is performed mainly for three cavities of the ribbon profile. Moreover, the extruded volumes in shaping the various hollows are significantly different. In this regard, accordion folding of the ribbon is possible, which requires experimental verification.

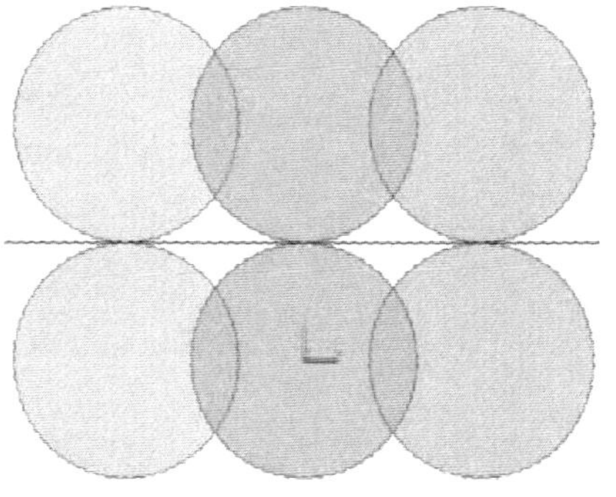

Figure 14. Modeling of ribbon shaping by two rollers.

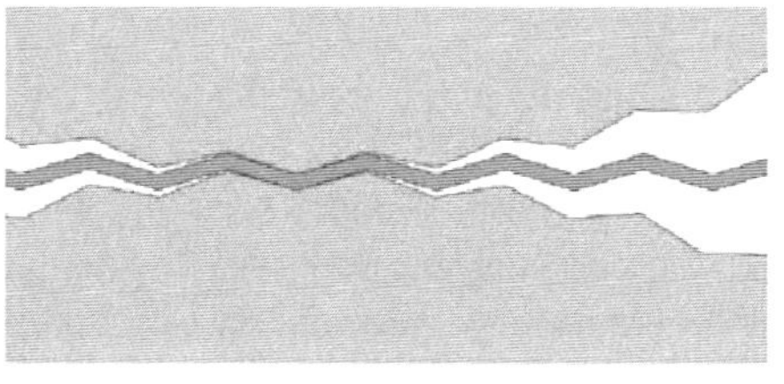

Figure 15. Scaled-up fragment of ribbon shaping modeling by two rollers.

Thus, the implementation of 3D modeling of direct and inverse shaping problems, the analysis of the results obtained at different stages of the shaping, as well as possible experimental verification, let us define optimal parameters of the roller profile and its installation relatively to the ribbon workpiece.

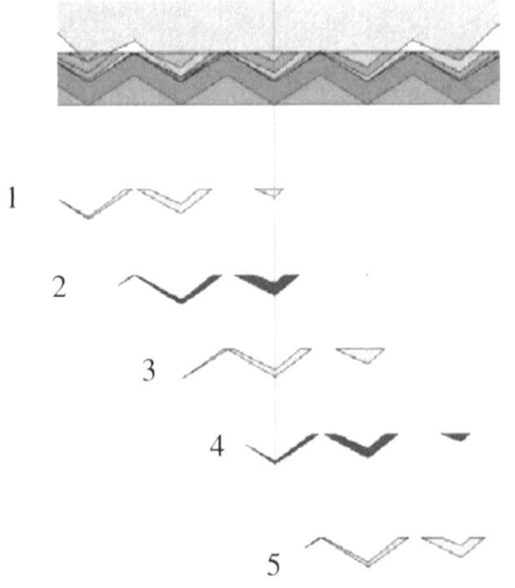

Figure 16. Modeling of extruded volumes under the family parameter change $\varphi = 3^{\circ}$.

4. Experimental results

The most laborious operation of manufacturing finned elements of the gas heater is rolling the periodic corrugated profile.

The rolling pattern has dominant influence on manufacturing efficiency and quality. Two rolling patterns are possible for rolling the finned surface of the ribbon with the angle of the profiles inclination: by straight-toothed rollers with feeding of the workpiece at the angle the profile inclination to the rotation axis of the rolling rollers (Figure 17a) or by helical rollers with feeding of the workpiece at an angle of 90 ° to the axis of the shaping rollers (Figure 17b).

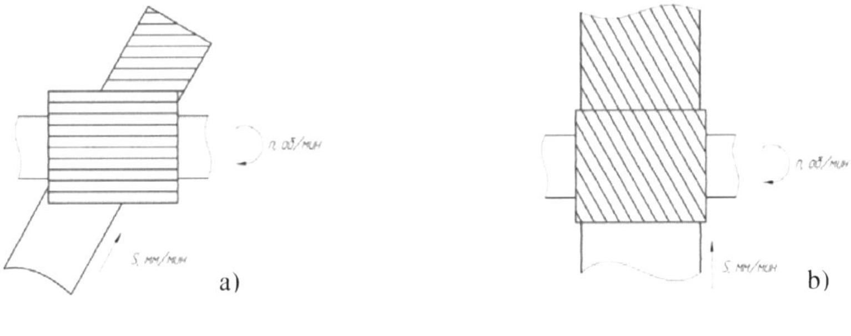

Figure 17. Schemes of rolling the finned elements of the nozzle with rollers: a) with straight teeth; b) with inclined teeth.

For rolling the finned elements of the nozzle between the rollers, a gap is set that corresponds to the thickness of the rolled ribbon, and the ribbon itself moves along the reference diameter. In the scheme shown in Figure 17a, the reference diameter has the shape of an ellipse, which leads to uneven rolling of the profile and "bending" of the ribbon (Figure 18). In this regard, the most optimal from the point of view of manufacturing efficiency and accuracy could be the scheme shown in Figure 17b.

IOP Conf. Series: Journal of Physics: Conf. Series **858** (2017) 012017 doi:10.1088/1742-6596/858/1/012017

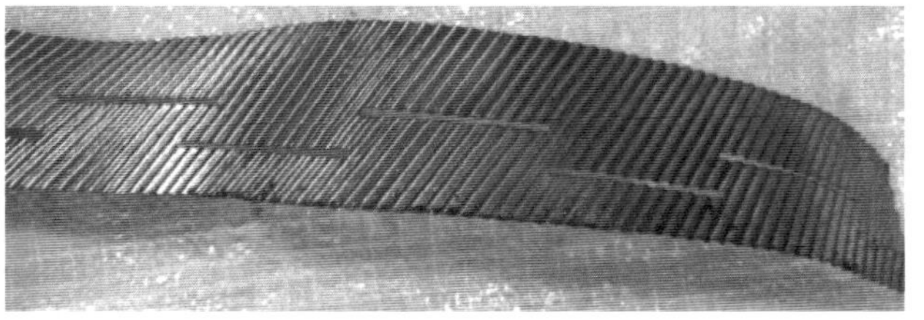

Figure 18. Distortion of the rolled ribbon configuration.

In accordance with the obtained scheme of ribbon manufacturing, adjustment for rolling the finned profile with helical rollers was developed - Figure 19. In this scheme, one shaping roller is fixed in the tool and has free rotation, while the second is attached onto the spindle of the machine and is fixed from axial rotation. The ribbon is fixed into in the catching-guiding device and is fed to the working surface of the rolling rollers. The spindle of the machine starts to tighten the ribbon between the rollers. Then the process continues in an automatic mode. Between the teeth of the rolling rollers, a guaranteed clearance f is provided equal to the thickness B of the rolled ribbon. With a decrease in the clearance size, plastic deformation of the ribbon surface occurs, which leads to unacceptable hogging (Figure 20).

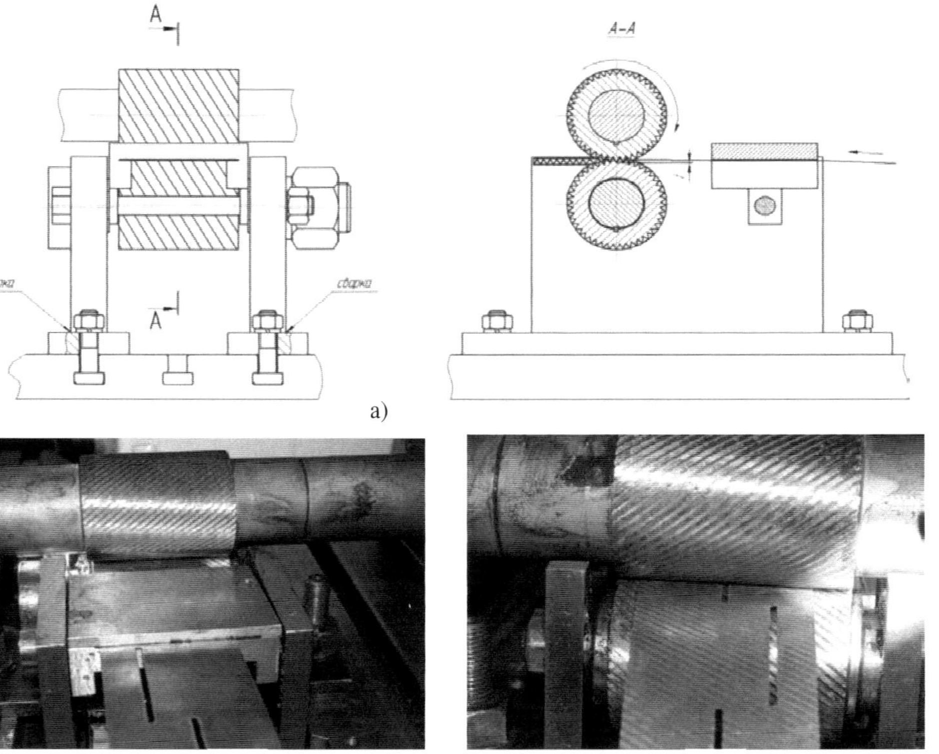

a)

b) c)

Figure 19. Adjustment for the rolling of the finned profile with helical rollers: a) general view of adjustment; b) feeding the ribbon into the catching-guiding device; c) ribbon after rolling.

The optimum range of roller speed, which determines the increment of the motion parameter, is from 10 to 20 rpm. Figure 21 shows the ribbon after rolling on the machine.

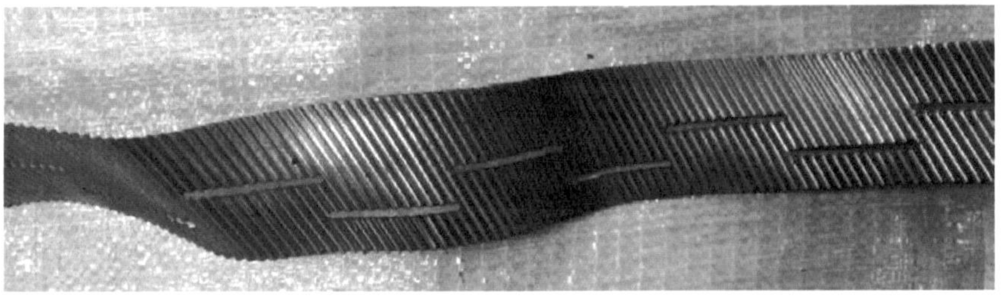

Figure 20. Ribbon surface hogging.

Figure 21. Ribbon after rolling operation.

5. Discussion

In the article, when solving the direct and inverse problem of shaping, modeling of the profile of the tool and then the profile of the ribbon according to the corrected profile of the tool was performed. During 3D modeling, models of extruded volumes were also obtained for different values of the tool motion parameter increment. The preconditions for the effect of this parameter on the possible ribbon hogging are revealed. The information obtained from 3D modeling concerning geometric parameters of the ribbon and tool profiles, as well as the extruded volumes, is necessary for further research, such as determining the shaping forces, tool stresses and ribbon performs at their contact points, as well as the development of tool wear.

6. Conclusion

Making corrugated ribbons of considerable length is technologically complicated process which requires the accuracy of the profile and straightness of the ribbon. It is proposed to solve this problem by the rolling method, similar to the rolling method used in the shaping of workpieces by cutting. To implement this method, a technique was developed that provides 3D modeling of the process of shaping the surface of a workpiece with a tool.

On its basis, geometric and computer modeling of the profile of the rolling tool was carried out. Based on the modeling results, the theoretical profile of the ribbon was replaced by a straight line (taking into account technological requirements). After replacing the roller profile, the inverse shaping problem was solved - a real ribbon profile was obtained. Then, a comparative analysis of the ribbon initial profile and the profile, designed theoretically and corrected (a real one) based on technological requirements was carried out. According to the results of the analysis, it was concluded that the obtained profile was consistent with the required result. A software package was developed that perform computer modeling of extruded volumes in the process of shaping. The analysis of the qualitative and quantitative parameters of the extruded volumes made it possible to give recommendations on setting the increments in the tool motion parameter. Based on the results of 3D modeling, a technology was developed for manufacturing finned ribbons from heat-resistant steels. It allows to carry out the rolling process characterized by high efficiency, accuracy and stability of the obtained dimensions.

References

[1] Litvin F L and Fuentes A 2004 Gear Geometry and Applied Theory (Cambridge University Press, Cambridge) 800
[2] Radzevich S P, 2001 Formoobrazovanie poverkhnostei detalei: osnovy teorii (Surface Shaping of Machine Parts: Theoretical Principles), Kiev, Rastan 593
[3] Argyris John, Litvin Faydor L, Qiming Lian and Lagutin Sergey A 1999 Determination of envelope to family of planar parametric curves and envelope singularities. *Computer Methods in Applied Mechanics and Engineering* **75(1–2)** pp 175–187 Doi: 10.1016/S0045-7825(98)00367-3.
[4] Dimitriou Vasilis, Vidakis Nectarios, 2006 Antoniadis Aristomenis. Advanced Computer Aided Design Simulation of Gear Hobbing by Means of Three-Dimensional Kinematics Modeling *Journal of Manufacturing Science and Engineering* **129(5)** pp 911–918 Doi:10.1115/1.2738947.
[5] Dimitriou V, 2012 Antoniadis A. CAD-Based Calculation of Cutting Force Components in Gear Hobbing, *Journal of Manufacturing Science and Engineering*. **134(3)** DOI: 10.1115/1.4006553
[6] Lyashkov A A 2012 Shaping of harts with a helical surface by means of a disk mill *Russian Engineering Research, Allerton Press, Inc.* **32(4)** pp 404–406
[7] Lyashkov A A, Panchuk K L 2015 Computer modeling of the pump screw and disc tool cross shaping process. *Procedia Engineering* **113** pp 174–180 Doi:10.1016/j.proeng. 2015.07.314.
[8] Lyashkov A A 2013 Geometrical and computer modeling of shaping surfaces of the parts 89

IOP Conf. Series: Journal of Physics: Conf. Series **858** (2017) 012018 doi:10.1088/1742-6596/858/1/012018

The PTFE-nanocomposites mechanical properties for transport systems dynamic sealing devices elements

Y K Mashkov, V A Egorova, O V Chemisenko and O V Maliy

Omsk State Technical University, 11, Mira ave., Omsk, 644050, Russia

E-mail: omgtu_physics@mail.ru

Abstract. The mechanical properties study results of polymer nanocomposites based on polytetrafluoroethylene with modifiers in the form of micro- and nanoscale cryptocrystalline graphite and silicon dioxide powders are determined. The nanocomposites mechanical properties determined values provide high sealing degree of transport systems dynamic sealing devices elements. When the temperature changes from cryogenic to high positive then the elastic modulus, tensile strength decrease significantly and nonlinearly, the latter limits the composite usage in heavily loaded tribosystems operating at elevated temperatures.

1. Introduction

The sealing devices reliability largely determines transport systems and machine building equipment operation durability and stability. Specific friction units features, especially in the transport system undercarriage elements operating under significant dynamic loaded conditions, require continuous improvement both as sealing devices design as materials used. Increased reliability and machine life ensuring raise the problem of structural materials development with higher physical and mechanical properties performance, retained for long operation period in a loading variable conditions at different temperatures.

The materials based on polytetrafluoroethylene (PTFE) are the most suitable for extreme conditions in a wide range of positive and negative temperatures among antifriction usage polymeric composite materials (PCM) [1, 2].

Composite materials with fillers-modifiers nanoparticles have properties that differ significantly from the material with micro-sized particles properties due to the higher surface energy and nano-sized filler particles activity [3, 4]. The combined micro-and nano-sized fillers usage effectiveness for PTFE-based PCM properties complex improvement is shown in works [5].

2. The problem setting

There is the same type of structural phase change in the samples and machine parts volumes for uniaxial and other forms of static and dynamic load and deformation as well as frictional interaction in the solids surface layers that lead to mating surfaces micro defects and destruction (wear) gradual accumulation [6]. During operating construction materials have different types of stress and significant deformation leading to defects and structural phase transitions occurrence and accumulation accompanied with change in the physical and mechanical properties. Therefore, one of new PCM development main tasks is to improve the mechanical and tribological properties performance to increase the polymer components reliability and durability at work in friction units. We solve the

IOP Conf. Series: Journal of Physics: Conf. Series **858** (2017) 012018 doi:10.1088/1742-6596/858/1/012018

problem of studying the fillers concentration and the ambient temperature influence on the PTFE-based PCM mechanical properties.

3. Methods of research

The study subject is PTFE based nanocomposites (Ftoroplast-4PN GOST 10007-80) widely used in metal polymer machinery friction units [1, 2]. According to the preliminary studies results [5, 7] to modify the structure and to improve PTFE properties we used complex content of the polymer with nanodisperse silicon dioxide SiO2 grade BS-120 powder (GOST 18307-78) in the concentration range from 1.0 to 3.0 wt. % and micro-sized powder cryptocrystalline graphite (CCG) at 8.0 wt. %. Samples for studies were prepared on the cold pressing and free sintering conventional technology.

The PCM mechanical properties study: tensile strength σ_p and elasticity modulus E in tension was carried out according to GOST 11262-80 on tensile testing machine "Zwick / Roell" completed with a heat chamber. PCM samples were made from flat slabs of rectangular cross section by stamping in cutting dies. Research methods included the testing of at least four samples for each PCM composition at tensile strain speed 50 mm/min. The study was conducted in the temperature range from 20°C to 180°C. The test results automated processing provided to obtain data with an error of not more than 5.0%.

4. The experimental results

The concentration dependence of the tensile strength and elasticity modulus is shown in Fig.1.

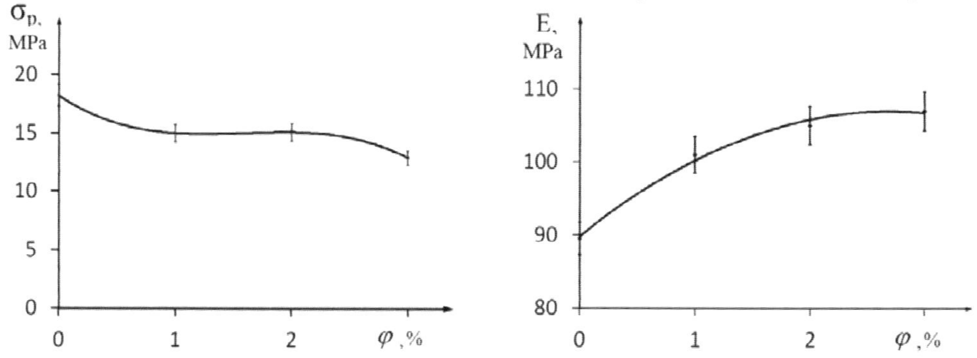

Figure 1. Concentration dependence of tensile strength σ_p and elasticity modulus E in tension for PTFE-based composite with 8 wt.% CCG and 3 wt.% BS.

Temperature dependence of the elasticity modulus in tension and tensile strength is shown in Fig. 2.

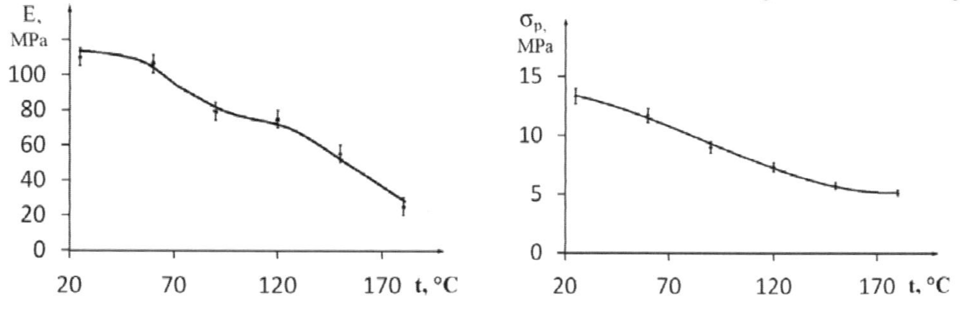

Figure 2. The temperature dependence of the Young's modulus E and the tensile strength σ_p for PTFE-based composite with 8 wt.% CCG and 3 wt.% BS.

Increasing nano-sized filler concentration results in decrease in tensile strength to 28% and increase in the elasticity modulus by 12% (Fig. 1). In terms of elastic modulus at room temperature the studied PCM should be classified as moderately soft materials (E ~ 20…500 MPa) [8, 9] used in sealing technology with average values of contact pressures and moderate speeds [8]. The combination of moderate hardness with a tensile strength close to the unfilled PTFE value corresponds to materials used in metal polymer tribosystems and allows using them in the reliable sealing elements development that provide high degree of tribo matching tightness when dynamic loads change.

Increasing the temperature from 20°C to 180°C leads to significant non-linear reduction in elasticity modulus and tensile strength – to 4.4 and 2.7 times respectively (Fig.2). Reduced mechanical properties characteristics both as PTFE as PCM on its basis are consistent with the data [10].

The relatively low elastic modulus values contribute to the sealing devices tightness degree rise, it makes it advisable to use PTFE-nanocomposites for making sealing elements of such devices. At the same time the strength characteristics limit the use of these materials in high-tribosystems composites, operating at elevated temperatures.

Reducing the PTFE-nanocomposites mechanical properties characteristics with temperature increasing may be associated with changes in molecular mobility in the investigated composites polymer matrix due to the filler energy impact. As a consequence, the phase transition change of the PTFE elementary triclinic cell type into the hexagonal in the temperature range from 20°C to 45°C is accompanied with the structure reorganization determining the PCM mechanical properties. With further temperature increase above 80°C there is transition slow physic relaxation process connected with the ordered micro areas mobility consisting of a large number of segments ('defrosting' relaxation transition of polymer matrix molecular chains micro Broun movement). To identify the filler influence on the polymer amorphous phase supramolecular structure it is necessary to study molecular mobility in the matrix and the considered nanocomposites viscoelastic properties.

5. Conclusion

The study results analysis leads to the following conclusions.

1. Increase in nano-sized modifier concentration in the complex filler content increases the elasticity modulus and decreases the composite tensile strength to 12% and 28% respectively.
2. Raising the ambient temperature from +20°C to +180°C has a significant impact on the PTFE based polymer nanocomposites mechanical properties performance and leads to monotonous and non-linear decrease in the elasticity modulus and tensile strength of the PTFE-nanocomposites in tensile into 4.4 and 2.7 times respectively.
3. The developed PTFE- nanocomposites mechanical properties characteristics make them to be used for the transport systems dynamic sealing devices elements manufacture.

References

[1] Okhlopkova A A., Struchkova T S, Alekseev A G and A.P. Vasiljev 2015 *Bulletin of the North-Eastern Federal University named after M.K. Ammosov* **4(48)** pp 51–63

[2] Mashkov Y K., Kropotin O V, Shil'ko S V and Pleskachevski Y M 2013 *Self-organization and structural modification in metal tribosystems: monograph* 232

[3] Morokhov I D., Trusov L I and Lapovok V N 1984 *Physical phenomena in the ultrafine metal environments* 224

[4] Avdeichik S.V. et al 2012 *Machine-building phtorcomposites: Structure, Technology, Application: monograph* 339

[5] Mashkov Y K, Kropotin O V, Egorova V A, Chemisenko O V and Shil'ko S V 2015 *Inorganic Materials: Applied Research* **6(4)** 289–292 DOI: 10.1134/S2075113315040176

[6] Mashkov Y K 1990 *Durability of friction machine parts* **4** pp 219-244

[7] Kropotin O V, Mashkov Y K, Egorova V A and Kurguzova O A 2014 *Technical Physics. The Russian Journal of Applied Physics* **59(5)** 696–700 DOI: 10.1134/S1063784214050156

[8] Kondakov LA et al 1986 *Seals and sealing technology* 464

[9] Markov A V and Vlasov S V 2004 *Polymer materials. Products. Equipment Technologies* **6-8** pp 17–19; 26, 28, 29; 20, 22-24

[10] Mashkov Y.K., Zyablikov V S and Kazantsev V M 1991 *Mechanics of Composite Materials* **27(1)** 15–19. DOI: 10.1007/BF00630712

IOP Conf. Series: Journal of Physics: Conf. Series **858** (2017) 012019 doi:10.1088/1742-6596/858/1/012019

Technology of wear resistance increase of surface elements of friction couples using solid lubricants

A P Morgunov[1], V B Masyagin[1], V V Derkach[1], N A Matveev[2]

[1] Omsk State Technical University, 11, Mira ave., Omsk, 644050, Russia
[2] Federal State Unitary Enterprise "Federal Research and Production Center "Progress", 4, 5st Kordnaya ave., Omsk, 644018, Russia

E-mail: nikitamatveev55@gmail.com

Abstract. Based on the results of experimental investigations in wear resistance increase using lamellar solid lubricants the technology of wear resistance increase of surface elements of friction couples by applying solid lubricants is developed with the following surface plastic deformation providing enough bond strength of solid lubricant with an element surface and increasing operational life.

1. Introduction

Films can be formed with submission of particles in the loaded contact rubbing in of particles in a surface or pressing of firm lubricant materials. Adhesion connected of a film produced by dispersion on a substrate of firm lubricant materials with addition organic or metal binding.

Savege has found out, that the film of graphite, drift on copper by a graphite brush, has the basic (base) crystal planes almost parallel a surface. Is shown, that this orientation makes 5... 10° concerning a surface of a substrate in a direction of sliding. In the basic results are received for MoS_2. The more detailed research with the help x-ray diffraction films MoS_2, generated on copper, has shown, that they contain a monocrystal layer by thickness 2...5 microns with the basic planes parallel surfaces of sliding.

Figure 1. Structure of a superficial film MoS_2.

It is interesting, that this guided, layer settles down on unguided (not focused) (figure 1). The authors believe, that the high-energy connections of edges of crystals provide coupling a guided, layer with a surface and increase cohesive durability of a guided, film. The lubricant film by thickness 2.4 microns consists approximately of 300 separate layers S-Mo-S. With deterioration of a film the guided, layer approaches with a metal substrate. This rather thick (2 microns) the layer testifies that the process of sliding is accompanied by the advanced plastic deformation.

Film received rubbing, for the first time were studied Johnston and Moore [1]. The cylinder covered sated MoS_2 with a fabric, sated rubbing about a copper surface with a different roughness under various. After the first 100 passes the space near to rough nesses was filled by lubrication, that made a surface more smooth. With the subsequent passes the carry (MoS_2) was carried out already on MoS_2, and the thickness of a film continued to be increased even the after 7000 passes. However other researchers have shown, that there is a limiting thickness of a film, which depends on loading. Certainly, the rougher the surface is, the more material it is required for its covering. A completely different film were formed in a dry and damp atmosphere. In a normal atmosphere the more dense packing of particles was realized, that is caused by influence adsorbed of a moisture ensuring the best coupling of the basic planes MoS_2/MoS_2. Lancaster has shown, that on smooth surfaces fragments of a lubricant material by the size 10 microns were transferred on a film by thickness 0.05 microns. If the film formed rubbing of a surface, simultaneously is exposed frictional to influence, at the end it collapses as a result of wear process. Though solid lubricant the core wears out with rather large speed, but on a film is transferred of very poorly worn out material. Frequently meet difficulties of restoration of a film with the help of ;free particles; the degree of restoration depends, apparently, on character of distribution of pressure. Such impossibility to keep a film specifies that has a place superficial sliding, as shear can be realized only with strong adhesion in interphase area MoS_2/MoS_2.

Thus, at the first stage of functioning of firm lubrication on both surfaces the thin guided, film is formed. Then has a place sliding either between films, or between a substrate and film. The subsequent sliding results in gradual wear process of a film up to a complete exhaustion of a firm lubricant material, or to destruction of a film, that depends on atmospheric conditions and parameters of sliding. Fusaro in detail has studied wear process and destruction films MoS_2 and of fluorinated graphite. The deterioration is displayed as gradual reduction of thickness of a film caused radial and tangential by its replacement from the zone of contact. Last is caused normal and by frictional loadings.

The second kind of wear process is cracking and chipping of a film (similarly of weariness). It is interesting to note, that the same character of behavior is found out in soft materials with moderate deterioration (Ag/Fe). Deterioration of a film, in essence, same, as well as for volumetric materials. Moreover, is shown, as the intensity of wear process similar. Under these conditions more isotropic materials (metals, organic substances, the glasses) collapse by current, while MoS_2 and the layered materials tend to fatigue to destruction of a guided, external layer.

The process of wear process connected and untied films proceeds in two stages. The process described above, has a place until uncover of roughness of a substrate. In the further durability of a film depends on ability of a firm lubricant material in a vicinity of roughness to cover its top.

The destruction of a continuous film occurs for the different reasons. Conducting are two basic processes. First, the heat generated during sliding, softens a film and promotes its oxidation. Secondly, the interaction with an environment can change structure of a film. It, in particular, is fair for MoS_2, which oxidation in steam of water, on air or in environments of oxygen results to deformation of a film.

2. Formulation of the problem

Already in the first works on inorganic firm lubrication were mentioned layered screen structures serving with the basic criterion with selection of lubricant components. However soon it was revealed, that not structure in itself, and nature of connections is important. The materials with hexagonal «layered» by structure appeared effective, if the connection between layers were weak, and within the limits of a layer - rather strong. The materials of a type borazon, having strong interlayer of connection, are not effective as firm lubricant materials. So, Holinski and Gansheimer and other authors connect lubricant action MoS_2 to strong polarization of atom of sulfur giving an opportunity to generate layered structure. The graphite has not weak, border, if its layers are not covered with a lubrication moisture. The early researches PTFE connected its low friction to the minimal force of molecular interaction caused by shielding by a large ion of fluorine a charge on atoms of carbon. Thus,

IOP Conf. Series: Journal of Physics: Conf. Series **858** (2017) 012019 doi:10.1088/1742-6596/858/1/012019

the mechanism of lubricant action PTFE, in essence, same, as well as at MoS_2, behind that exception, that PTFE consists of the poorly connected among themselves circuits, instead of layers. There is, a formation on the initial stage of the transferred layer PTFE on counterspecimen rather, it is essential for efficiency of lubrication. Then actually there is a sliding PTFE on PTFE. With sliding PTFE (volumetric) on PTFE two modes of friction are observed: high friction with characteristic intensive carry and low friction with guided films of shift in interphone of area. Further will be shown, that MoS_2 and other firm lubricant materials (for example, Ag/Fe) behave similarly.

The approach based on representation weak interlayer of connection, was deeply advanced Jamison, come to a conclusion, that the efficiency of firm lubrication is caused by this weak connection. However MoS2 has unique structure among layered lubricant materials, that makes it especially effective. In essence, the lubricant ability depends on distance between the basic crystal planes being function of electronic structure of metal. In MoS2 the atoms of molybdenum settle down above and below «holes» in the nearest layer, but not above or under other atoms of molybdenum. Such specific structure connect with spin coupled electrons, that at the end causes absence residual interlayer of connections. Further is shown, that such type of a structure can be received by implantation of atoms copper and silver in layered structures with rather strong interlayer by connections of a type NbS_2 and $NbSe_2$. In such cases factor of friction decreases with 0.30 up to 0.10. The implantation in graphite chlorides and metals also raises it wear resistance and loading ability.

The researches in a little bit other direction have shown, that of a film MoS_2 received dispersion, have not by lubricant ability, if are used in amorphous a condition (418 K). Besides it is revealed, that of a lubricant film are effective with thickness 200 nm, that makes approximately 300 layers MoS_2. However if the dispersion was carried out with higher temperature (423 K), wear resistance was reduced even with constant factor of friction. This result connect to presence of a porous irregular film with the low contents of sulfur.

Fleischauer has carried out detailed research films, received by dispersion, and has found out, that they can have two versions: crystalline grains with base planes parallel a surface or perpendicular to it. The serviceability such films is various, and this distinction is caused by different chemical potentials of crystal planes and their edges. The weak connections and low chemical activity are characteristic of base planes, while the edges of planes form strong connections and actively oxidize. Thus, the lubricant ability is essentially connected to orientation of crystals

3. The study of the structure of the modified lead-tin-base bronze

Bowden and Tabor for the first time have developed the theory of lubricant action thin films, being based on experiences with films India, lead and copper on substrates from steel, nickel, copper and lead [2]. Were used of a film of different thickness, and counter-specimen served steel indenters of different radius. They have found out, that the force of friction in all cases depends on width of a path of friction (contact platform). It, certainly, serves confirmation that, that

$$F=AS_f, \tag{1}$$

where F - force of friction; A - area of contact; S_f - durability on shift of a material of a film. Varying such parameters, as thickness of a film, geometry of a sample, loading and hardness of a substrate, they achieved change of the area of contact and accordingly force of friction. Their concept of lubricant action thin films was retied that

$$f=\frac{S}{P}=\frac{S_f}{P}=\frac{S_f}{H_S}, \tag{2}$$

where f - factor of friction; P - pressure; Hs - hardness of a substrate. In other words, the area of contact depends on hardness of a substrate, while the durability of a film on shift determines specific force of friction S. Friction of a metal film was much lower, than volumetric sample, as for last

$$f= \frac{S_f}{H_f}, \tag{3}$$

where Hf - hardness of a material of a film.

For example [2], with friction of spheres (r = 8 and 3 mm) on leaden films (thickness 1... 12 microns) on a steel substrate the area of contact first of all are defined by elastic deformation of steel. Thus,

$$f= \frac{S_f}{P}=\frac{\alpha S_f}{L}=\pi\left(\frac{3LR}{4E}\right)^{2/3}\frac{S_f}{L}, \tag{4}$$

where a - area; L - loading; R - radius of a sphere; E - module of elasticity. The plastic deformation of a film begins to influence the area of contact, when radius of contact is less, than five-multiple thickness of a film. Thus, the area of contact can depend both from elastic, and on plastic deformations.

The experiences of Bridgeman and other researchers have shown, that the durability grows by shift with increase of pressure. Therefore equation (4) should be modified, by replacing Sf, on Sp.

$$S_p=S_f+\alpha P, \tag{5}$$

Where Sp - durability on shift of a material of a film with the given pressure P; a - constant. The equation (4) will look like

$$f=\pi\left(\frac{3LR}{4E}\right)^{2/3}\frac{S_p}{L}=\pi\left(\frac{3LR}{4E}\right)^{2/3}\frac{S_f+\alpha P}{L}, \tag{6}$$

With plastic deformation of a film

$$f= \frac{L}{H_f}\cdot\frac{S_p}{L}=\frac{S_f}{H_f}+\alpha, \tag{7}$$

here P is equal to hardness of a film.

However usually deal with surfaces, flat initially or generated during wear process of curvilinear indenter, thus the area of real contact can be defined by a film [see equation (7)] or geometrical area of contact. Then factor of friction becomes function of loading (or pressure) (figure 2), and, as shown numerous researches, frictional the behavior is best described by the equation

$$f= \frac{S}{P}+\alpha, \tag{8}$$

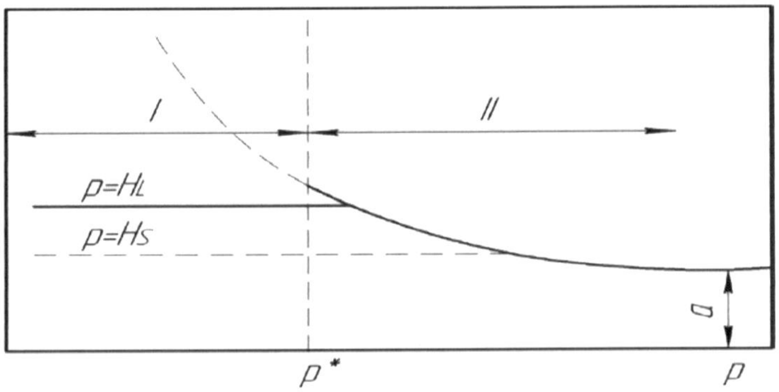

Figure 2. Dependence of factor of friction f of firm lubricant materials on pressure p: I - zone of sliding; II - zone of shear.

From the equation (8) follows, that factor of friction decreases with growth of pressure, S and α remain constant. If the material is used as a film, in accounts are accepted its durability on shift and parameter a. In quality S the durability on shift of a film Sf or interphase of area Si can be accepted, if in the latter case has a place sliding. The pressure P also accepts different meanings: with low loadings P the equation (7) can be equalled of hardness of a film see. In this case surface of contact (Ar) is discrete and is limited to tops of roughnesses. Thus, the friction should be same, as well as with sliding the friend on the friend of volumetric samples of a lubricant material. With very thin films elastic or plastic deformation of a substrate the influence on the area of contact up to limiting meaning P = Hs can render. In this case friction should be much lower, as shown in a figure 3.

With growth of loading the area of contact is increased, and the friction remains constant up to some critical pressure P = P*. Above than this pressure Ar = Aa (nominal area of contact) and P = L/Aa. Then the friction is reduced, while (with rather high pressure) will not reach size f = a. Thus, for flat surfaces it is possible to use three equations:

$$f=\frac{S_f}{H_L}+\alpha_L \text{ at } P < P^* \text{ and } A_r \approx L, \tag{9a}$$

where H_L - hardness of a lubricant material.

$$f=\frac{S_f}{H_L}+\alpha_L \text{ at } P < P^* \text{ and } A_r \approx L, \tag{9b}$$

(for thin films) and

$$f=\frac{S_f}{P}+\alpha_L \text{ at } P > P^* \text{ and } A_r = A_\alpha, \tag{9c}$$

(Aa - constant determined by geometry of system).

4. Conclusion

So, it is possible to assume, that there are two modes of friction of firm lubricant materials; a mode of sliding with low pressure and shift mode with high. In the latter case, the mechanism of viscous or plastic current is realized. Factor of friction is directly proportional to the area of contact. This area is defined by the nominal area for flat samples or elastic deformation for curvilinear of contact. In a mode of sliding the area of contact is defined or hardness of a lubricant film, or (for very thin films)

IOP Conf. Series: Journal of Physics: Conf. Series **858** (2017) 012019 doi:10.1088/1742-6596/858/1/012019

hardness of a substrate. The mechanisms of sliding are described by models 3, 4 or 5 (see tab. 1). And as durability on shift Sf it is necessary to use adhesional durability Si. If the above mentioned reasons are applied to real materials, for silver, MoS_2, lead and model viscous material (200 N/m2 s) it is possible to receive curves shown in a figure 3. The appropriate data on hardness and durability on shift are taken from several sources. The meanings of parameter and are received from work of Bridgeman.

These curves are based on simple reasons and do not apply for severity. However they illustrate the tendency in behavior of firm film lubricant materials. Let's notice, that with small loading the friction should be very high, if has a place pure shift or current

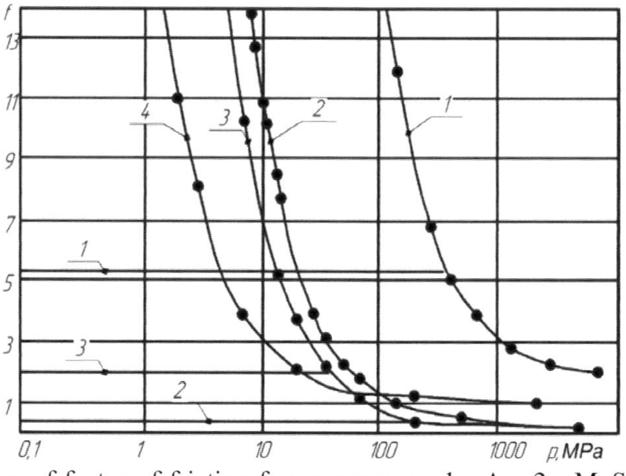

Figure 3. Dependence of factor of friction f on pressure p: 1 - Ag; 2 - MoS_2; 3 - Pb; 4- model material by viscosity 2000 poise (200 Pa×s).

Obviously, in this case will occur grip or transition to other type of sliding. Usually has a place transition to slipping or essential reduction of the area of contact caused by decrease of normal pressure owing to deformation of a material.

Numerous literary given for films MoS_2 (experiments in conditions of a dry atmosphere) it is enough precisely approximate by the equations (9a) and (9b). Factor of friction makes 0.04 down to pressure ~ 560 MN/m^2. Then it begins to decrease. It is interesting to note, that for guided films MoS_2 the hardness makes 600 MN/m^2, so for MoS_2 P* = Hf (hardness of a film). Both Barry and Binkelman, and Reed and Shaw it is revealed, that with small pressure the friction, does not depend on hardness of a substrate. Thus, the equation (9b) is not applied. All experiments with small pressure are carried out for the different nominal areas, so the friction does not depend and on the nominal area of contact. It serves confirmation that, that the hardness of a film determines the actual area of contact. Last circumstance predetermines subsequent tribologycal behavior.

The film MoS_2 is not ideally smooth. With small loading of roughness of a film perceive this loading and, being deformed, form the actual area of contact. The effort of shear individual spot also makes measurable force of friction. With growth of loading the actual area of contact is proportionally increased. Thus, factor of friction remains constant. At the end, when the pressure becomes equal to hardness films MoS_2, the contact is entered by the whole area of a contact. Then the friction begins to decrease, as now and Ar and Sf constant:

$$f = \frac{A_r S_f}{L} + \alpha$$
$$where\ \frac{A}{L} \approx \frac{1}{H}\ and\ A_r = A_\alpha\ at\ P^* = H. \quad (10)$$

If the pressure exceeds P*, the friction decreases up to size equal a, owing to ArSfL-1 —>0. When the durability on shift is not increased with growth of pressure, and should be equalled to zero and factor of friction will aspire to zero. Thus, low factor of friction MoS_2 is caused by high hardness of a guided film.

For very soft backs the friction can grow, as shown in a figure 4 (on the data Barry and Binkelman). If the hardness of a substrate is less, than the hardness MoS_2, friction grows, probably, because of deformation of a substrate. The thicker film, the effect will be less significant.

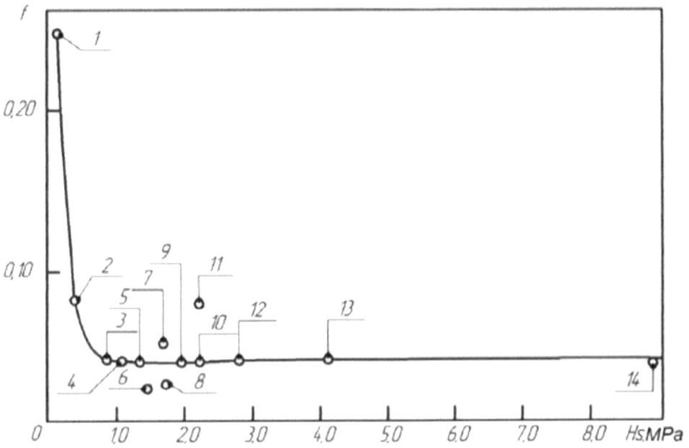

Figure 4. Influence of hardness of a substrate on factor of friction MoS_2 ("+" — Spherical particles MoS_2) 1 - lead; 2 - babbit; 3 - silver; 4 - copper; 5 - silver plate; 6 - brass; 7 - aluminium; 8 - bronze; 9 - steel 1020; 10 - molybdenum; 11 - titanium; 12 - TZM; 13 - tungsten; 14 - rigid steel.

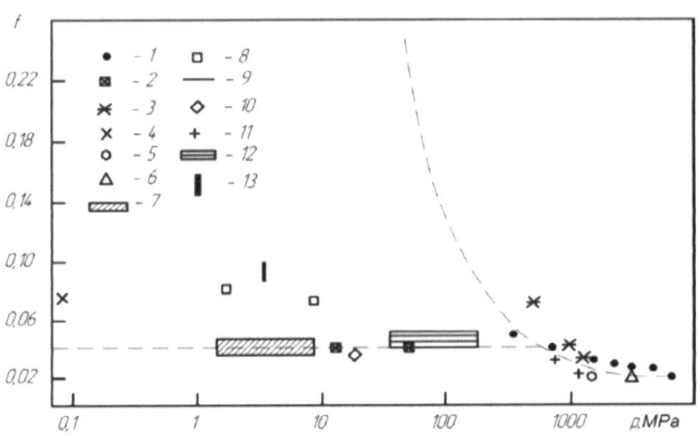

Figure 5. Friction MoS_2 in a wide range of pressure: 1 - film; 2 - film; 3 - film; 4 - balls; 5 - film; 6 - film; 7 - film; 8 - balls; 9 - balls; 10 - film; 11 - Film; 12- balls; 13 - balls.

From the data submitted in a figure 5, follows, that factor of spherical particles MoS_2 sliding on metal. Such behavior is clear in light of the data of a figure 4. The volumetric hardness of particles MoS_2 makes, as show measurements, 180 MN/m^2. If to accept this size as hardness of a substrate, the meaning of factor of friction 0.08 is possible to consider quite acceptable.

For very firm backs (more than 8 GN/m^2) was not observed of essential reduction of friction, though a brass and bronze give lower, and titanium the higher debate, than predicts dependence submitted in a figure 4. With pressure exceeding 560 MN/m^2, continuous film MoS_2 shearing. Factor of friction is directly proportional to the area of contact with the given loading. The contact is defined by the nominal area of a sample or elastic deformation of the concentrated contact. For softer backs there can also be their plastic deformation, which increases factor of friction a little. However, when f comes nearer to $\alpha = 0.02$, the influence of pressure becomes insignificant.

The small meaning of factor of friction (0.02) with high pressure (2.8 GN /m^2) is fixed Peterson and johnsori for very thin films, roughnesses, generated at tops, of a substrate. In this case pressure is equal to hardness of a substrate and the actual area of contact is small. For continuous films other meanings turn out. Thus, in a mode of shear of adhesional connections the friction first of all is defined by pressure; for the given pressure it remains constant.

The data for films of tin are submitted in a figure 6. There is a satisfactory conformity between the equations (9a) and (9b) and experimental results. With small pressure factor of friction equal turns out 0.40. This size is close to received Rabinowich with friction of a volumetric sample of tin on steel (0.29... 0.51).

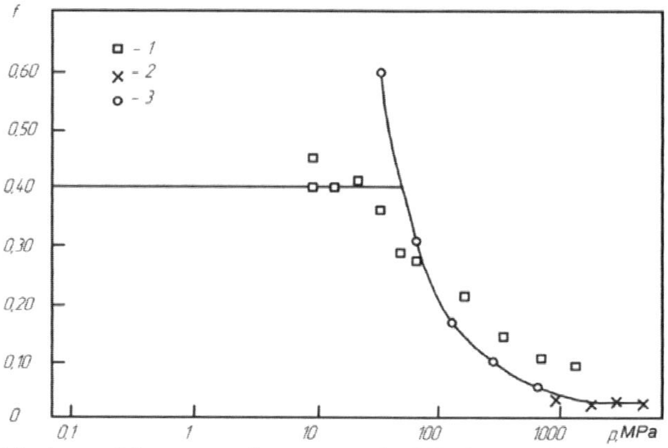

Figure 6. Friction of tin in a wide range of pressure: 1 - [3]; 2 - [4]; 3 – on the data of a Figure. 4.

The data for lead are submitted in a figure 7. Though as a whole the described above tendency is appreciable, but there are some distinctions, especially with small pressure. Factor of friction lays in a range 0.40... 0.70, instead of is equal, as was predicted, 0.20. However it nevertheless is lower, than meaning 1.30, received in experiences Tsus with sliding lead on steel. This fact becomes clear if to proceed from of the adhesional theory of friction. The strong coupling of lead with steel results in increase of the area of contact for the account of tangential effort. The same behavior observed Kato, working with thick films. Also has noticed, that of a thicker film give higher factor of friction close to friction of a volumetric sample of lead.

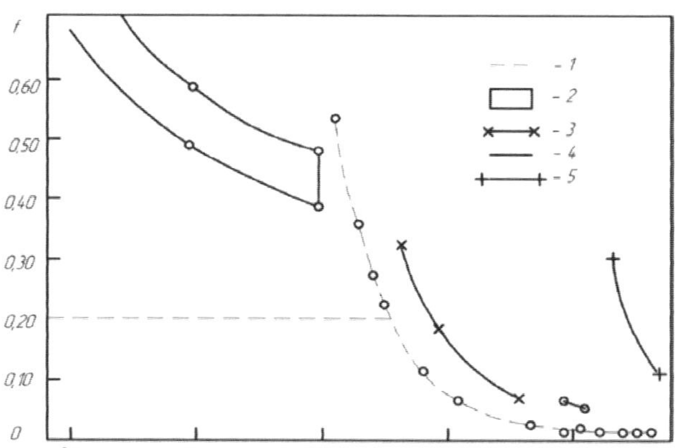

Figure 7. Friction of lead in a wide range of pressure: 1 - [4]; 2 - film by thickness 0.1... 32 microns, small influence of thickness up to 10 microns; 3 - [3]; 4- films by thickness 0.1 microns; 5 - thick film.

As the growth of adhesional connections is not characteristic of tin and MoS_2, they frictional behavior can be predicted. It, probably, is caused low adhesion (Sn/Fe and MoS_2/MoS_2) or that the deformation of these materials does not conduct to growth of friction. Some researchers have fixed high friction (0.30) with sliding MoS_2 on films MoS_2 in a damp atmosphere [5]. Thus the carry of large fragments was observed, that connect to increase adhesion MoS_2/MoS_2. Other examples of such influence adhesion are known also. The silver behaves similarly to tin or MoS_2, when is used as a film for lubrication steel or nickel, but with friction on an aluminium surface its behavior is similar to behavior of lead. For the same reason inefficient lubricant materials are of an aluminium film.

5. Acknowledgments

The analysis of dependences of friction from pressure for solid lubricant films allows better to understand them frictional behavior. The assembled data enable to offer the simple theory agreed to the numerous literary data. According to this theory friction and deterioration solid lubricant films are simple adaptation of their behavior in the volumetric form. Such adaptation is limited to growth of connections, which can take place for thin films.

Certainly, this concept is speculative and is based on the limited data received under different conditions. These results should be reproduced on one equipment in a wide range of loadings with the control and measurement of the actual area of contact. The measurement of micro hardness of surfaces is necessary for carrying out after sliding for definition of hardness thin films. The concept of growth of connections is not suitable for a number deformative of processes, for example, with reference to contact of materials with hexagonal and cubic structures.

It is offered five different models solid lubricant films. The extensive review of the literature with the purpose is carried out to determine, what from models will be coordinated with the basic knowledge about tribologycal behavior of firm lubricant materials. The behavior films with the account adhesion of a substrate, features of formation films, their deterioration and destruction, crystal structure, influence of the atmospheric factors and of frictional the characteristics is discussed. On the basis of the limited data the conclusion about applicability standard of the adhesional theory of friction to thin films is made. It is offered to distinguish two modes of friction: of sliding with small loadings and shear with large. The transition from one mode to other occurs, when the pressure becomes approximately equal to hardness of a film. In a mode of sliding the actual area of contact is caused nominal, which depends on elastic deformation of a material of a substrate. Exhalation and oxides of a surface change frictional behavior solid lubricant films.

References

[1] Johnston R R M, Moore A J W 1964 The burnishing of molybdenum disuiphide on to metal surfaces *Wear* **7** 498-512

[2] Bowden F P and Tabor D 1943 The Lubrication by Thin Films and the Action of Beerings of Metals *App. Phy* p 141

[3] Peterson M B, Cheng H S and Ling F F 1958 Investigation of Friction and Shear in Hot Deformation *MIT* pp 9–20

[4] Bridgeman P W 1936 Shearing Phenomena at High Pressures *Proc. Amer. Acad. Arts Sci.,* **71** pp 387

[5] Mansha M, Holl Y and Gauthier C 2014 Stude of friction coefficient of colloidal polymer films *Proceedings of 11th IBCAST* pp 22–32

IOP Conf. Series: Journal of Physics: Conf. Series **858** (2017) 012020 doi:10.1088/1742-6596/858/1/012020

Research of lamellar lubricants mechanical-and-physical properties and structure

A P Morgunov[1], V B Masyagin[1], V V Derkach[1], N A Matveev[2]

[1] Omsk State Technical University, 11, Mira ave., Omsk, 644050, Russia
[2] Federal State Unitary Enterprise "Federal Research and Production Center "Progress", 4, 5st Kordnaya ave., Omsk, 644018, Russia

E-mail: nikitamatveev55@gmail.com

Abstract. in the article ascertaining the applicability of lamellar solid lubricants is considered for the purpose of wear resistance increase of metal surface elements by their contact interaction in different operational conditions. Forms of binding in the structure of solid lubricants are described through the example of molybdenum disulfide MoS2, as well as change properties at standard and high (350-480°C) exposure temperature. After solid lubricant application the additional treatment is made on the work surface of flanged wheels in the wheel set of a locomotive using plastic deformation for getting of multilevel micro relief resulting in working service increase of the wheel set in comparison with control set from 1.6 to 3 times.

1. Introduction

The large distribution, which was received last years by layered firm materials as lubrication substances, has involved in them attention of the researchers. However nowadays still there are no uniform reasonable representations about a nature of friction of layered lubrication

At the present it is impossible of the unequivocal and settling answer to a question that determines lubricant ability of layered film lubrication.

However already now it is possible to name some characteristics, common for materials with the brightly expressed lubrication ability.

To such characteristics it is possible to attribute the form of a crystal lattice; the form of surfaces of layers limiting a planes of sliding; a degree of heterogeneity of connections between atoms of a crystal lattice; size of work spent on splitting of a crystal of layered lubrication on a plane of sliding; high adhesion to metal surfaces.

The film layered materials having lubrication ability have in the majority a lattice hexagonal of a structural type.

2. Formulation of the problem

The layered structure is characteristic of all firm lubrication, with which the atoms which are taking place in one plane, are enough densely packed, and their crystal lattice consists of a number such mutually of parallel layers. In a lattice of graphite these unary layers consist of homogeneous atoms of carbon, in a lattice at of disulfides both diselenides of molybdenum and of tungsten there are threefold layers: the layers of atoms of molybdenum or of tungsten settle down between two layers of atoms of sulfur (disulfides) or between two layers of atoms selenium (selenides). In a lattice CdJ_2 the layer of atoms cadmium settles down between two layers of atoms iodine.

The layered firm lubrication belong to a class so—called anisodesmal of connections, at which the relative durability of connections between atoms is sharply various in various directions [1]. This distinction is caused by that for anisodesmal of structures the sharp difference in intemuclear distances is characteristic: shortest distance up to one — Three nearest atoms much more differs from the following on size of distance between similar atoms. For example, in a lattice of graphite distance between the nearest atoms of carbon in a layer is equal 1.42 °A (shortest distance between atoms in parallel layers — 3.44 °A).

In structure of firm lubrication two are usually combined such as connections: residual or Van—Der—Vaals connection with ion—covalent by a type of connection. So, in a crystal lattice MoS_2 the connection between atoms of sulfur located in parallel planes, is Van—Der—Vaals, and connection between atoms Mo — S — covalent. The forces connecting atoms of carbon in one layer of a lattice of graphite, concern to covalent to connections, and interlayer of connection are connections of a metal type.

The durability of Van—Der—Vaals connection is extremely small; besides it back is proportional to the sixth degree of intemuclear distance and in tens time more poorly purely ion of connection.

The presence of much weaker forces of connection between separate layers in comparison with similar forces inside layers is common for all materials with layered structure. In case of sharp anisodesmic of connections in a crystal, for example, such film layered lubrication, as sulphides both selenides of molybdenum and of tungsten, and also selenides niobium, under influence of external forces have sliding, i.e. Displacement of one blocks of a crystal concerning others in a plane, normal to a direction of action weakest of forces. At considered crystals this plane usually is a plane of cleavage.

So, in crystals of disulfides or diselenides the sliding occurs accordingly between parallel layers of atoms of sulfur or selenium, removed from each other on much large distances, than atoms of sulfur or selenium, located in one plane. The crystal structure of firm

layered lubrication though does not determine entirely their lubrication ability, however, essentially influences plasticity of a material and its behavior in conditions of friction. It proves to be true by existence of some correlation between factor of friction and parity of parameters of a crystal lattice. Last size (c/a) in any measure qualitatively reflects distinction of forces of connections between atoms located in a plane of shift and in a plane, perpendicular to it, and, hence, can characterize resistance to shift.

The mechanism of the considered phenomenon is not still completely opened, two possible processes — Diffusion and formation of electrical charges — Influence increase of energy of inteijunction — type connection of graphite and mica in vacuum.

In the first case the water steam contained in air, diffuse in depth of a firm body (mica or graphite) also influence forces of connection of these layered substances, as the molecules of water weaken ion of connection between atoms. In vacuum diffusion, which is directly proportional to a gradient of concentration of gas, is considerably reduced, so the influence of molecules water pair on connection practically is excluded. Therefore energy of connection of a crystal, which is characterized by size of work spent on its splitting, grows in some times.

According to the second assumption with splitting a crystal on planes of cleavage there are forces of an attraction between planes. On air this effect will be quickly neutralized by atoms of gases, adsorbed on planes of sliding. In ultra high vacuum this opportunity is almost excluded, as results in increase of work spent on splitting of a crystal in these conditions.

The hypothesis which is put forward by Bryant with the co—authors, deserves attention and requires experimental check with reference to such known and perspective layered lubrication, as sulphides and selenides of molybdenum, of tungsten, niobium.

From positions of this hypothesis it is good explainable behavior of graphite with friction on air and in vacuum with moderate temperature. In the latter case factor of friction of graphite grows up to 0.5 and above depending on a material counterspecimen and considerably surpasses factor of friction

IOP Conf. Series: Journal of Physics: Conf. Series **858** (2017) 012020 doi:10.1088/1742-6596/858/1/012020

on air. In the literature, accessible to us, there are no data on factor of friction of mica in vacuum, and consequently it is inconvenient to establish precisely, whether there is for mica, as well as for graphite, a correlation between size of work spent on splitting of a crystal on planes of cleavage, and factor of friction. However, if to compare factor of friction on air for latest splitting of a surface of mica, when the various pollution from air had no time(was in time) yet adsorbed onto surface of friction, with factor of friction of mica after long sliding on air, it is possible to notice the certain connection between the data of Bryant and by frictional behavior of mica. So, in the first case factor of friction achieved size 0.35, and in second — 0.48.

The works of Bryant with the co—authors represent interest in the respect that they show an inconsistency of a hypothesis "easy shear" on planes of cleavage, explaining good lubrication property of graphite on air proceeding only from its layered structure and ability it is easy to be splitted on planes of cleavage. The theoretical size of work spent on splitting of a crystal of graphite on planes of cleavage, as well as certain experimentally in vacuum is rather high that, specifies presence significant cogesion between planes of sliding of graphite.

Concerning the mechanism of lubrication action MoS_2 there are various assumptions and hypotheses. So, Johnson and Van in 1956 have put forward a hypothesis, according to which high lubrication ability and low factor of friction MoS_2 in vacuum are caused by formation during sliding a thin layer amorphous of sulfur having extremely low resistance of shear. The authors believed, that during friction MoS_2 occurs original «tribo—cracking», which result is formation free sulfur.

Against this hypothesis speaks, first, that fact, that factor of friction of sulfur on sulfur is much higher, than MoS_2 on MoS_2, and also research on friction MoS_2 in ultra high vacuum and nitrogen executed by Haltner.

The special point of view on a nature of friction MoS_2 has stated Braiwaite, which considers, that the good lubrication properties MoS_2 are caused by its layered structure and ability to be stratified with heating arising at the expense of friction.

Agrees Braiwaite with friction MoS_2 on air there is its oxidation, thus will be formed as nonvolatile oxides of molybdenum (MoO_2, MoO_3), and volatile oxide of sulfur SO_2. With heating MoS_2 during friction the formed oxides of sulfur are liberated earlier. Last, as believes Braiwaite, considerably weakens connection between planes of sliding MoS_2 and by that provides their easy shift and low factor of friction.

Braiwaite has shown, that the partial oxidation of a powder MoS_2 at the presence of oxygen can occur with temperature 423 — 443 K. It connects lubrication ability MoS_2 to process of its oxidation with friction. It is impossible to consider reasons of Braiwaite rather convincing, as they are in the contradiction with results of experiments on friction MoS_2 in vacuum of the theory, executed in laboratory, of friction IES (institute of Engineering Science) and in USA by Haltner. The numerous tests, earned out by us, have shown, that with friction in high vacuum, where the opportunity of oxidation MoS_2 is shown up to a minimum, factor of friction it falls in comparison with air in 2 times and more than [2].

From various hypotheses and assumptions which have been put forward by the scientists for an explanation of the lubrication ability MoS_2, most correct are the representations of Bryant with the co-authors in detail advanced by them with reference to graphite. Proceeding from these representations it is necessary to assume, that for splitting a crystal MoS_2 on planes of cleavage it is necessary to spend considerably smaller work, than in case of splitting a crystal of graphite.

As against graphite, which is a rather inert material. MoS_2 chemically is considerably more active. With friction MoS_2 on air its oxidation on separate sites of contact even is possible with temperature of the order 353 K. Such oxidation is considered as mechano—chemical reaction typical for juvenile of surfaces formed during oxidation. Oxidated the sites MoS_2 hydrophilic and easily are absorbed by water steam always contained in air.

The atoms of oxygen and water, adsorbed between layers MoS_2, can become chemically connected, that will cause increase of resistance to shift along planes of sliding and as a consequence it growth of factor of friction. The study of kinetics of destruction films MoS_2 during friction on air [N =

IOP Conf. Series: Journal of Physics: Conf. Series **858** (2017) 012020 doi:10.1088/1742-6596/858/1/012020

5 kg (50 N), v = 1.6 m/s], executed with use electronic microscope and filming of a surface of friction, has allowed to establish. that the destruction of a film MoS_2 (earlier put on a metal surface by frictional a way) begins with formation on its surface of microswellings (size 400 °A), which then are increased up to ~ 3 mm; the film under swellings becomes flat and flake. The authors consider, that such little bubbles will be formed on sites of concentration of pressure, as a result of which action arise fatigue of the phenomenon in superficial layers of a film MoS_2 and the chemical activity (interaction with a moisture and oxygen of air) amplifies. The combination of such processes results to fatigue to destruction of a film.

3. Theory

In work Savege is shown, that the efficiency of use of graphite as a lubricant material depends on presence water steam. Water and other exhalations, adsorbed on base planes, reduce factor of friction with 0.80 up to 0.18. The strong wear process named «pulverization», occurs in vacuum. Later was shown, that the graphite has rather strong interjunction — type connections. Pairs work as a lubricant material. These experiments were carried out with graphite cores sliding on copper, but film of graphite order how and with room temperature. The damp film provides factor of friction 0.12, which is reduced with its drying by heating. Such behavior confirms, that has a place infilm sliding model 4 tab. 1. The surfaces of graphite are greased with steam in the same way, as surfaces of diamond by a moisture, and metalin conditions of boundary friction — organic films.

These results have induced to research of influence of humidity on MoS_2. It is revealed, that friction it is not enough in a dry atmosphere, however with growth of humidity is increased approximately by 85 %, and then is reduced. Growth of friction connect to presence of metal contacts through a film, while subsequent its reduction explain by chemical reactions between MoS_2 and surface of metal reducing metal contact. It was confirmed during the further researches. However friction grew even with low humidity and with absence of metal contacts.

Mataunaga [3] studied friction of spherical particles MoS_2 on steel and has received the following results. In a dry atmosphere the thin continuous film ensuring low meanings of factor of friction (0.15) and intensities of the wear process (2×10^{-6}) was formed. In a damp atmosphere factor of friction made 0.30, and intensity of wear process – 90×10^{-6}. These results similar received for graphite, confirm, that MoS_2 slides on MoS_2, however adsorbed the moisture increases friction. Such increase of friction causes carry MoS_2 per itself in the form of separate fragments

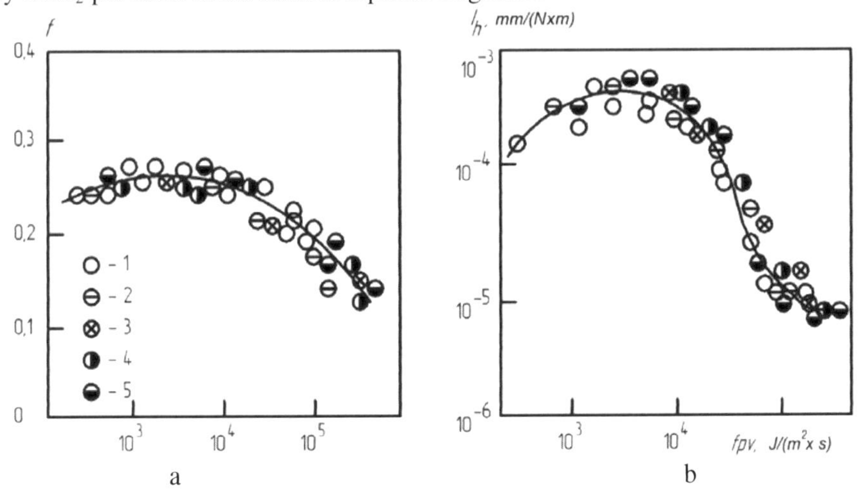

Figure 1. Influence of humidity of a film on factor of friction f (a) and intensity of wear process I_H (b) with contact pressure, MPa: 1 – 0.082; 2 – 0.165; 3 – 0.247; 4 – 0.330; 5 – 0.

4. Experimental results

The subsequent researches have shown, that features of sliding with higher loadings, speeds and temperatures same, as well as in a dry atmosphere. The increase of temperature of a film results in downturn of its humidity and reduces friction and deterioration. For example, in a Figure 1 the data, illustrating influence «predrying» of a film MoS_2 on friction and intensity of wear process are submitted. Haltner studied influence different steams on lubricant ability MoS_2 in vacuum. It has come to a conclusion, that the presence steams is not an obligatory condition of lubrication; with vacuum 1.3×10^{-7} Pa factor of friction was equalled 0.08. The increase of factor of friction fixed, with lower vacuum (1.3×10^{-4} Pa), coincides with the beginning of adsorbtive pollution of surfaces MoS_2. Thus, adsorbed of pair increase adhesion. Similarly layers of talc, connected by forces of Van—der—Vaals, at the presence of a moisture appear connected strongly, as their weak connections are replaced hydrogen.

Haltner investigated also behavior WS_2. With high vacuum 8.5×10^{-8} Pa factor of friction essentially did not depend on structure of environments and was equated approximately 0.17, however with low vacuum 1.3×10^{-3} Pa exhalation hydrocarbons (isoamyl alcohol) provided small friction ~ 0.09. Thus, exhalation of some organic substances grease layered structures, however pressure should be rather low, that could be formed of a strong film. The similar results are received for MoS_2 with research films, received by dispersion, and individual crystals [3]. Such behavior, first of all, testifies to an opportunity superficial of sliding, but thus it is necessary also to take into account chemical and physical effects. For example, anybody from the listed researchers did not consider, how exhalation can influence hardness of a film. The softer, the there should be a factor of friction more. Fleischauer has found out, that the chemical activity of a film is influenced also by its orientation. As have shown Johnston and Moore, it is quite possible, that the orientations of a film are promoted by humidity.

The stratification was studied some by the researchers. Bryant has measured interlayer cohesion of graphite in high vacuum and has found, that it makes 1.75×10^{-4} J/cm^2. It is higher the order, than appropriate size on air. The authors came to a conclusion, that MoS_2, talc and pirophyllite lubricant ability internally is inherent. These materials are easily stratified without influence of gas environments. For others of materials the presence steams is necessary, that they have displayed lubricant ability. A question in this connection is of interest, why talc it is much worse as lubrication, than MoS_2. The obvious answer consists that, first, it chemically is not linked to a surface. Mechanically it should be linked so well, as well as MoS_2, but the chemical reactions are not found out. Secondly, cohesion the durability of its film is lower. The experiment is of interest which would allow to find conditions or materials, for which talc would be an effective lubricant material. It would ensure essential progress in understanding of criterion of lubrication.

Mataunaga stratified individual crystals MoS_2 and has found out surplus of sulfur on planes of cleavage. Frictional of test the similarresults have given. With sliding diamond on an individual crystal MoS_2 in a mode of faltering friction it is revealed, that the contents of sulfur is increased with coupling, and subsequent sliding of surfaces has a place, when superficial density of sulfur grows up to some appreciable size. Johnson and Vaughn the hypothesis about a layer MoS_2 with adsorbed by sulfur is offered, but Haltner refute it, as this effect could not observe with higher vacuum. It has come to a conclusion, that the faltering friction is caused by any pollution of vacuum system. However it is necessary to note, that in the experiences on shift with high pressure Bridgeman very much frequently observed free metal, in particular for $CuCl_2$, $CuBr_2$, PbO, PbI_2, Bi_2S_3 and $A1_2SO_4$. All this – effective lubricant materials. If under action of high pressure the substance is decomposed, should be observed certain friction effect. This idea requires the further research.

MoS_2 (a — MoS_2) crystallizes in hexagonal to system. The atoms of molybdenum are located between two layers of atoms of sulfur. Distance between the nearest atoms of molybdenum and sulfur makes 2.41 °A, and nearest distance between atoms of sulfur in parallel layers — 3 °A.

Natural MoS_2 — MoS_2 receive with production of copper ores from collateral products, which contain so—called molybdenum shine. The first mentions of application $_{MoS2}$ as a lubricant means concern to XVI centuries.

Synthetic MoS_2 also has structure of a layered type, but or rhombohedral (β – MoS_2) with

IOP Conf. Series: Journal of Physics: Conf. Series **858** (2017) 012020 doi:10.1088/1742-6596/858/1/012020

parameters a = 3.15 °A, c = 18.38 °A, or intermediate between $\alpha - MoS_2$ and $\beta - MoS_2$.

Table 1. The basic properties MOS_2

Property	Meaning
Densities	4,8 g/cm3
Hardness on Moos	1,0—1,5
Temperature of a beginning sublimation	723 K
Temperature of melting	1458 K
Coefficient of linear expansion in the field of temperatures 323 — 973 K	$7{\times}10^{-6}$ l/K
Electroconductivity	Dielectric or semiconductor
Thermal conductivity	$4{,}9{\times}10^{-3}$ kcal/cm s K
Magnetic properties	Diamagnetic
The chemical resistance	Is dissolved in imperial vodka, oxidize in concentrated H2SO4 and HNO3
Shine	Metal
Color	Bluish—gray up to black
Transparency	Weak in seen area, high in infrared area of spectrum.

The electrical resistance MoS_2 large with a low pressure, but falls with increase of a pressure. Partially occurrence conductivity in MoS_2 is caused by heating at the expense of passage of an electrical current and partially at the expense of action of an electrical field. With temperature about 1123 — 1173 K it becomes a rather good conductor.

Light falling on the surface MoS_2, reduces its electrical resistance. In this respect MoS_2 reminds selenium, but to change of light exposure reacts more quickly. The greatest influence is rendered by beams concerning red area of a spectrum (with length a waves up to 10000 °A).

On air MoS_2 oxidize up to MoO_3 and sulfur or SO_2. Such oxidic of a film begin to be formed with 623 K, and with temperature the fast oxidation MoS_2 is higher 753 K occurs. In vacuum MoS_2, up to temperature 1373 K is stable.

Fluorine vigorously reacts with MoS_2, the chlorine with heating transforms MoS_2 in $MoCl_5$, and bromine practically does not react with MoS_2.

The hydrogen restores firm MoS_2 directly up to metal without formation intermediate connections:

$$MoS_2 + 2H_2 = Mo + 2H_2S \quad (1)$$

MoS_2 has high radiating resistance: any damages are absent with a doze $5{\times}10^9$ rad.

The moderate heating MoS_2 in the electrical furnace without access of air results in formation Mo_2S_3 on reaction:

$$2MoS_2 = Mo_2S_2 + S \quad (2)$$

Crystals Mo_2S_3 is grey—steel colors have a needle structure of density 5.9 g/cm^3 ($5{,}9 \times 10^{-4}$ kg/m^3).

With friction on air MoS_2 in some modes (temperature on a surface of friction is higher 343 —373 K) there is a partial oxidation MoS_2 up to MoO_3.

MoO_3 represents a white powder, frequently with a greenish or dark blue shade. Sublimation MoO_3 begins with 973 K, melt it with 1068 K. Firm MoO_3 — dielectric, but in melted a condition becomes a good conductor.

MoO_3 is restored up to MoO_2 with heating in hydrogen (573 — 743 K) and up to metal of molybdenum with temperature above 773 K. The acetylene (C_2H_2) will form with MoO_3 carbide of molybdenum — Mo_2C; methane (CH_4) restores trioxide of molybdenum up to dioxide and, probably, up to metal; oxide of carbon restores MoO_3 up to oxides of violet colouring.

Such metals as Na, To, Hg, A1 u Si are capable to restore MoO_3 up to metal. MoO_3 is well dissolved in alkali, forming hydrate oxide of molybdenum MoO_3—H_2O and MoO_3—$2H_2O$.

WS_2 crystallizes in hexagonal to system [5]. The crystal lattice WS_2 is similar to a lattice MoS_2, in which the atoms of molybdenum are replaced by atoms of tungsten. In comparison with MoS_2 WS_2 has greater heat resistance (is stable on air up to temperature 783 K) and resistance to oxidation. Its bearing ability in 3 times is higher, than at MoS_2.

WS_2 chemically is inert, is insoluble in all environments, including water, oil, alkali and almost all acids. Is sensitive only to influence free gaseous fluorine, hot chamois and hydrofluoric of acids. WS_2 is non—toxic a material and does not cause corrosion on metals.

5. Conclusion

Use WS_2 as the additive to lubricant oils for formation colloidal suspensions is complicated because of its high densities ($\rho = 7.4$ g/cm^3), exceeding almost in 8 time densities of mineral oil a little.

So, in suspension on the basis of mineral oil and 50 % (on weight) graphite, MoS_2 or WS_2 will accordingly be on volume 36 % of graphite or 15.5 % MoS_2, or 11 % WS_2.

For work in a usual atmosphere with temperature use WS_2 is higher 673 K is recommended, whereas with lower temperatures it is better to use cheaper MoS_2. With use in vacuum WS_2 and MoS_2 display almost identical properties and have lubrication ability down to 1593 K.

In vacuum of 1.3×10^{-13} Pa WS_2 is stable up to 1373 K, and the decomposition it occurs with temperature, exceeding 1673 K.

In more late research executed by the soviet scientists, was established, that in vacuum dissociation WS_2 occurs completely only with 2273 K. With heating WS_2 vacuum up to 673 K from it the part of sulfur is allocated.

In England to units of dry friction working with temperature up to 1853 K in vacuum or in an inert atmosphere, apply paste containing 50 % (on weight) WS_2. Such paste keeps the lubrication properties and with low temperatures (up to — 328 K) [6], it is insoluble in water and has high resistance to the concentrated acids, alkali and gases.

Selenides of molybdenum, of tungsten, niobium successfully are applied last time in quality ingredients self—lubricating of materials intended for work on friction in ultra high vacuum. Despite of it, about properties of these connections there are only very poor isolated items of information.

Diselenide of molybdenum — $MoSe_4$ for the first time was received by heating of molybdenum acid with selenium in a current of hydrogen. Crystallochemical and the electrical properties MoSe2 remained unexplored till 1961, when diselenide of molybdenum was received by sintering a powder of molybdenum with 973 K during 100 hour (cleanliness — 99 %) with stochiometric by quantity selenium in the quartz ampoules, sealed — in under vacuum.

The received preparation $MoSe_2$ represented a grey powder with metal shine, quite steady on air. X—ray research of a powder $MoSe_2$ has allowed to establish, that diselenide of molybdenum has hexagonal an elementary cell with parameters: a = 3.28 °A and c = 12.84 °A, density 6.9 g/cm3 (6.9×104 kg/m^3).

The crystal lattice $MoSe_2$ is similar to a lattice MoS^2, in which the atoms of sulfur are replaced by atoms selenium.

Diselenide of molybdenum is the semiconductor, its specific electroconductivity makes: with 295 K – 1.23×10^{-4}/(Ohm × cm), and with 333 K – 2.3×10^{-4} 1/(Ohm × cm).

Diselenide of tungsten — WSe_2 has hexagonal a lattice of a type MoS_2. The ways of reception WSe2 are similar to ways of reception MoSe2. Diselenide of tungsten has semi—conductor properties [4].

IOP Conf. Series: Journal of Physics: Conf. Series **858** (2017) 012020 doi:10.1088/1742-6596/858/1/012020

Diselenide niobium. In the literature there are only data on crystal structure $NbSe_2$ [4] With temperature 1073 K on air diselenide niobium begins sublimate. The possible updatings of an elementary crystal cell are considered in work.

The graphite, as well as other layered firm lubrication, has a lattice hexagonal of a structural type. Distance between atoms of carbon in parallel layers (3.44 °A) exceeds similar distance between atoms in a lattice borazon. In this connection the connection between planes of cleavage at graphite should be more poorly, than at borazon. The indirect proof it is served by lower size of factor of friction of graphite in comparison with borazon, observable in identical conditions of test.

Such carbonaceous of substance, as stone coal, coke, soot have crystal structure of graphite and usually are used for reception from them by thermal processing (temperature 3173 K) of polycrystalline graphite with a high degree of perfection separate crystallites.

Some physico—mechanical properties of graphite used in reactor building (table 2) are resulted below

Table 2. The basic properties of graphite

Property	**Meaning**
Density	2.21 – 2.25 g/cm^3
Porosity	20 – 32%
Temperature of melting	4073 – 4173 K
Thermal conductivity	30 – 32 kcal/m hour K
Factor of linear expansion in an interval 273 — 1273 K	3.5–8×10^{-6} 1/K
Strength on a stretching	50 – 95 kg/cm^2 (5×10^6 – 9.5×10^6 N/m^2)
Strength on compression	160 – 300 kg/cm^2 (16×10^6 – 30×10^6 N/m^2)
The module of elasticity	5,6×10^4 kg/cm^2 (56 × 10^8 N/m^2)

Borazon crystallizes in hexagonal to system. The crystal lattice borazon is similar to a lattice of graphite, in which one atom is replaced on boron, and other with nitrogen. The atoms of boron and nitrogen are located in horizontal layers alternately. Distance between the next layers at borazon (3.34 °A) is less a little, than at graphite (3.44 °A), therefore connection between layers at BN is stronger, than at graphite.

As against graphite at borazon there is no metal connection between layers. The connection between atoms of boron and nitrogen has ionic character. BN has low electroconductivity, and also is steady in neutral and reducing atmospheres.

Oxidation BN largely depends on temperature of its reception or preliminary of calcination. So, with increase of temperature preliminary of calcination with 1273 up to 1373 K speed of its oxidation decreases in 4 times. In the form of dense products borazon begins oxidize with temperature 973 K, the amplified oxidation occurs with 1273 — 1373 K. With heating in vacuum up to temperature 1273 K is not decomposed. With carbon borazon reacts with 2273 K, forming carbide of boron and nitrogen. With metals in an atmosphere of nitrogen borazon will form borides. In boiling water and diluted acids is decomposed with formation of ammonia NH_3 and boric acid. The amplified decomposition begins with temperature 403 K. The Concentrated acids on work poorly. For example, H_2SO_4 decomposes borazon only after 6 — 10 hours influence.

Besides considered above, lubrication ability other materials (CdJ_2, PbJ_2, BiJ_3, Sb_2S_3, AgJ) find out also. In firm lubricant coverings are used also PbO, CaF_2 and organic materials phthalocyanine and PTFE.

6. Acknowledgments
Experimental studies carried out on the surfaces of flanges tires wheelsets and necks motor — axial bearings with the application of solid laminated lubricants followed, proved the possibility of increasing the wear resistance of flanges surfaces by 40 percent in the initial period of operation, and the surfaces of the necks of motor — axial bearing not less than 6 times in comparison with surfaces treated according to current technology.

References
[1] Chijikov D M and Schastlivii V P, 1964 Selenium and Selenides (Moscow: Science) pp 165–171.
[2] Wainstein V E, Suchkova O A and Memelov V L 1965 The Effect of Friction Conditions on the friction characteristics of desulfide of molybdenium (Moscow: Knowledge) pp 108–114
[3] Matsunaga M., Nakagawa T. 1978 Frictional Behavior of Various Kinds of MoS2 *Proc. Second International Conference on Solid Lubrication, ASLE SP-6* 45.
[4] Burwell F T, Miln A A 1953 Grease solids (Moscow: Eds. Of Foreign literatures) pp 220–230
[5] Gramels J 1953 Selenium rectifiers – Factor in their application *The Bell System Technical Journal* **32** pp 1469–1492
[6] Katano T, Oka M, Nakazawa S, Aramaki T and Kusakawa K 2003 Characteristics of dual lubbricant layers *IEEE Transactions on Magnetics* pp 2489–2491

IOP Conf. Series: Journal of Physics: Conf. Series **858** (2017) 012021 doi:10.1088/1742-6596/858/1/012021

The influence of ultrasonic exposure on polytetrafluoroethylene structure modified with boron nitride

D A Negrov[1], **E N Eremin**[1]

[1]Omsk State Technical University, 11, Mira ave., Omsk, 644050, Russia

E-mail: Negrov d a@mail.ru

Abstract. The purpose of study is to state influence patterns of ultrasonic exposure on the density and porosity change of the synthesized polymer composite material based on polytetrafluoroethylene, modified with boron nitride. The density of 5 % boron nitride samples manufactured by ultrasonic pressing is 9% higher than the similar samples manufactured by cold pressing have and this confirms the structure defectiveness results received by electron microscopy. As the result of carried studies it was stated that polymer composite material pressing with ultrasonic vibration is an active technology increasing efficiency of matrix structure modification and affecting significantly the structure formation processes in it.

1. Introduction

Structural modification is the basis of different methods to increase polymer composite material (PCM) mechanical properties. The most widely used structural modification method is introducing different type fillers in polymer matrix: dispersed, fiber, ultradispersed [1–4].

The development of PCM obtaining technology in order to increase external energy deposition level allows one to get significant improvement of PCM properties.

One of the advanced active energy deposition methods is imposing ultrasonic vibrations (UV) on pressed powder mixture. This increases powder flow time, levelness of polymer particle lay-up, arched structures are destroyed, the development of powder particle viscous deformation is rather simplified, the porosity is decreased and the density of pressed material is increased. This has a positive influence on friction changing between particles and mould walls, enables one to form complex products with rather small efforts of pressing.

2. Statement of problem

The purpose of study is to state influence patterns of ultrasonic exposure on the density and porosity change of the synthesized polymer composite material based on polytetrafluoroethylene, modified with boron nitride.

To make research, the following concentrations of boron nitride in polytetrafluoroethylene were chosen: 1.0, 3.0, 5.0, and 10.0%.

To produce samples the special equipment based on the hydraulic press MT-50 was assembled [5]. Ultrasonic generator UZG 2-4M with 2.5 kW input operating at frequency from 17.5 to 23 kHz was used. The magnetostrictive transducer PMS 15-A-18 with 17.8 kHz resonance frequency was taken as the source of ultrasonic vibrations. The effort of pressing was 54 MPa and the range was 16 μm. After pressing the samples were subjected to baking at 360 °C.

Composite material structure defectiveness was studied on the samples obtained with and without ultrasonic vibrations.

3. Theory

To clarify processes of polytetrafluoroethylene structural modification as the result of ultrasonic treatment, the polymer composite material (PCM) substructure was studied by electron-microscopic analysis.
To examine structure defectiveness, in this study JEM–6460 LV microscope was used (resolution is 3nm) and in order to make the conductive coating cleave on the surface, silver spraying in conditions of high vacuum was used.
The density of samples was studied by hydrostatic weighing on the 0.0001 g/cm^3 high-accuracy weighing machine VIBR-HTR.

4. Experiment Results

Microscopic studies found differences in composite morphology. Electron microscope data showed that modified composite obtained without ultrasonic treatment has non-uniform structure. It consisted of various shape nanoobjects. Besides, in the sample there can be noticed agglomerates of particles less than 2 nm covering lots of crystallites by dense layer.

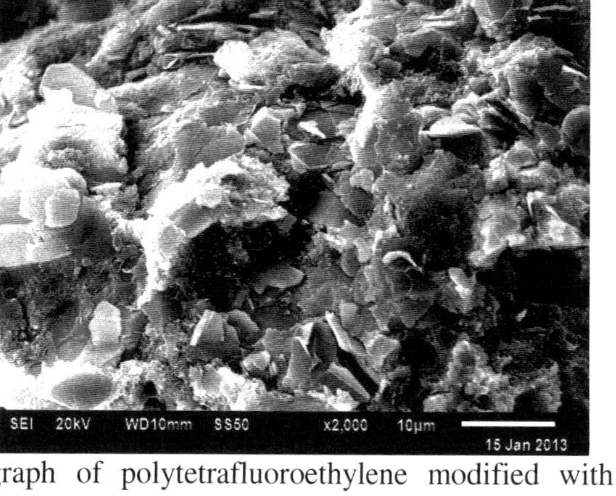

Figure 1. The micrograph of polytetrafluoroethylene modified with boron nitride and manufactured by static cold pressing.

In the modified composite obtained with ultrasonic treatment crystallites have specific morphology. They are characterized by complex structure-layer extended ones consisting of 50 nm blocks and 1–5 nm nanoparticles. In the PCM sample fracture nanoparticles are noticed to be denser than the others forming contaminations in crystallites.

Figure 2. The micrograph of polytetrafluoroethylene modified with boron nitride and manufactured by ultrasonic pressing.

The second type of objects observed in such composite is chain structures consisting of blocks with 50–100 nm width (Fig. 3). The chain structures, observed on this sample morphology study, are suggested to be parts of PTFE supramolecular structure formed as the result of ultrasonic treatment. These parts contain the abundance of carbons. In the PTFE molecular composition carbon is twice less than fluorine.

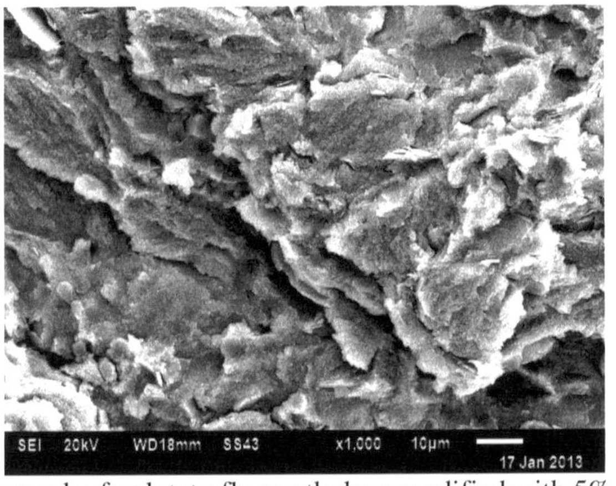

Figure 3. The micrograph of polytetrafluoroethylene modified with 5% boron nitride and manufactured by ultrasonic pressing.

Also it is important to state that sample surface of PCM obtained by static cold pressing is more spongy and porous; the defined boundaries of formations that can be identified as spherulites are not observed (Fig. 4). Crystalline formations near filler particle surfaces are not observed.

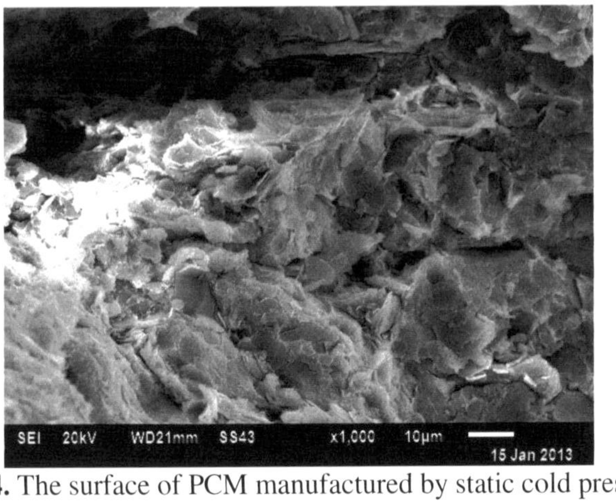

Figure 4. The surface of PCM manufactured by static cold pressing.

It should be said the increase of PCM wear resistance in result of ultrasonic treatment is defined by setting up spherulites of different shapes and sizes in the matrix.

In composite material obtained by ultrasonic pressing along with supramolecular formations similar to PTFE there can be found sites of polymer with structure not typical for pure PTFE, which can be identified as strong defective irregularly shaped spherulites (Fig. 5).

The boron nitride particles decrease polytetrafluoroethylene layer mobility that in its turn can result in wear resistance increase of synthesized composite material based on it. In composite material synthesized by ultrasonic pressing at optimum performance a high-volume contact between boron nitride particles can be observed and this defines the main way of obtaining indicated structure and its interference with physical and mechanical properties.

Figure 5. The cleave surface of polytetrafluoroethylene modified with boron nitride and manufactured by ultrasonic pressing.

When filler concentration is increased, the volume of polymer interreacting with boron nitride is decreasing. Therefore, the rate of more solid intermediate layer becomes higher up to polymer self-reinforcement effect followed by strengthening molecular interaction. As a result, the polymer acquires the structure similar to network polymer where strengthening phase acts as crosslinks preventing polymer chains from moving (Fig. 6).

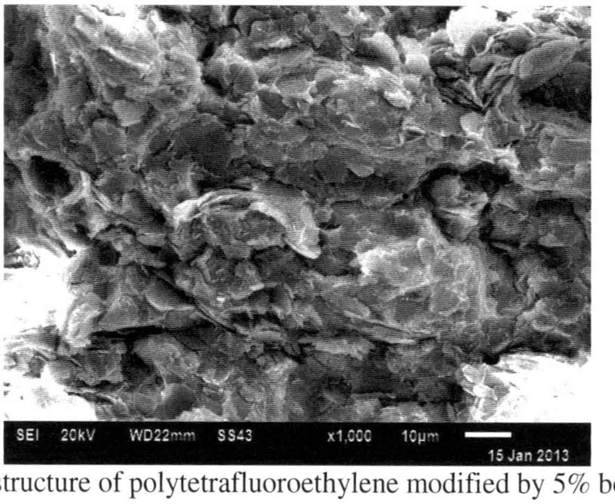

Figure 6. The structure of polytetrafluoroethylene modified by 5% boron nitride.

The stated difference between structures results from large volume changes of PTFE practically not connected by additional adhesion bond with boron nitride on melting crystalline phase in static pressed materials. This confirms again the best interaction of PTFE and boron nitride upon ultrasonic treatment and polymer monolitization with less volume changes and, therefore, small shrinkage upon baking that is very important when composite products with boron nitride are manufactured, and also the formation of the base while being treated by ultrasound on its being filled with 5% and its primary influence on material properties (Fig. 8).
The result of carried studies of cold pressing sample cleave showed on the surface of it small shells and voids are observed (Fig. 7).

IOP Conf. Series: Journal of Physics: Conf. Series **858** (2017) 012021 doi:10.1088/1742-6596/858/1/012021

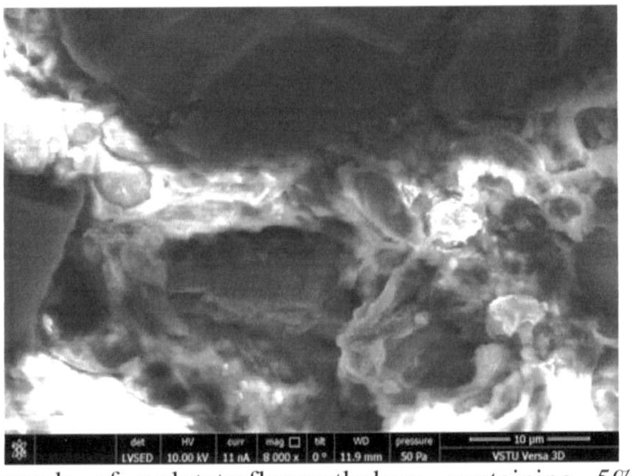

Figure 7. The micrograph of polytetrafluoroethylene containing 5% boron nitride and manufactured by static cold pressing.

On the surface of ultrasonic pressing sample cleave the shells and voids are not observed (Fig. 8).

Figure 8. The micrograph of polytetrafluoroethylene containing 5% boron nitride and manufactured by ultrasonic pressing.

The decrease of composite structure defectiveness should result in increase of its density after ultrasonic treatment. For this purpose the study of ultrasonic treatment influence on changes of received PCM density was performed.

The results of synthesized composite material density are given in Fig. 9.

The density of 5 % boron nitride samples manufactured by ultrasonic pressing is seen to be 9% higher than the similar samples manufactured by cold pressing have and this confirms the structure defectiveness results received by electron microscopy.

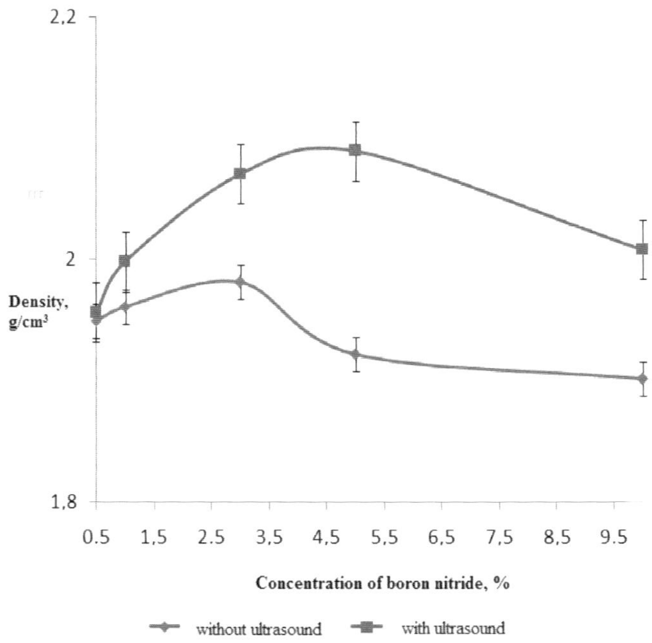

Figure 9. The influence of ultrasonic treatment and filler concentration on composite material density.

The density of polytetrafluoroethylene and boron nitride is practically similar – 2.18 g/cm^3, therefore, composite material based on polytetrafluoroethylene modified with boron nitride should have the same density theoretically (Fig. 10).

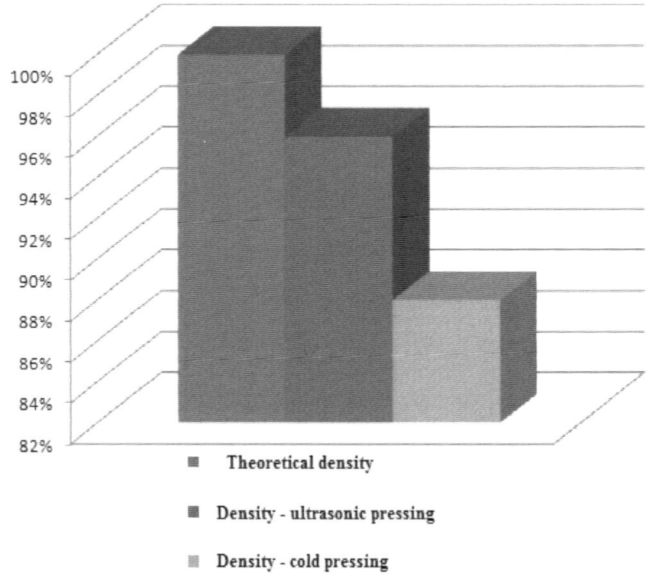

Figure 10. The density rate of composite material.

The study of quality of pressed samples from composite based on polytetrafluoroethylene and 5% boron nitride showed that 96% maximum density rate is available in material manufactured by ultrasonic pressing. PCM manufactured by static cold pressing has 88% density rate.

IOP Conf. Series: Journal of Physics: Conf. Series **858** (2017) 012021 doi:10.1088/1742-6596/858/1/012021

5. Results and discussion

The received results confirm the low adhesion strength of PCM with filler synthesized without ultrasonic exposure which acts as defect and promotes composite mixture breaking in contrast to PCM having high adhesion strength of compositions after ultrasonic treatment. On baking static pressed samples the rapid change of strain-stress polymer state is likely to occur that results in weak adhesion polymer-boron nitride bond breaking and after cooling along boundaries of metal particles pores can be formed. This confirms their lower density during baking in free condition in contrast to materials after ultrasonic treatment.

Ultrasonic treatment leads to noticeable increase of studied composite density. This result coincides with the data of supramolecular structure studies before and after ultrasonic exposure. The ultrasonic treatment of composites based on PTFE leads to fundamental change of polymer morphology, puts into operation the process of structure reconstruction. The supramolecular structure becomes more homogeneous, the porosity decreases significantly and filler particles act as structure-forming factors - centers of spherulite nucleation.

6. Conclusion

The study of ultrasonic exposure on the structure of pressed materials from polymer compositions informs of molecular interactions in powder particle surface layers promoting the process of further baking. In contrast to static pressing the promotion of phase transition in PTFE treated by ultrasound is confirmed by more advanced change of its density, strength and hardness, density of monolithic material.

The behaviour of polymers upon baking is the result of structural changes occurred during ultrasonic treatment and caused by decreasing thermal stability of crystalline formations and their lower defectiveness, as well as molecular interaction strengthening up to cross-linking during ultrasonic exposure.

The carried studies confirm that polymer composite material pressing with ultrasonic vibration is an active technology increasing efficiency of matrix structure modification and affecting significantly the structure formation processes in it.

References

[1] Mashkov Yu K, Kropotin O.V and Shilko S.V. et. al. 2015 *Mater.* **1** pp 22-25.
[2] Petrova P N, Okhlopkova A A and Fedorov A L 2015 *Friction and Wear* 36, **1** pp 14–20
[3] Negrov D A and Eremin E N 2012 *Russian Engineering Research* 32, **1** pp 42–44
[4] Eremin E N and Negrov D A 2014 *Chemical and Petroleum Engineering* **49** pp 9–10
[5] Negrov D A and Eremin E N 2014 *IEEE Dynamics of Systems, Mechanisms and Machines (Dynamics)*

Scale Control and Quality Management of Printed Image Parameters

O A Novoselskaya[1], V L Kolesnikov[1], T V Solov'eva[1], I V Nagornova[2], E B Babluyk[2], O V Trapeznikova[3]
[1] Belarusian State Technological University, 13a, Sverdlova st., Minsk, 220006, Belarus
[2] Moscow Polytechnic University, 38, Bolshaya Semenovskaya str., Moscow, 107023, Russia
[3] Omsk State Technical University, 11, Mira ave., Omsk, 644050, Russia

E-mail: novolochka@yandex.by

Abstract. The article provides a comparison of the main valuation techniques for a regulated parameter of printability of the offset paper by current standards GOST 24356 and ISO 3783: 2006. The results of development and implementation of a complex test scale for management and control the quality of printed production are represented. The estimation scale is introduced. It includes normalized parameters of print optical density, print uniformity, picking out speed, the value of dot gain, print contrast with the added criteria of minimizing microtexts, a paper slip, resolution threshold and effusing ability of paper surface. The results of analysis allow directionally form surface properties of the substrate to facilitate achieving the required quality of the printed image parameters, i. e. optical density of a print at a predetermined level not less than 1.3, the print uniformity with minimal deviation of dot gain about the order of 10 per cents.

1. Introduction

Offset paper has several features, i. e. anisotropy sheet structure, complex composite and chemical compound. Overview the modern technologies on studying of multiscale structure of the printing paper in [1] showed the importance and urgency of this issue.

Forecasting and management of properties of printing paper is now receiving more attention. Paper printability is studying by J. Aspler, A. Mackay, D. V. Dunayev, V. N. Leontiev, P. V. Lukanin. J. Aspler in the works [2, 3] connected the quality of prints and paper, has shown that the structure of its surface significantly affects the print image parameters. A. Mackay has studied the correlation between print quality and test results according to standard methods using proofing devices. He showed that data correlation of the actual print close to the threshold value determined for the laboratory tests that indicated an indirect relation [4]. D. V. Dunayev, V. N. Leontiev, P. V. Lukanin [5, 6] during studying papermaking technology for printing have introduced the concept of "the cloud" as a parameter of paper characterizing its irregularity of structure and properties, but their researches haven't embraced the joint effect of composition and the quality of printed image.

Questions on quantitative evaluation of size of the filler and its distribution in paper using SEM-techniques are discussed in [7]. Despite the results achieved, to control the quality of offset paper need to know exactly how its properties change by the composition. The papermaking process involves many factors, i. e. temperature, concentration of the compound at the inlet / outlet flow, feed rate of the mixture, the amount and type of additives that can cardinally change the properties of the final material. On the other hand, paper requirements include such parameters as breaking length, opacity, whiteness, Kobb test, tightness, breaking strength with special demand of picking strength that characterizes the printability of paper.

2. Experimental setup

Comparison of estimation techniques of the main regulating parameter of printing properties of the offset paper the picking out speed in accordance with GOST 24356 [8] and ISO 3783: 2006 [9] shows the differences in method of its determining. For example, according to GOST 24356 regulated thickness of specimen on form is 4.3 micron, and by ISO 3783: 2006 is given 8.0 ±0.5 microns. Printing pressure is recommended about 650 N / m2 in accordance with GOST 24356, and 350 N/m^2 by ISO 3783: 2006. There is a difference in normalizable value of picking strength (i. e. according to GOST 24356 it is 2.2 m/s, and according to ISO 3783: 2006 it is 4.0 m/s). Also, there is a discrepancy in number of experiments. All above leads to difficulties in comparing the properties of printed paper by regulated parameters and the need of more detailed studying the printing paper properties, including by used in printing industry scale control methods for forecasting paper behavior during the printing process and the management its quality.

It should be noted that most of standardized parameters characterize the strength of printing paper but not its behavior in the printing process. Interaction with ink is described with a very limited number of parameters determination of which is carried out by indirect methods. It caused the developing of a complex test scale. The main difference between a complex test scale and the scales used in printing industry is its high resolution, which allows showing changes in printed image from the paper furnish. A number of additional graphic elements is also introduced in it and allow determining the machine and cross direction of paper sheet formation and the front side of the grid. The main purpose of this work is to develop a method of quality management of the printed image with taking into account the composition of the offset paper compound. To solve this aim it is necessary to work out a set of criteria to estimate and control the quality of outputs, carry out tests on selected criteria, develop recommendations for quality managing of printing paper.

3. Material and methods

A proofing machine Korex and laboratory proofing device IGT have been taken for the process equipment. Printing was performed under standard conditions. The test samples have been taken with a mixed stack of paper of varying composition to obtain prints with the image of a complex test scale, which made it possible to evaluate such print image parameters, as the absolute optical density of a print, print uniformity, dot gain value, print contrast, microtext, paper slip, resolution and effusing ability of paper surface. In addition, we evaluated the surface resistance to picking by the ISO 3783: 2006. The minimum number of replicates was 10 prints for each paper type determined by its compositional structure.

The main problem in the evaluation of the experimental data was the large amount of data (more than 400 types of paper and about 15 criteria for each). Therefore, to predict the behavior of paper in printing process and decision management the technology of complex quantitative quality estimation was applied. Algorithm of solution includes two main stages, i. e. the estimation of separate parameters and properties, and complex assessment of quality overall. Evaluation object property reduces to determining the location of a separate property on a scale. The analysis is performed on the basis of calculating the geometric mean of expert evaluation:

$$D = \left(\prod_{u=1}^{p} d_u^{\delta_u} \right)^{\frac{1}{\sum_{u=1}^{p} \delta_u}}, \tag{1}$$

where δ_u – statistical weight (importance) u-criterion.

IOP Conf. Series: Journal of Physics: Conf. Series **858** (2017) 012022 doi:10.1088/1742-6596/858/1/012022

In (1) d_u – private desirability function for each of the parameters calculated by the formula:

$$d_u = \exp\left[-\exp\left(-y'_u\right)\right].$$

where y'_u – coded parameter value.

As the rating scale the normalized parameter values of optical density of a print, print uniformity, surface resistance to picking, the value of dot gain, print contrast, and also added the criteria of minimizing microtext, the value of paper slip, resolution and effusing ability of paper surface have been taken.

The results of tests carried out by us have shown that the limited values not always fit into the recommended limits for print contrast index, calculated by Schirmer index K_{Sh}, and corresponded with the worst characteristics of paper printability. Therefore, it was decided to bring the indicator to the intervals within a scale experimental data, $K_{Sh} = 0.3 \div 0.5$, and the best value was on the point of 0.4.

Processing of the experimental data with the calculation of desirability private functions and the generalized quality criterion have been carried out using software tools of program FUZZY developed in BSTU. Screen copies on assign the desirability private functions curves of offset paper are shown in Figure 1.

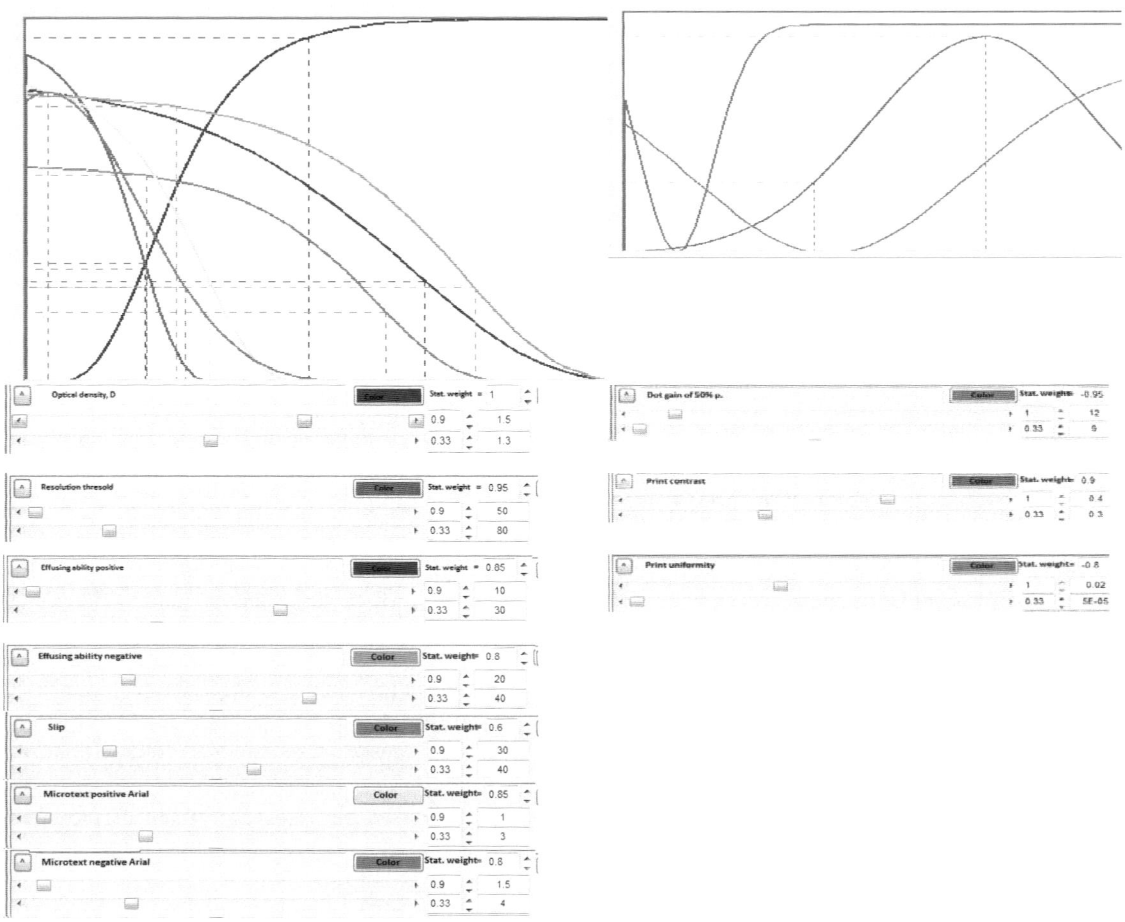

Figure 1. Shape of desirability private functions curves: a – function type – sigmoid; b – function type – Gauss.

The value of dot gain is normalized by ISO 12647-2 [10]. Practice shows that for the production of paper with improved printing properties it should be limited the allowable range of dot gain for the 50% raster field to 12%. For negative microtexts and effusing ability in comparison with the positive ones the weight have been reduced on 0.05 as they are used relatively rare.

4. Experimental results

To evaluate the paper printing properties a complex test scale has been developed (Table 1) [11, 12], and the basic requirement was a frequency of no less than 300 lpi, which guaranteed a minimum size of microdots on the level of 5 micron that matches to the thickness of paper sheet fibers.

Table 1. Desirability scale for quality parameters of the offset paper.

Parameter Name	**Values on a scale of desirability metrics**				
	Discard	*Satisfactory*	*Good*	*Very Good*	*Perfect*
Optical Density, D	менее 1.1	1.1	1.3	1.5	более 1.5
Printing Contrast, K_{Sh}	менее 0.25	0.25	0.3–0.5	0.35–0.45	0.4
Print Uniformity, D	более ±0.1	±0.1	±0.02	±0.001	0
Dot Gain of 50% p., %	более ±15	±15	±12	±9	0
Resolution threshold, μ	более 100	100	80	50	менее 50
Microtext, pt.:					
– positive	более 3.5	3.5	3.0	1.0	менее 1.0
– negative	более 5.0	5.0	4.0	1.5	менее 1.5
Effusing ability, μ:					
– positive	более 40	40	30	10	менее 10
– negative	более 50	50	40	20	менее 20
Slip, μ	более 50	50	40	30	менее 30

The values of parameters were normalized with the recommendations of various normative documents. The statistical weight for a set of standardized parameters was on the level of 0.9, and for non-standardized – 0.65.

For example, for the optical density of a print according to the technological instructions for printing of Institute of printing for the offset paper the standardized value for the black ink is 1.2 D, and by the requirements of ISO 12647-2 it is 1.0 D, and according to recommendations of X-Rite company the optical density for uncoated paper is 1.55 D for black ink.

Therefore, the minimum acceptable value of optical density was taken on 1.1 D based on the range of variation of the experimental data of 0.95÷1.6 D, and to assess perfect metric the optical density must be greater than 1.5 D.

The results of approbation on Goznak "Paper mill" of Borisov (Byelorussia) with optimized composite structure are shown in Tables 2 and 3.

Analysis of experimental data in Tables 2 and 3 shows that the management of a composite compound of the paper by the use of fuzzy sets and generalized desirability functions for processing the results of approbation suggested the increase not only the physical and mechanical properties of the offset paper but its printing properties for all quality parameters of the printed image.

In particular, increasing ink sensitivity to the surface by the surface sizing, increase the homogeneity of the surface as a result of changes in the composition be the fiber filler leads to increase of printing uniformity and optical density, minimizing microtexts, reduces dot gain and slip.

SEM-image of the sample paper layer surface under the magnification of $1000^{\times}$ is shown in Figure 2*a*. In the sample the formation of modified inorganic filler obtained by the technology of Goznak "Paper mill", containing $CaCO_3$, is revealed (Figure 2*b*).

Table 2. Physical and mechanical properties of the offset paper obtained in an industry.

Parameter Name	Offset Paper (control)	Paper With Optimized Composite Structure
Fracture Strength, n. d. f.	149	264
Breaking length, m		
– in machine direction	6700	6900
– in a transverse direction	3900	4000
Resistance to tearing, mN	630	640
Kobb test, g/m^2	18	18
Smoothness, s	50	67
Linear deformation, mm	2.9	2.8
Sizing degree, mm	2.0	2.0
Ash content, %	6.2	9.7

Table 3. Paper printing properties obtained in an industry.

Parameter Name			Offset Paper (control)	Paper With Optimized Composite Structure
Optical density of print, D			1.17	1.45
Uniformity Print, D			±0.0924	±0.0531
Microtext, pt.	Negative	Arial	2.0	1.9
		Times	2.5	2.0
		Script	4.0	2.5
	Positive	Arial	1.4	0.9
		Times	1.7	0.9
		Script	2.5	1.7
Paper resolution threshold, μ			80	50
Effusing ability, μ			30	10
Printing Contrast K_{Sh}			0.324	0.350
Slip, μ			60	40
Dot Gain of 50% p., %			+12.8	+10.8

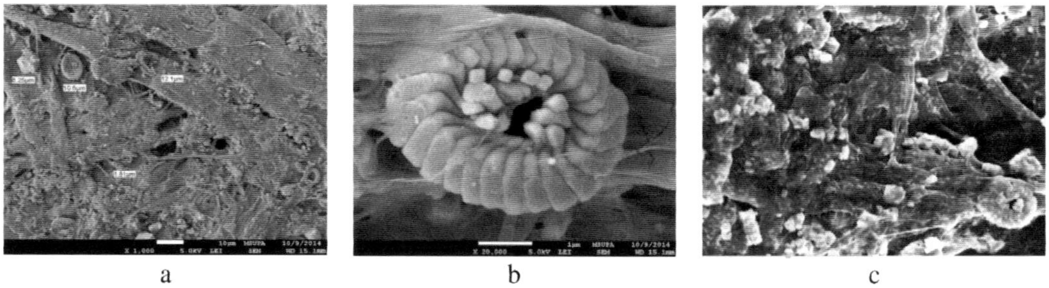

a b c

Figure 2. SEM-images of the structure of the paper (a) and the filler (b) overlaid with EDS map the distribution of elements (c).

It is clearly visible forming the grid of paper sheet from fibers, microfibrils and filler particles on Figure 2*a*. Figure 2*c* shows the distribution map of the elemental composition of a paper sheet with carbon C by blue, oxygen O by red and Calcium Ca by green.

5. Results and discussion

Printing uniformity according to the recommendations of the Institute of printing for black ink needs to be no more than ±0.1 D. The experiment carried out with the paper surface sizing has showed that printing uniformity does not exceed the level of ±0.05 D. Therefore, as a criterion for this parameter the average value is taken ±0.02 D. Parameter should show function with two-sided contingencies with a minimum at 0. For the rest parameters the logistic sigmoid with 0.33-0.9 terms in corresponding percentages are used as a membership function, which are evaluated as "good" and "very good".

To determine the criteria by which parameters of paper printability change the decision tree is modeled by the data set processing in the JMP program for each of identified by the scale parameters. The example of the decision tree for the optical density is shown in Figure 3.

Detailed analysis is shown in an example for the surface sizing of the offset paper by compositions based on various types of starches with additives recommended by manufacturers of chemicals for the pulp and paper industry to improve paper printability.

Parameter of optical density of a print obtained on the offset paper with surface sizing is greatly affected by the concentration and type of the starch composition (Figure 3).

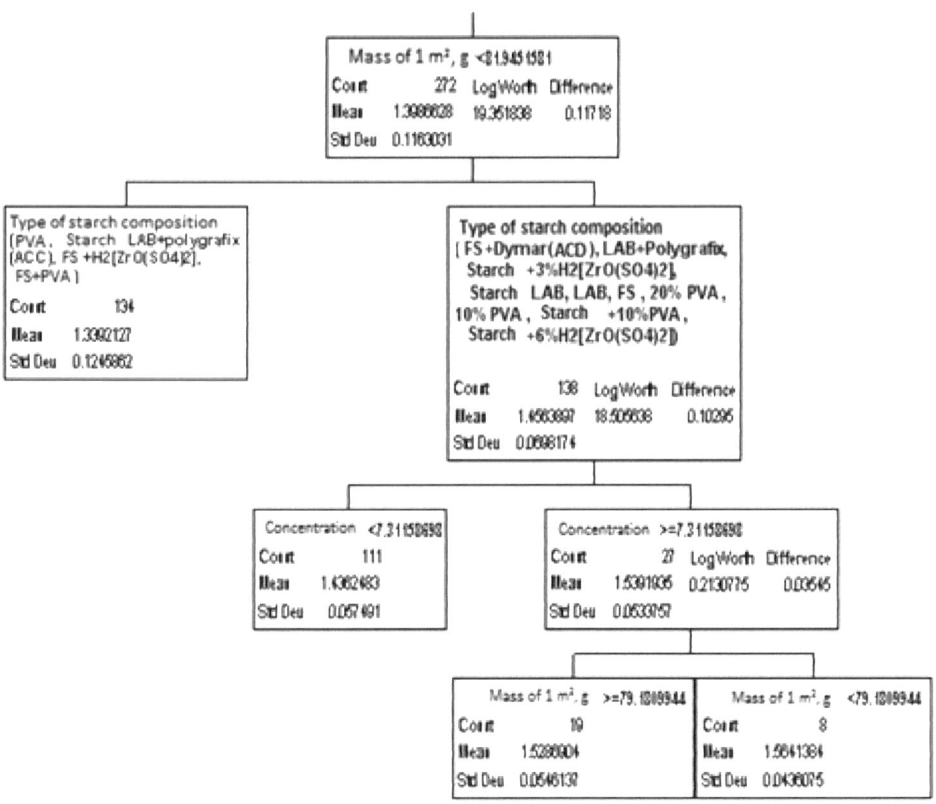

Figure 3. Example of decision tree in JMP.

For maximum of optical density it is necessary to have mass of 1 m^2 not exceed to 80 g. Further analysis showed that to improve the resolution and effusing ability of the paper surface it is necessary

to change the type of starch composition and starch expiry speed. Slip parameter also has the greatest effect by the type of starch composition and its concentration.
In order to increase the reproducibility of fonts by parameter of microtext it is prefered to vary the type of the starch composition, starch flow discharge and its velocity. To reduce dot gain value it is needed to modify the starch digestion temperature, the concentration and its discharge velocity. Printing Contrast is within optimal limits when changing the type of composition and starch concentration. For the parameter of printing uniformity it is important to change all the dependent factors.
It was found that in the aggregate for the offset paper with its surface sizing parameters vary from changes in type of the composition for surface sizing, discharge velocity and starch concentration.
Overall analysis of SEM-images showed a significant increase in the uniformity of the paper sheet due to a more equal distribution of the filler particles in the structure of a sheet and its retention by fibrils and microfibrils of cellulose fibers, changes in its shape and dispersion.
In general, quantity of filler particles distributed over the surface in the relative atomic units is low and assists to increase of the smoothness, resolution and effusing ability of the paper surface, reduces the amount of dot gain.

6. Conclusion and summary

Correlation of quality parameters of the printed image with the condition parameters of the technology in papermaking process by complex test scale is shown in the work. This allowed predicting and directionally controlling final printing paper properties in order to improve the uniformity of paper sheet structure, change the resolution and effusing ability, which is particularly important for special types of paper.
The use of fuzzy sets, linguistic variables, membership functions and formulas for calculation the complex quality criterion of image reduces the dimension of the problem for the analysis of printing paper properties.
From the viewpoint of reproducibility fonts improve by microtext parameter, increase of prints saturation by enhancement of ink sensitivity with the compositions for surface sizing, rise of paper structure uniformity, its smoothness and ash content due to the introduction of the compositions with modified fillers that higher the natural paper whiteness, the most important criteria for predicting and quality control of the printed image are revealed.
The results of the analysis allow directionally improve surface properties to give a paper with improved quality of the printed image, i. e. optical density at a given level of not less than 1.3 D, print uniformity with minimal deviation $\sim 10^{-5}$, dot gain for offset paper at a level of coated paper about 10%, microtext reproducibility, resolution and effusing ability and minimum slip. This procedure can be applied not only to paper, but, for example, and to ink, the blanket and the plate.

References

[1] Chinga-Carrasco G 2009 Exploring the multi-scale structure of printing paper – a review of modern technology *J. Microscopy* **234** 211–242
[2] Aspler J, Jordan B and O'Neill M 1999 Offset print quality of uncoated mechanical printing papers: Results of a commercial heatset offset trial *TAGA Proc.* pp 302–327
[3] Donderi D C and Aspler J 2000 Proscale: a fast new way to measure subjective print and paper quality *Test methods* pp 43-45
[4] Mackay A and Wright A 1996 Correlation between laboratory printability testing and commercial printing *50th Appita Annual General Conf.* pp 405-414
[5] Dunayev D V, Leontiev V N and Lukanin P V 2007 Method for estimating the impact of paper on the heterogeneity of density fluctuations in the optical impression *Pulp, paper, cardboard* **1** 50-51
[6] Leontiev V N 2009 System approach to the problem of improving printability of paper *Forest magazine* **3** 125-128
[7] Bennis H, Benslimane R, Vicini S, Mairan A, Princi E and Bennis H 2010 Fibre width

measurement and quantification of filler size distribution in paper-based materials by SEM and image analysis *J. of electron microscopy* **59(2)** 91–102

[8] GOST 24356. Method for determination of printability [Electronic resource], access mode: http://vsegost.com/Catalog/78/7845.shtml

[9] ISO 3783:2006. Paper and board. Determination of resistance to picking. Accelerated speed method using the IGT-type tester (electric model), 2006

[10] ISO 12647-2:2004. Graphic technology – Process control for the production of half-tone colour separations, proof and production prints – Part 2: Offset lithographic processes, 2012

[11] Novoselskaya O A, Temruk V I, Penkin A A and Solov'eva T V 2012 A method for controlling properties of paper printed deviations from nominal: Pat. Rep. Belarus № 15509, MPK8 G 01 N 21/89 (2006.01), G 01N 33/34 (2006.01) – № a20091585; stated 2009.11.10; published 2012.02.28. // Official bulletin / National Intellectual Property Center **1** p 150

[12] Novoselskaya O and Kulak M 2009 Modelling the printing properties of the offset paper *Proc. of Chemnitz University of Thechnology* pp 171–176

IOP Conf. Series: Journal of Physics: Conf. Series **858** (2017) 012023 doi:10.1088/1742-6596/858/1/012023

Geometrical and computer modeling of mechanical engineering surfaces products intersection line

K L Panchuk[1], N V Kaygorodtseva[1], T N Kaygorodtseva[2]

[1]Department of Engineering Geometry and CAD, Omsk State Technical University, prospect Mira, 11, Omsk 644050, Russia
[2]Radio Technical Faculty, Omsk State Technical University, prospect Mira, 11, Omsk 644050, Russia

Kaygorodtceva@gmail.com

Abstract. In the design and manufacture of engineering products geometrical problem is known by shaping the surface of the product. An important element of general solution of this problem is to define the lines of surfaces intersection, forming the shape of designed product. Existing possibilities of modern CAD systems do not allow achieving fullness of the result in this direction. For example, control points and change point of visibility is difficult to identify in the product drawings. In addition, there are no possibilities of detecting imaginary points which are necessary for a complete analysis of intersection surfaces, and mapping these points in the drawing. The aim of the study is to develop a geometric algorithm of constructive determining the intersection line and is devoid of these shortcomings. The objectives of the study are testing the obtained algorithm by experimental verification with geometric modeling solutions to specific problems by tools CAD. This study adopted the method, which is based on quotient of geometric sets, which are regarded as intersecting surfaces in space E^3. One area of practical use of surface engineering products geometric algorithm - shaping is based on their intersection line.

1. Introduction

At all stages of new engineering products creation - from design to manufacture, have to solve geometric problems of forming surfaces of the product. The geometric form of modern engineering products is a set of merged into one elementary and compound surfaces: polyhedron, spherical, conical, cylindrical, toroidal, helical, spline and others (Fig. 1). When shaping geometric shape products of mechanical engineering we must take into account technological and technical (choice of equipment) the possibility of their production - metal casting, stamping, forging, machining, pressure, welding and others. It is also necessary to take into account the pragmatic (functional) and esthetical requirements for the form of the designed product, which in some cases may reach full unity. For example, in perfect shape airliner or supersonic fighter. Often there is the predominance of one requirements over the other. Accounting for of technological and technical possibilities, pragmatic and aesthetic requirements leads eventually to formation of a specific geometric shape certain engineering products. At the same time, some of the combined in one form and geometric shape of the surface can

IOP Conf. Series: Journal of Physics: Conf. Series **858** (2017) 012023 doi:10.1088/1742-6596/858/1/012023

form edge - intersection of the line representing the boundary between two or more surfaces (Fig. 1, a). In another case, it may move one surface to another at the contact line (special case line of intersection) (Fig. 1, a), or on the surface rounding formed along the intersection line of surfaces (Fig. 1, b). Therefore, whenever it is possible, to synthesize the geometric shape of the product in the form of a set of elementary surfaces has to deal with the lines of intersection of these surfaces.

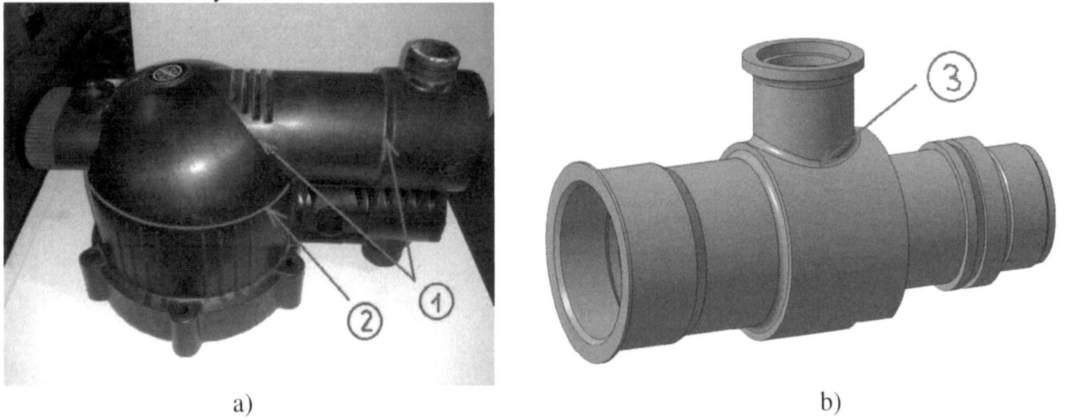

a) b)

Figure 1. Examples of the creation of complex surfaces of engineering products:

a) 1 – the line of intersection, 2 – the contact line; b) 3 – the surface rounding.

There are plenty of methods for determining the line of intersection of surfaces, each of which solves the positional problem of intersection for a certain limited class of surfaces [1-4].The purpose of this work is to develop a common and universal of geometric algorithm of definition constructive line of surfaces intersection with the possibility of its implementation by means of modern CAD.

It should be noted that the problem of constructing the intersection line of surfaces is basic and, at the same time, one of the most complex functions in the geometric modeling by means of modern CAD. It is noted, for example, by the authors: Liang Sun, Christopher M. Tiemey, Cecil G. Armstrong, Trevor T. Robinson [5], V. A. Korotkiy, E. A. Usmanova, L. I. Khmarova [6]. Depending on the mathematical content of the geometric modeler the considered problem has different levels and degrees of complexity solutions. Solving such problems by means of CAD begins with the creation of models of surfaces, some of which are simple primitives CAD. The result of the intersection of surfaces, depending on the choice of the way of modeling (the surface or the solid), can be a line, as a geometric definition of the object, or the boundary between the surfaces, which is a certain vicinity of the intersection line. In the latter case the line is determined only approximately. In modern a CAD it is possible to allocate the line of intersection from this boundary. In the transition from 3D-model of the intersecting surfaces to drawing it is required the positions that are defined intermediate and anchor points of the line of intersection. For example, these are the points of change of visibility. These points form the projection of lines of intersection on projection planes. Furthermore, in some cases it is necessary to consider the presence an imaginary component of the intersection line. It is necessary for completeness of representation of the complete picture of the surfaces intersection. The thing is that the imaginary part of the required line of intersection should be considered and have a real figurative representation on the planes of projections. The modern CAD-systems do not have yet the means and opportunities for electronic geometric realization of the imaginary part of the solution.

2. Formulation of the Problem

Based on the above information it is necessary to develop a general geometrical algorithm for determining the intersection line of surfaces. The possibilities of this algorithm should describe, on the one hand, accurately and completely the geometry of the final result - the line of intersection, and on the other hand, should provide various approaches. Under the variability the existence of several directions in solving the problem of intersection with the opportunity of selection of optimal solutions should be understood.

From practical demand of the completeness and accuracy of solving the problem on the intersection of the surfaces it is necessary to perform a theoretical rationale and develop an algorithm for a constructive solution of this problem, implemented by means tools 2D and 3D geometric modeling in CAD.

3. Theory and Methods

In the work on the set theory, the authors K. Kuratowski and A. Mostowski [7] described how to perform the operation of intersection of two sets A^m and B^q over the set Γ^p at the space of intersection E^n, where *m, q, p* and *n* – the dimension of nonintersecting sets, the set of intersection and Euclidean space respectively:

$$p = m + q - n \tag{1}$$

where $m + q \geq n$.

Then, for the space E^3 formula of intersection will be:

$$p = m + q - 3. \tag{2}$$

It is obviously, that $m + q \geq 3$ and, hence for the two surfaces ($m = q = 2$) has the equality $p = 1$, that is the intersection of surfaces happens on a line or a finite number of lines.

According to the theory of sets [7] between the elements of sets *A* and *B*, a binary equivalence relation Δ can be installed. Then, for every element α_1, α_2 and α_3 of set *A* the conditions will be performed:

1) reflexivity: $\alpha \, \Delta \, \alpha$, for any $\alpha \in A$;
2) symmetry: $\alpha_1 \, \Delta \, \alpha_2 \Rightarrow \alpha_2 \, \Delta \, \alpha_1$;
3) transitivity: $(\alpha_1 \, \Delta \, \alpha_2, \alpha_2 \, \Delta \, \alpha_3) \Rightarrow \alpha_1 \, \Delta \, \alpha_3$.

Elements of the set *A*, which are equivalent to an element *α*, form a class of equivalence. Thus, the equivalence attitude Δ allows dividing, the set *A* into nonintersecting classes of equivalence. All the set of equivalence classes, which fills the set *A*, is called factor set of the set *A*. Operation of the division of set *A* classes of equivalence is called factorization of this set. Factorization of set can be expressed as follows:

$$A^m = \Sigma^{r(t)} = \bigcup_r \sigma^t, \tag{3}$$

Where $r + t = m$, r – dimension of the set Σ of equivalence classes, t – the dimension of the class σ of equivalence. For surface *A* of space E^3 formula (3) takes the form:

$$A^2 = \Sigma^{1(1)} = \bigcup_1 \sigma^1 . \tag{4}$$

Therefore, the surface may be divided into a plurality of lines σ^1, that is a one-parameter set, where $\Sigma^{1(1)}$ – the factor-set of surface.

For the space E^3, both for the set, there are the following factorization:

$$E^3 = \Sigma^{2(1)} = \bigcup_2 \sigma^1, \tag{5}$$

$$E^3 = \Lambda^{1(2)} = \bigcup_1 \lambda^2. \tag{6}$$

In the first case, the space is filled with a two-parameter set of equivalence classes – nonintersecting lines, in the second case – it is the one-parameter set of nonintersecting surfaces.

Division (5) of the space E^3 allows installing different conformity between the sets A^2 and B^2, which are the surfaces. Data conformity is such that a finite number of elements from A^2 and B^2 will belong to the same equivalence class σ^1. If in the case of entering the third set C^2 in the space E^3 (surface or plane), then the factor-set

$$\Sigma^{2(1)} = \bigcup_2 \sigma^1$$

can be taken as projecting and review the operation (5) as an opportunity to perform the projection (mapping) the surface A^2, as well as the B^2, to the set C^2.

Thus, known in Descriptive Geometry the projection operation takes the new interpretation, which can be the basis for solving the problem of the definition of line of intersection surfaces A^2 and B^2.

We considered the kinematic surface known in descriptive geometry, i.e., surfaces, which are formed by the movement of the lines of constant or variable forms. This surface may be factorized, i.e. it could be represented by the formula (4). We have introduced the concept «the expansion of kinematic surfaces». The operation of expansion based on the possibility of «propagation» of each equivalence class (lines) σ^1 from the factor-set $\bigcup_1 \sigma^1$ of surface A^2 in a the classes of one type. For example, one-parameter set of circles of a cylindrical surface of revolution «propagated» in the two-parameter set of circles centered on the axis of the surface and in planes perpendicular to the axis.

The operation of «propagation» can be represented symbolically as follows:

$$A^2 = \Sigma^{1(1)} = \bigcup_1 \sigma^1 \Rightarrow \bigcup_2 \sigma^1 = \Phi^{2(1)}. \tag{7}$$

Obviously, the original surface A^2 belongs to the set $\Phi^{2(1)}$. In addition, the set $\Phi^{2(1)}$ can be viewed on the dimension of as a factor set of the space E^3.

On the basis of the given information we can propose the following algorithm of constructive determine intersection line of surfaces A^2 and B^2:

1) Perform a factorization of one of the kinematic surfaces A^2 or B^2, for example, A^2: $A^2 = \Sigma^{1(1)} = \bigcup_1 \sigma^1$.

2) Perform the extension of the surface A^2: $\bigcup_1 \sigma^1 \Rightarrow \bigcup_2 \sigma^1 = \Phi^{2(1)}$.

3) The set should be taken as projecting factor-sets of space E^3: $E^3 = \Phi^{2(1)}$.

4) Assign a plane of mapping (the plane of projection) C^2.

5) Perform a factorization of surface B^2: $B^2 = \Lambda^{1(1)} = \bigcup_1 \lambda^1$.

6) Determine a confluent projection of the surface A^2 (its factor-set is included in the projecting the factor-set) on a the projection plane C^2: $\bigcup_1 \sigma^1 \rightarrow \delta_C^1$.

7) Determine the set of projections of equivalence classes (the lines λ^1 of surface B^2) in the plane C^2: $\bigcup_1 \lambda^1 \rightarrow \bigcup_1 \lambda_C^1$.

8) Determine the set of projections ψ_C^0 - the images of points at the required lines of intersection: $\delta_C^1 \cap \lambda_C^1 = \psi_C^0$; $\bigcup_1 \psi_C^0 = \delta_C^1$.

9) Determine the required line of intersection Ψ^1 as a set of points of inverse images: $\psi_C^0 \rightarrow \psi^0$; $\bigcup_1 \Psi^0 = \Psi^1$.

The proposed algorithm can be used for finding of intersection of surfaces positions of the physical object that simulates the value of radio navigation parameters. It is known that when determining the location of the physical object according to the set of radio navigation parameters in radiolocation and radio navigation to such surfaces, as noted in this work [8], belong spheres, ellipsoids, hyperboloids.

In the first example (Fig. 2) the intersection of the sphere $A^2(0, R)$ and the hyperboloid of rotation $B^2(j, g)$ are defined on the basis of 2D-modeling tools.

According to the proposed algorithm, we adopted one surface, for example a hyperboloid B^2 as the base for the factorization of space E^3. According to the first stage of the algorithm we have presented the surface B^2 as a continuous set of circles with variable radius, each of which is disposed in the plane which is perpendicular to the axis j (j_1, j_2) of the surface. According to stage 2 of the algorithm we have done the expansion of the set of circles. In expanding each circle is transformed into the bundle of circles, and all the set of circles of surfaces B^2 is converted into congruence (two-parameter set) circles. We accept the congruence as projecting factor-set of space E^3. Then we appoint the plane of projection C^2. As a sample, we chose the frontal plane - the overall plane symmetry of surfaces A^2 and B^2. Next, we perform the factorization of surface of the sphere A^2.

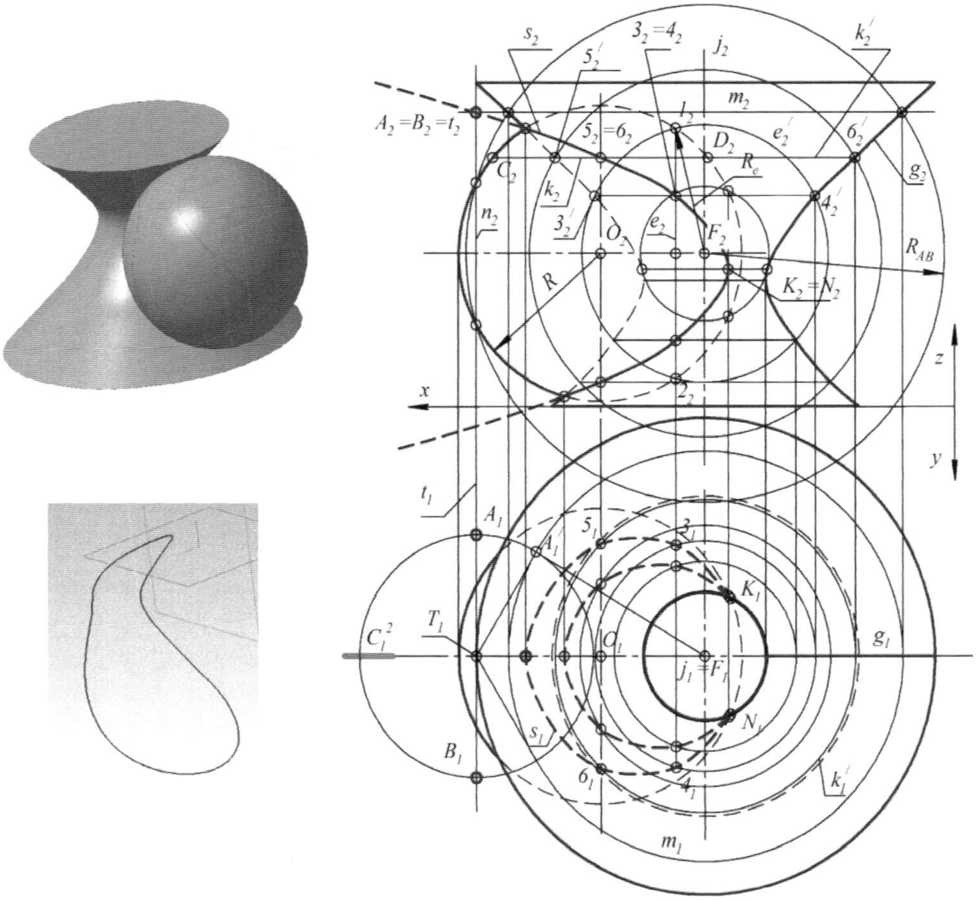

Figure 2. 3D- and 2D-modeling of the intersection of the sphere and the surface of the hyperboloid.

One option of the factorization of the surface is the presentation of the sphere as a continuous set of circles in the profile planes. One of these circles is a circle e (e_2) with a diameter 1_22_2. Then, on the sixth stage of the algorithm, we have drawn the confluent of the hyperboloid B^2 on the plane of the projection C^2.

This is hyperbole g (g_1, g_2). Then on the seventh step of the algorithm the set of projections on the plane C^2 of the circles in the selected variant the factorization of sphere A^2 is defined. Construction of projections is based on the following obvious fact: the rotation of the circle about the point on the perpendicular to the plane of the circle passing through center forms a sphere. For example, circle $e(e_2)$ while rotating about the point $F(F_1, F_2)$ on the perpendicular $OF(OF_1, OF_2)$ forms the sphere for which the contour circle e'_2 will be the projection of the circle $e(e_2)$ on the plane C^2 under mapping by the projecting factor-set (the congruence of circles). As a result, all the circles of factor-set of the sphere A^2 is displayed on the plane C^2 in the form of a bundle (F) of circles centered at the point F (F_1, F_2). A bunch of circles (F) is the image factor-set circles of sphere A^2. The intersection points $3'_2$, $4'_2$ the image e'_2 of circle $e(e_2)$ and contour hyperbole g_2 represent images of points $3(3_1, 3_2)$ and 4 $(4_1, 4_2)$ of the required intersection line s (s_1, s_2). Performing such actions for each circle of factor-set of sphere A^2 allows to get a discrete set of points. Interpolation of the set allows us to construct the intersection line s. Factorization of sphere A^2 and projecting the factor-set for given conditions of the problem are basis and evidence base of the method of spheres, which is known in descriptive geometry. This method, as applied to the present problem of the surfaces intersection, is the only one which allows to determine pivots points $K(K_1, K_2)$ and $N(N_1, N_2)$ of the intersection line s and the points its imaginary of continuation, which are real projections, for example, points $A(A_1, A_2)$ and $B(B_1, B_2)$.

For the sphere A^2 you can select another variant of factorization in the form the continuous set of circles in horizontal planes. In this case, any circle of a new factor-set of the sphere, for example, the circle k (k_2) with the diameter C_2D_2 is displayed on plane C^2 in the form of a horizontal line that intersects with frontal contour g_2 of surface B^2 at points $5'_2$, $6'_2$ - the images of points $5(5_1, 5_2)$ and $6(6_1, 6_2)$ of the required intersection line s. Moving from the images $5'$ and $6'$ to the prototype of 5 and 6 is performed in the circle k' (k'_1, k'_2) to the intersection with the circle k (k_2). The new version of factorization of sphere A^2 with the same projected factor-set is the basis and the evidence base of other known method in Descriptive Geometry, the method of auxiliary planes. In the example in Fig. 3 the intersection line of two ellipses of the private location by the method auxiliary planes is determined using 2D-modelling.

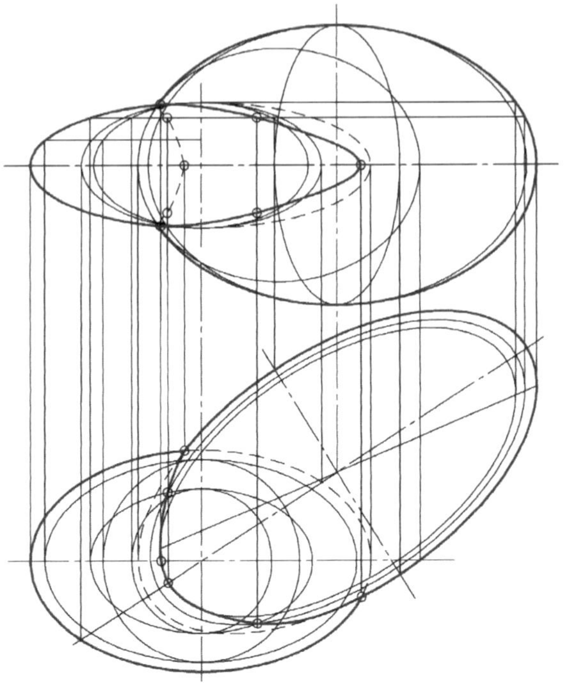

Figure. 3 2D-modeling of the surfaces of intersection of ellipsoids private location.

If the intersecting surfaces (ellipsoids, hyperboloids) have more complex geometry and general location, then the search of the intersection line by means of 2D-modeling becomes difficult. In this case, it is advisable to use 3D-modeling tools that are available in CAD (Fig. 4, 5).

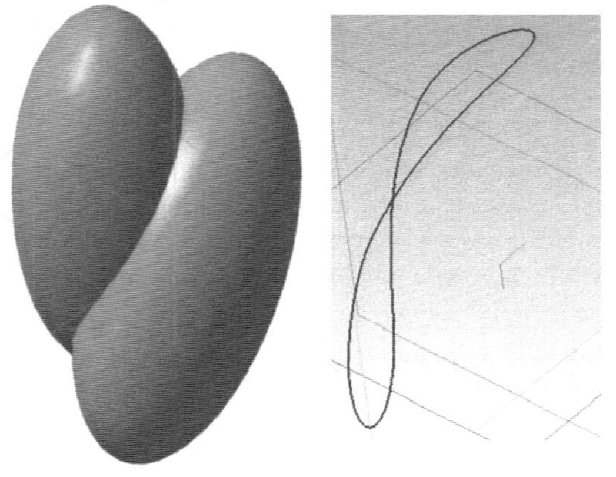

Figure 4. 3D-models of the intersection of the surfaces of the two ellipsoids general arrangement.

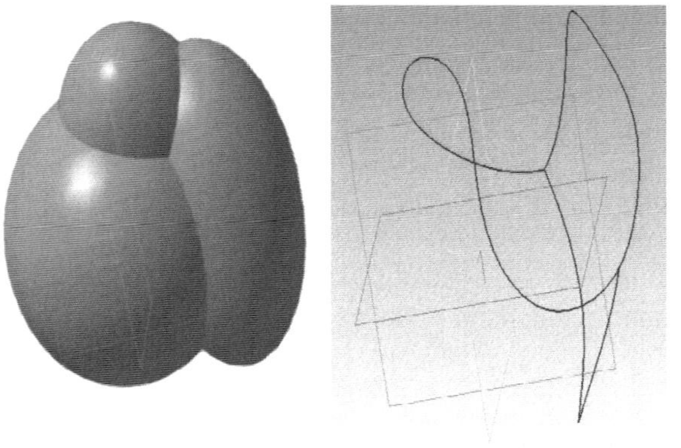

Figure 5. 3D-models of the intersection of the three surfaces of ellipsoids general arrangement.

Wherein there is the possibility of surface or solid modeling to obtain the form and coordinates of the points of the intersection line with the necessary accuracy.

4. Results and Discussion

The line of intersection of two surfaces of a set of spherical, ellipsoid and hyperboloid, which are used in forming products engineering, may consist only of the real points (for pairs of a set of spheres and ellipsoids) and may have other than the real part of the imaginary extension (for the pairs of surfaces, one of which - hyperboloid).

Obviously, the spheres method provides a more complete picture of the intersection as opposed to the auxiliary planes method, which is used to solve the same problem. The auxiliary planes method is more universal and allows determining the intersection line of surfaces of simple and complex geometric forms in the most general case their location. Both methods are a consequence of the proposed factorization algorithm. They can be used with the use of CAD means on the basis 2D- and 3D-modeling for the tasks at the intersection, which is required to obtain the most complete picture of the intersection of surfaces, forming a geometric shape of the product engineering.

5. Conclusion

We propose a factorization algorithm for determining the intersection line of surfaces, forming a geometric shape of the product engineering. The algorithm is based on the quotient of surfaces and can be implemented by means of a CAD on the basis 2D- and 3D-modeling.

Methods of surface and solid modeling in CAD should be used to get an idea of the line of intersection of surfaces as a whole. A more complete and accurate picture of the formation of the line of intersection can be obtained using the factorization algorithm, allowing to make the choice of optimal solutions of the problem of the intersection.

References

[1] S. Ambrosimov, G. Harlamov, A. Cherepenko, Kinematic Model of Complex Surfaces Formation with Three None Linear Movements Forming, Procedia Engineering, ICIE-2016, 150 (2016)1089-1097.

[2] Brigham Bagley, Shankar P. Sastry, Ross T. Whitaker, A Marching-tetrahedra Algorithm for Feature-preserving Meshing of Piecewise-smooth Implicit Surfaces, Procedia Engineering, 25th International Meshing Roundtable, 163 (2106) 162-174.

[3] G. Rachkovskaya, Geometrical Modeling and Graphics of Kinematical Ruled Surfaces Based on Triad of Contacting Axoids, Geometriya i grafika, 2016, V. 4, I. 3, pp. 46-52. DOI: 10.12737/21533.

[4] V. Korotkiy, L. Khmarova, E. Usmanova, Computer Simulation of Kinematic Surfaces, Geometriya i grafika, 2016, V. 3, I. 4, pp. 19-26. DOI: 10.12737/17347.

[5] Liang Sun, Christopher M. Tiemey, Cecil G.Armstrong, Trevor T. Robinson, Automatic Decomposition of Complex thin Walled CAD Models for Hexahedral Dominant Meshing. Procedia Engineering, 25th International Meshing Roundtable, 163 (2106) 225-237.

[6] V. A. Korotkiy, E. A. Usmanova, L. I. Khmarova, Surface as an Object of Computer Geometric Modelling. Procedia Engineering, ICIE-2015, 129 (2015)775-780.

[7] K. Kuratowski, A. Mostowski, Set theory, With an inroduction to descriptive set theory. North-Holland Publishing Co., Amsterdam-New York-Oxford; PWN—Polish Scientific Publishers, Warsaw, 1976, p. 509.

[8] B. Hofmann-Wellenhof, K. Legat, M. Wieser, Navigation –Principles of Positioning and Guidance, Springer-Verlag, Wien, New York, 2003, P. 427.

IOP Conf. Series: Journal of Physics: Conf. Series **858** (2017) 012024 doi:10.1088/1742-6596/858/1/012024

The mathematical and computer modeling of the worm tool shaping

K L Panchuk[1], A A Lyashkov[1], T V Ayusheev[2]

[1]Omsk State Technical University, Mira ave., 11, Omsk, 644050, Russia
[2]East Siberian State University of Technology and Management, Klyuchevskaya ave., house 40, building 1, Ulan-Ude, 670013, Russia

E-mail: Panchuk_kl@mail.ru

Abstract. Traditionally mathematical profiling of the worm tool is carried out on the first T. Olivier method, known in the theory of gear gearings, with receiving an intermediate surface of the making lath. It complicates process of profiling and its realization by means of computer 3D-modeling. The purpose of the work is the improvement of mathematical model of profiling and its realization based on the methods of 3D-modeling. Research problems are: receiving of the mathematical model of profiling which excludes the presence of the making lath in it; realization of the received model by means of frame and superficial modeling; development and approbation of technology of solid-state modeling for the solution of the problem of profiling. As the basic, the kinematic method of research of the mutually envelope surfaces is accepted. Computer research is executed by means of CAD based on the methods of 3D-modeling. We have developed mathematical model of profiling of the worm tool; frame, superficial and solid-state models of shaping of the mutually enveloping surfaces of the detail and the tool are received. The offered mathematical models and the technologies of 3D-modeling of shaping represent tools for theoretical and experimental profiling of the worm tool. The results of researches can be used at design of metal-cutting tools.

1. Introduction

The means and the methods of three-dimensional modeling in modern CAD represent the powerful tool for designing metal-cutting tools, in particular at the stage of their profiling. The determination of geometrical form of the surface of the tool at the set form of the detail and parameters of the positioning of the detail concerning the tool is understood, first of all, under profiling.

Now in the field of profiling the set of theoretical approaches are created. The kinematic method considered in the known works of the founder of the theory of gear gearings F. L. Litvin [1, 2] and method of envelope surfaces which has been reviewed in scientific works, for example [3, 4, 5] prevail among them.

Among the set of tasks of profiling of the cutting tools the incomplete one in respect of theoretical study and implementation by means of CAD is the task of determination of the surface of the worm tool for handling of the screw surface of the detail. In the existing mathematical models of profiling of this tool the first T. Olivier method of formation of the mutually envelope surfaces is used, according to which there is an intermediate surface at models (the surface of the making lath) [1]. It complicates the process of mathematical profiling and its implementation by means of CAD. Therefore, receiving of mathematical model with simpler algorithm of profiling is an actual task. Another actual task accompanying the first is the accomplishment of virtual computer modeling of technological process

IOP Conf. Series: Journal of Physics: Conf. Series **858** (2017) 012024 doi:10.1088/1742-6596/858/1/012024

of shaping of the worm tool. In the case of such modeling, the important technological subtleties and features of profiling, hidden in case of mathematical modeling, are found.

2. Formulation of the Problem

The purpose of the given work is the improvement of mathematical model of profiling of the worm tool and its approbation with the use of tools of frame, superficial and solid-state modeling. Research problems are: receiving mathematical model of profiling with simpler, in comparison with known, algorithm of definition of the form-building surface of the tool; realization of this model based on methods of frame and superficial modeling; development of technology of solid-state modeling of shaping of the mutually envelope surfaces of the detail and the tool.

3. Theory and Methods

The basis for the mathematical model containing simple algorithm of profiling of the worm tools have served the following theoretical position [1, 2]:
1. Total normal at the point of contact of mutually envelope surfaces belongs to the complex of straight lines of the kinematic screw of the instantaneous relative motion of bodies with these surfaces.
2. The mutually envelope surfaces of the detail and the tool are formed on the first T. Olivier method. It means that the intermediate surface, which is provided in the method, forms two surfaces by two independent movements, one of these surfaces belongs to the detail, and another - to the worm tool. The contact of the mutually envelope surfaces in each from several couple of products: the detail – the making lath and the making lath – the worm tool, occurs on the line. The contact of the surfaces in the main couple the detail – the worm tool occurs in the point.
On the basis of the given theoretical positions the geometrical scheme of profiling of the worm tool is developed and its mathematical description in the form of algorithmic sequence of calculation formulas presented in list form is executed.

3.1. Computational formulas

1. Equations of a face profile of a screw surface of the detail

$$x_{d1} = a - r\,sin\,\sigma,\ y_{d1} = r\,cos\,\sigma, \sigma_0 \le \sigma \le \sigma_n \tag{1}$$

2. Module profile point radius vector

$$A = \sqrt{x_{d1}^2 + y_{d1}^2} \tag{2}$$

3. Angle between radius vector and a tangent to the profile in their common point

$$\sigma_f = arccos\frac{a\,cos\,\sigma}{A} \tag{3}$$

4. The equations of the contact line l_l of surfaces of the detail and the making lath, h_l – the screw parameter of the surface of the detail

$$\sqrt{R_1^2 - (A\,cos\,\sigma_f)^2} = C,$$

$$x_{p1} = \frac{A}{R_1}\left[A\,cos^2\,\sigma_f + C\,sin\,\sigma_f\right],$$

$$y_{p1} = \frac{A}{R_1}\left[C - A\,sin\,\sigma_f\right]cos\,\sigma_f, \tag{4}$$

$$z_p = h_1\left[arccos(\frac{A}{R_1}cos\,\sigma_f) - arctg\frac{y_{d1}}{x_{d1}} - \sigma_f\right]$$

5. The size of the current shift of the contact line l_l along the axis of the detail, β – an angle between axes of the detail and the tool

$$s_i = -(z_p + \frac{y_{p1}}{tg\beta}) \tag{5}$$

6. The equations 4–6 describe the profiling line l_P at mutually envelope of surfaces of detail and tool

$$z_{pi} = -\frac{y_{p1}}{tg\beta} = z_p + s_i \tag{6}$$

7. The size of initial shift of the contact line l_l along the detail axis

$$s_0 = -(\frac{y_{p1}(\sigma_0)}{tg\beta} + z_p(\sigma_0)) \tag{7}$$

8. Angles of rotation of the detail and the tool in the course of mutually enveloping

$$\varphi_1 = \frac{1}{h_1}|s_0 - s_i| \tag{8}$$

9. Angles of rotation of the detail and the tool in the course of mutually enveloping, u – the transfer relation

$$\varphi_2 = \frac{1}{u}\varphi_1 \tag{9}$$

10. Formulas of transformation of systems of coordinates of the detail and the tool, B – distance between axes of the detail and the tool

$$x_2 = (x_{p1} - B)cos\,\varphi_2 + (y_{p1}\,cos\,\delta + z_{p1}\,sin\,\delta)\,sin\,\varphi_2,$$
$$y_2 = (y_{p1}\,cos\,\delta + z_{p1}\,sin\,\delta)cos\,\varphi_2 + (B - x_{p1})\,sin\,\varphi_2, \tag{10}$$
$$z_2 = -y_{p1}\,sin\,\delta + z_{p1}\,cos\,\delta$$

11. The equations of an axial profile of the required surface of the tool, h_2 – the screw parameter of the surface of the tool

$$x_{t2} = x_2\,cos(\frac{z_2}{h_2}) + y_2\,sin(\frac{z_2}{h_2}),$$
$$y_{t2} = y_2\,cos(\frac{z_2}{h_2}) - x_2\,sin(\frac{z_2}{h_2}) \tag{11}$$

12. Equations of the face profile of the required surface of the tool

$$x_{02} = \sqrt{x_2^2 + y_2^2},$$
$$z_{02} = z_2 - h_2 arctg(\frac{y_2}{x_2}) \tag{12}$$

The mathematical model presented by calculation formulas (1) - (12) allows to make computer visualization (figure 1) of the contact line l_1 of surfaces of the screw detail and the making lath and the profiling line l_P (the line of gearing on terminology in the theory of gear gearings [1, 2]). From the lines of l_1 and l_P, on the basis of kinematics of the mutually envelope movements of a detail and the

IOP Conf. Series: Journal of Physics: Conf. Series **858** (2017) 012024 doi:10.1088/1742-6596/858/1/012024

tool, transition to the surface of the worm tool is carried out. In figure 2 the frame models of an initial surface of the detail and the surface of the tool which are in the position of the contact are presented. Superficial models of these contacting products are given in figure 3.

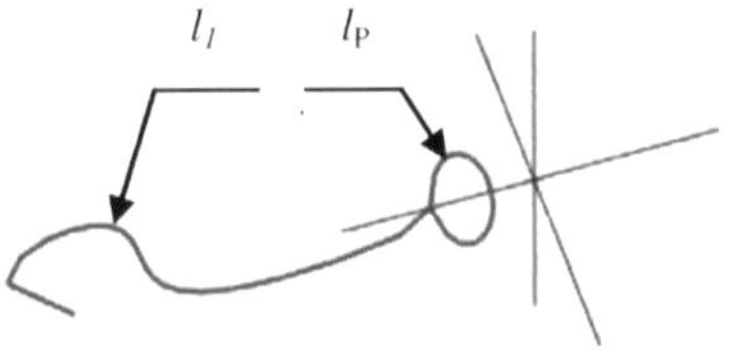

Figure 1. Line of the contact and line of profiling (closed).

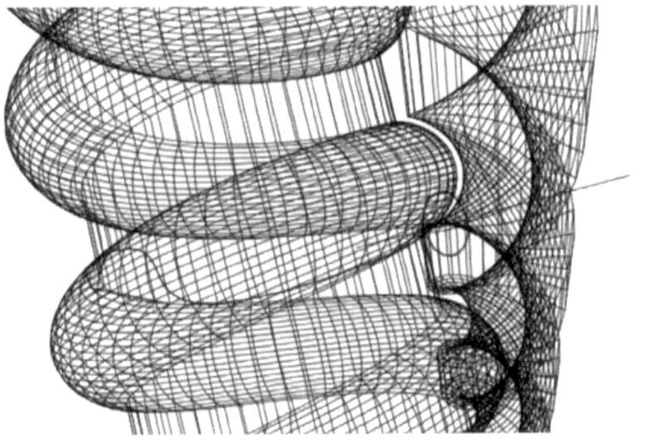

Figure 2. Frame model of the contacting surfaces.

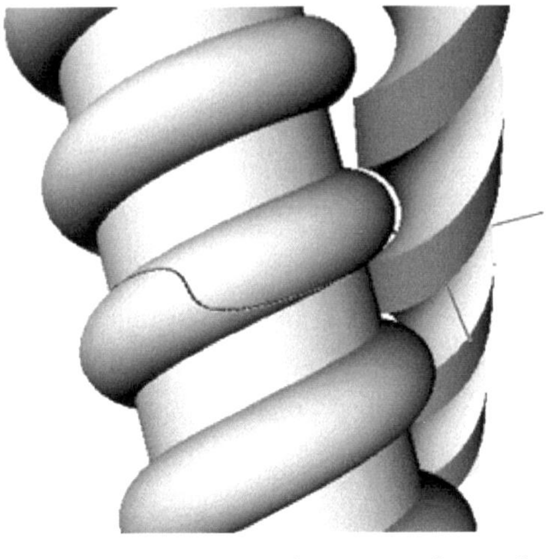

Figure 3. Superficial model of the contacting surfaces.

Among modern profiling techniques the method based on solid modeling is becoming increasingly common. It is regarded, for example, in scientific works of Dimitriou Vasilis, Vidakis Nectarios and Antoniadis Aristomenis [6], A.A. Lyashkov and K.L. Panchuk [7]. The following factors serve as motivation to it:

IOP Conf. Series: Journal of Physics: Conf. Series **858** (2017) 012024 doi:10.1088/1742-6596/858/1/012024

1. The need in the mathematical research and the solution of the transcendental equations or their systems connecting geometrical and kinematic parameters of the process of shaping on the basis of mutually envelope of surfaces of the detail and the tool disappears at solid-state modeling.
2. Solid-state modeling allows solving two interconnected shaping problems at the virtual imitating level. One of them is the task of geometrical synthesis including solutions of the direct and return task of profiling. In case of the direct task the determination of the surface of the tool is carried out, and in case of the return – a detail surface on the received tool surface for the quality evaluation of the solution of the direct task also is carried out. Solutions of both tasks of profiling are based on the virtual computer reproduction of relative movements of the detail and the tool provided on the practice by means of tuning of a kinematic chain of the machine equipment. The other task of the shaping is the task of technological synthesis, namely – consecutive removal by the cutting tool of amount of material of billet of the detail before receiving its necessary geometrical form.
Virtual imitating modeling allows receiving the quality characteristics of technological process of the shaping and numerical parameter values of the cut-off layers with the subsequent introduction of necessary geometrical and technological amendments to the scheme of shaping.
Solution of both tasks of synthesis is carried out in the mode of virtual experiment in which, if necessary, modification or adjustments of basic data for receiving of required results of profiling without accomplishment of material and finance costs is possible.
The algebraic logic of the theory of sets forms theoretical base of virtual technological experiment. The model of billet of the detail and model of the tool are considered as three-dimensional sets of geometrical elements for which transaction of the difference of sets of $M_1 = M_2 \backslash M_3$ is carried out where M_1, M_2 and M_3 – respectively a resultant set, a set of elements of billet of the detail and the tool. In the process of a cyclic shaping the set of M_1 after each iterative step accomplishment of transaction of the difference is updated and the cycle of the shaping is repeated for already updated set of M_1. As a result we receive discrete virtual modeling of the process of shaping.

4. Results of Experiments

In this work, proceeding from essence of solid-state modeling, the technological sequence of the solution of the considered profiling task is developed and virtually realized. The initial stage of solid-state modeling of shaping is the creation of initial models of the detail and billet. For the considered case the detail with the screw flute is the initial one. In the beginning model surfaces on its face section (figure 4) are created, and then they are combined with the model of billet (figure 5). These surfaces are used for cutting off its parts from the billet, the result of which is the detail model with a screw flute (figure 6).

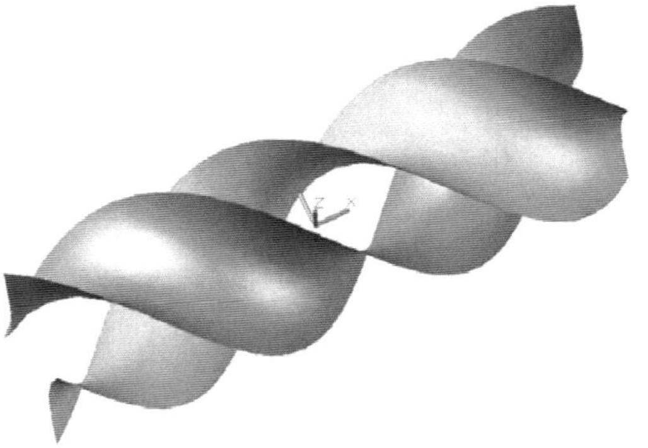

Figure 4. Initial screw surfaces of the detail.

IOP Conf. Series: Journal of Physics: Conf. Series **858** (2017) 012024 doi:10.1088/1742-6596/858/1/012024

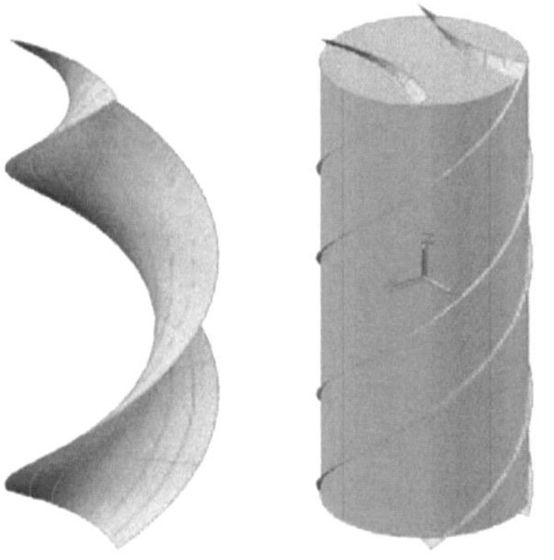

Figure 5. Cutting off the amounts of material from model of billet of the detail.

At the following stage the billet model (a cylindrical body) for the tool lath (figure 7) is added to the model of the detail.
The model of an instrumental lath is formed on the basis of the model of the detail which is given forward movement along forming a cylindrical body (figure 8).

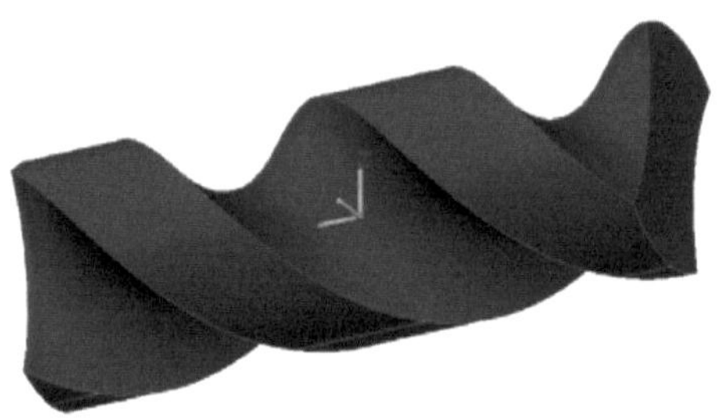

Figure 6. Result of cutting off.

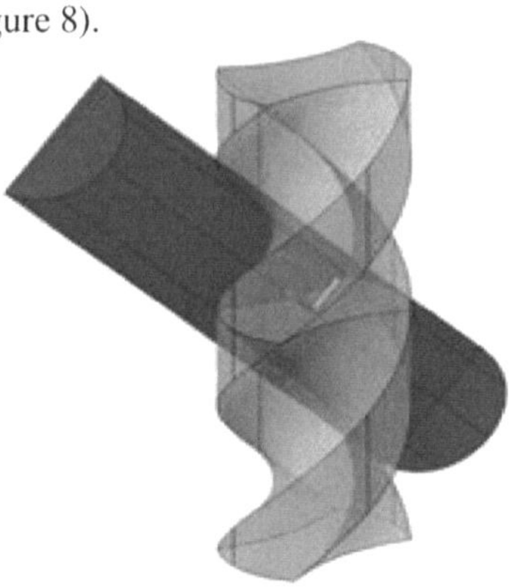

Figure 7. Addition of model of cylindrical billet.

The process of shaping is carried out in the automated mode on the base of the developed algorithms and programs.
The accuracy of modeling is defined by the size of an increment of the parameter of forward movement of the product. This parameter is set in the dialogue mode before modeling. The result of modeling and face section of an instrumental lath are given in figure 9. This section is used for receiving model of the cylindrical surface interfaced to surfaces of the detail and the instrument (figure 10). Then the cylindrical surface is used for creation of solid-state model of other instrumental lath used for shaping the instrument (figure 11).

IOP Conf. Series: Journal of Physics: Conf. Series **858** (2017) 012024 doi:10.1088/1742-6596/858/1/012024

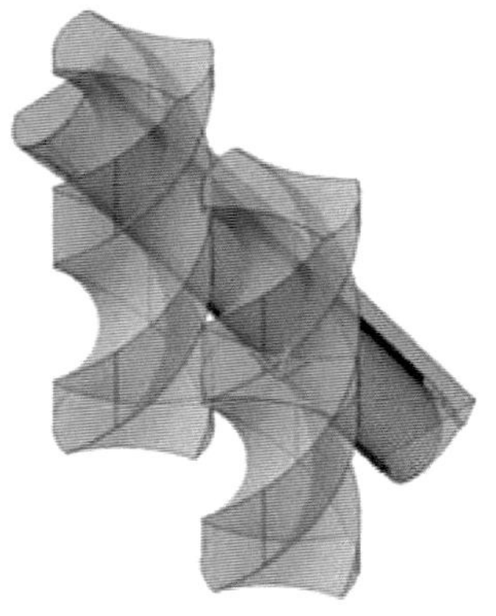

Figure 8. Shaping of the surface of the tool lath.

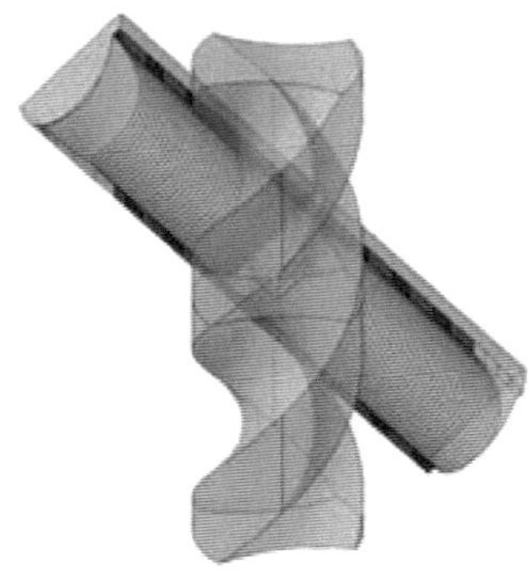

Figure 9. Result of modeling of the surface of an instrumental lath.

In figure 12 the models of the second lath and of the cylindrical billet for the tool are shown. The model of the tool is formed on the basis of the received model of the second instrumental lath making the screw movement with the set parameter (figure 13). The process of shaping is also carried out in the automated mode. Accuracy of modeling is determined by the size of the increment of parameter of screw movement of the product. This parameter is also set in the dialogue mode before accomplishment of the modeling.

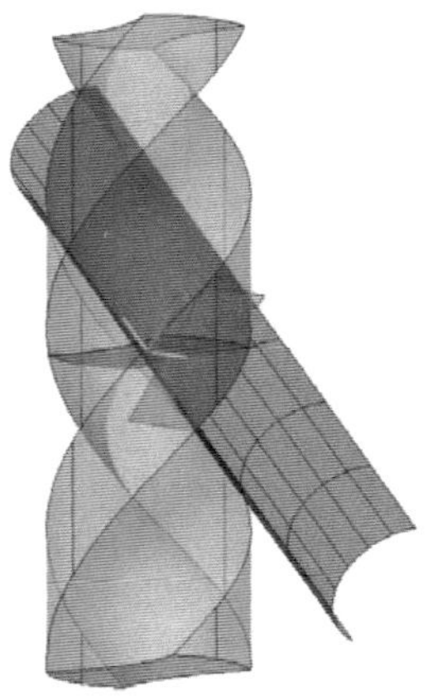

Figure 10. Surface of the tool lath.

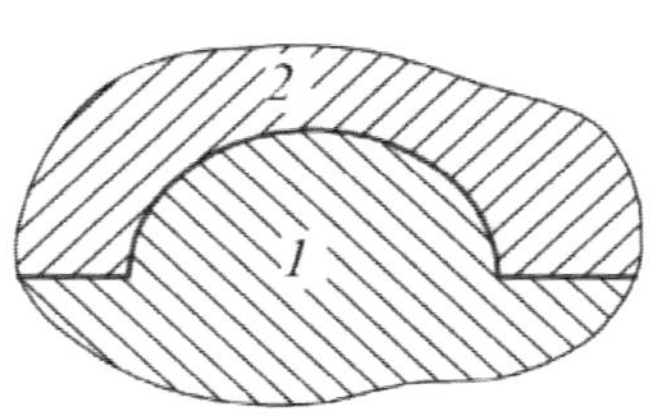

Figure 11. Two bodies with one surface

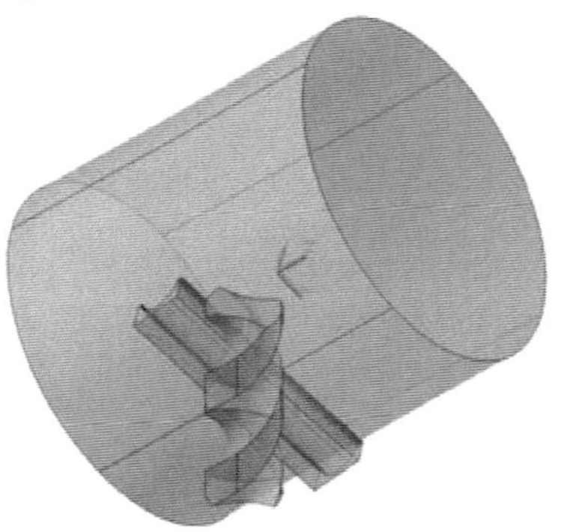

Figure 12. Accession of the surface of the tool lath to model of billet of the tool.

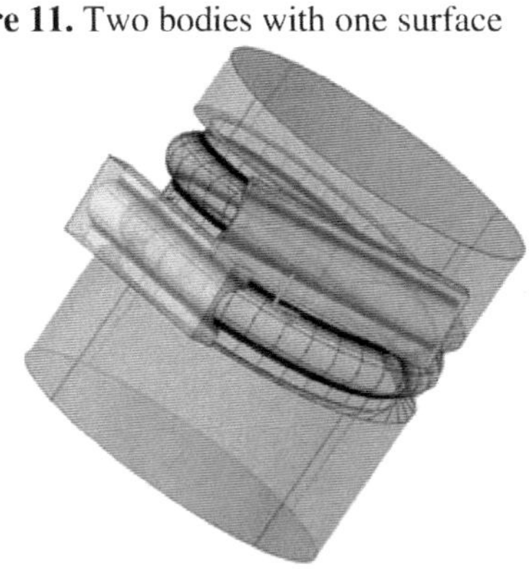

Figure 13. Instrument surface shaping by screw movement of an instrumental lath.

The result of modeling of the part of the instrument together with the detail and cylindrical surface of an instrumental lath is shown in figure 14. The same models are given in figure 15, but the instrument body is defined on its axial section from the previous stage of modeling.

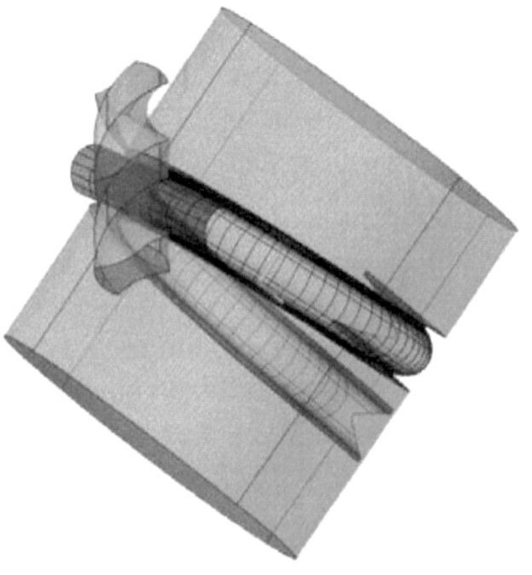

Figure 14. Result of shaping of the surface of the tool.

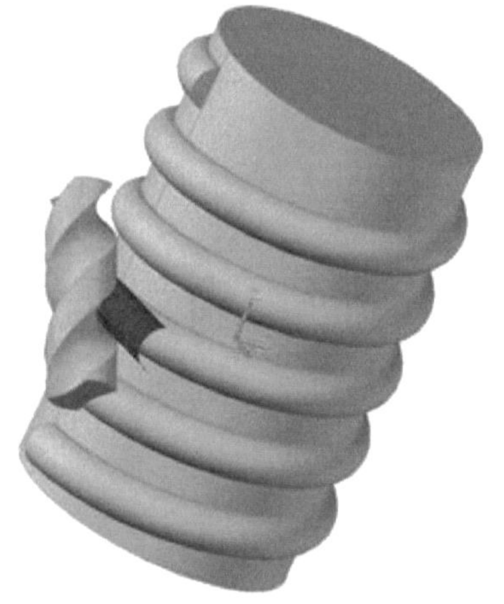

Figure 15. Result of the full shaping of the surface of the tool.

For the assessment of quality of the executed shaping of the worm tool, the received models can be used for comparison of results of solid-state modeling with the results of frame and superficial modeling. For this purpose it is possible to construct, for example, sections by various planes. So in figure 16 the section of models by the plane passing through the tool axis, and in figure 17 – through the detail axis is shown. The comparative analysis of the axial sections of the worm tool received by various ways of modeling shows their coincidence to within the fourth sign after the comma. Besides, from images in figures 16 and 17 it is possible to draw the conclusion that the practical contact of the surface of the detail and the tool occurs on some area though according to mathematical model they have pointed contact.

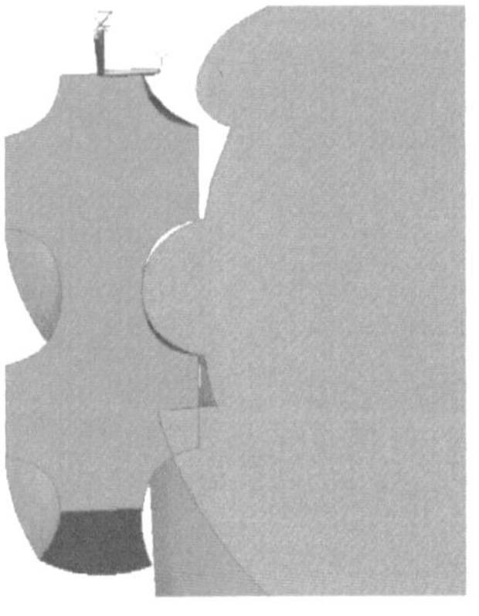

Figure 16. Visualization of the contact of surfaces of the detail and the tool in the axial section of the tool.

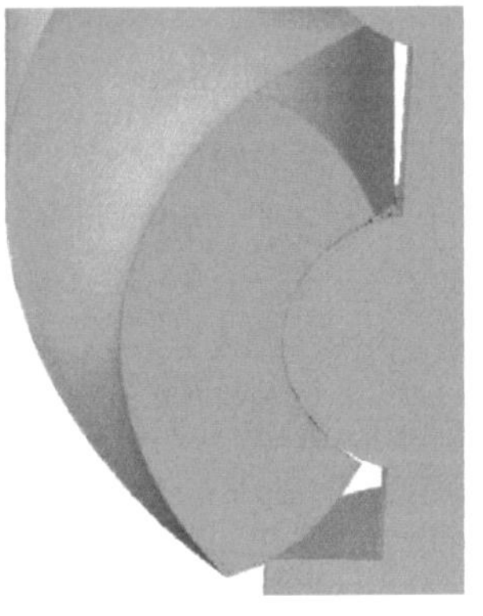

Figure 17. Visualization of the contact of surfaces of the detail and the tool in the axial section of the detail.

5. Discussion of Results

The design of the face profile of the screw detail is more difficult, the geometry of her screw surface is more difficult. The difficult design of the profile can include several natural lines joined on the first or higher order of smoothness. It can consist also of one or several spline lines, the points of the face profile constructed on the discrete set. For screw details of difficult geometry additional researches within the accepted mathematical model is realized by the frame or superficial modeling, for example, identification and research of transitional curves and cutting are required. The solid-state modeling realized at the virtual imitating level allows tracking emergence of transitional curves and cutting. It gives an idea of constructive and technological conditions of their emergence, allows to conduct the corresponding researches and to introduce necessary amendments in the scheme of shaping of the tool. Obviously, frame and superficial modeling of the problem of profiling of the worm tool are expedient for screw details with rather simple geometry, and solid-state modeling – for details of complex geometry. It should be noted also that for the considered problem of profiling the unexplored is the question of modeling of the shaping of the tool without participation of the making lath, i.e. directly, giving of an initial screw surface of the detail of the two-parametrical movement of enveloping.

6. Conclusion

1. The advanced mathematical model of the solution of the problem of profiling of a worm instrument for processing of the screw surface of the detail is offered. The model is based on the first T. Olivier method of shaping of the mutually envelope surfaces and unlike the existing mathematical models doesn't demand the use of an intermediate surface of tool rack. The model is simple concerning computing procedures and allows solving the problem of profiling with visualization of frame and superficial modeling without application of numerical methods.

2. The solution of the problem of profiling of the worm instrument on the basis of solid-state modeling is received. Profiling is carried out at the virtual imitating level in the form of sequence of the technological stages corresponding to the first T. Olivier method.

3. Proposed frame, superficial and solid-state modeling solutions of the problem of profiling of the worm instrument are submitted in total by the necessary set of design-technology instruments of profiling. For the screw details of simple geometry which is used, for example, in some designs of rotors of screw compressor cars for profiling of the worm instrument it is enough to use offered frame and superficial modeling, and for screw details of irregular shape – solid-state modeling. All three types of modeling supplement each other and allow reaching the qualitative solution of the problem of profiling.

References

[1]Litvin F L 1989 Theory of Gearing *NACA Reference Publication 1212 AVSCOM Technical Report 88-C-035* 490

[2]Litvin F L and Fuentes A 2004 Gear Geometry and Applied Theory (Cambridge University Press) 800

[3]Argyris John, Litvin Faydor L, Qiming Lian and Lagutin Sergey A 1999 Determination of envelope to family of planar parametric curves and envelope singularities *Computer Methods in Applied Mechanics and Engineering*. **175(1-2)** pp 175–187. DOI: 10.1016/S0045-7825(98)00367-3

[4]Raikhman G N 2002 Theory of conjugate surfaces formation by generating lines *Mechanism and Machine Theory* **37(9)** pp 999–1005 DOI: 10.1016/S0094-114X(02)00024-1.

[5]Litvin Faydor L, Donno M De, Lian Qiming and Lagutin S A 1998 Alternative approach for determination of singularity of envelope to the family of parametric surfaces *Computer Methods in Applied Mechanics and Engineering* **167(1–2)** pp 153–165 DOI: 10.1016/S0045-7825(98)00117-0.

[6]Vasilis Dimitriou, Nectarios Vidakis and Aristomenis Antoniadis 2006 Advanced Computer Aided Design Simulation of Gear Hobbing by Means of Three-Dimensional Kinematics Modeling *Journal of Manufacturing Science and Engineering* **129(5)** pp 911–918 DOI 10.1115/1.2738947

[7]Lyashkov A A, Panchuk K L 2015 Computer modeling of the pump screw and disc tool cross shaping process *Procedia Engineering* **113** pp 174–180 DOI: 10.1016 / j.proeng.2015.07.314.

IOP Conf. Series: Journal of Physics: Conf. Series **858** (2017) 012025 doi:10.1088/1742-6596/858/1/012025

Numerical Calculation of Total Radial Forces and Rotary Moments From the Cylinders Surface

A V Panichkin[1], **L G Varepo**[2]

[1] S L Sobolev Omsk branch of Institute of mathematics of the Siberian Branch of the Russian Academy of Sciences, 13, Pevtsova st., Omsk, 644043, Russia
[2] Omsk State Technical University, 11, Mira ave., Omsk, 644050, Russia

E-mail: panich@ofim.oscsbras.ru

Abstract. The determining of the acting moments of force when the cylinder is characterized by roughness after the deformation of contacting surfaces of the rotating cylinders and the substrate while interacting with incompressible viscous liquid is still one of the main problems in printing. The numerical calculation of the radial forces and rotary moments with integral formulas for deforming surfaces was carried out with regard to the discrete grid at the grid points of the upper and lower cylinders surfaces and their deformation speeds. The paper considers the results of the numerical modeling for the quantitative assessment of rotary moments of radial forces from the cylinders surfaces, occurring during the transfer of incompressible viscous liquid on the substrate between two rotating cylinders. The definition of dynamic forces (radial and tangential force, moment of force) in the discrete form on the example of one computational cell is presented. It is noted, that the consideration of boundary deformation results in the decrease of the calculated dynamic values by 3-5 times as well as to the reducing of the ink bleeding near the boundaries of the printing surface.

1. Introduction

The competitiveness trends in the modern printing equipment market dictate the terms of their continuous improvement. The main purpose of offset printing machines is the obtaining of multiple identical printouts by applying viscous incompressible liquid (ink) on the coated substrate by contact under pressure.

On the other hand, their purpose is also the implementation of the printing process with its inherent variety of deformations caused by contact phenomena in offset printing machines.

The presence of different deformations has a significant impact on the production quality and operation of various mechanical systems.

The deformations in the real printing process and the impossibility of their total elimination constitute an important scientific problem, i.e. their account and control, the solution of which requires comprehensive analysis of the phenomena, because the deformations are fundamental in assessing the quality of printed products and the equipment performance.

A prerequisite for achieving the desired quality of the final printed product [1] during the operation of offset printing machines is the consideration of the effect of the process of transferring viscous

incompressible liquid (ink) to the substrate on the cylinder dynamic characteristics while simulating and the following design of the printing units in offset printing machines.
The search for new approaches to solving the above problem, taking into account that the theoretical developments, in particular the development of mathematical models simulating the work of the printing machine functional units, are among the priorities and has not lost its relevance today.

2. Statement of the problem

In [2], there is a stress on the active control of the viscous incompressible liquid flow with respect to the rotating cylinder only, but not with account for the liquid transfer to the substrate, contacting with the rotating cylinder surface. Computer simulation of the non-monotonic distribution of the ink film during its transfer to the substrate in the printing machine is given in [3], where the issue of vibration compensation is considered.
The attention to this problem is also paid in the following papers [4, 5, 6, 7] where there is only information given on the questions of the liquid transfer between two parallel plates at the strain, rapture and liquid output stages. The influence of the surface tension and the contact angles on the liquid transfer is estimated.
In [8, 9] the results of numerical modeling of calculating the indices of viscous incompressible liquid transfer on a substrate, taking into account the deformation, are presented. However, the effect of the rotational torques is not considered by the authors
The definition of the acting moments of force occurring as a result of the cylinder raggedness when the cylinder surface deformation occurs while interacting with viscous liquid and the other cylinder is still in the focus of researchers.
In its turn, the need to select the projection of the tangential and the normal stress to the appropriate direction (the radius vector to the axis of the cylinder, or to the normal radius of the cylinder in the rotation plane) during the integration of these values has also both scientific and practical significance for the printing machines design and operation.

3. Methods

The numerical calculation for radial forces and moments with these integral expressions for the deformable surfaces was carried out based on the discrete grid at the points of the upper and lower cylinders surfaces and their deformation rates, which is essential in determining the shear stresses.
The simulation is performed using an algorithm developed by the authors of the numerical solution of the Navier-Stokes equations for viscous incompressible liquid on a two-dimensional grid with the help of finite-difference approximations of differential operators on a compact template.
The laminar type of viscous incompressible fluid flow at the point of surfaces contact and a constant cylinders rotation speed is considered.
The numerical calculation on defining the dynamic characteristics was performed on the grid 80×80 for the initial liquid size with thickness $\delta_H = 2.0\cdot10^{-6}$ m and width $\delta_S = 8.0\cdot10^{-6}$ m, for kinematic viscosity $\nu = 0.012$ m^2/s, density $\rho = 10^3$ kg/m^3, liquid surface tension coefficient $C_n = 0.03$ N/m.

4. Method implementation

Let us consider on the example of a cylinder the defining of the discreet values for stresses at the i-th point in the cylinder transversal direction with computational region $h_x(r)$ to L_z in size (Figure 1), where L_z is the length of the calculated element along the cylinder.
In Figure 1, on the example of the boundary point «04», let us consider the definition in a discrete form the dynamic values from one computational cell.
As the system of polar coordinates (r,φ) is considered with its transformation into the similar system $(x,y)=(R_\varphi,R-r)$, then the step size to the grid h_x will correspond to the dimension at radius $r = R$, u $h_x(r) = h_{x0} = \Delta_\varphi\cdot R$ where Δ_φ is the grid step to the angle φ.

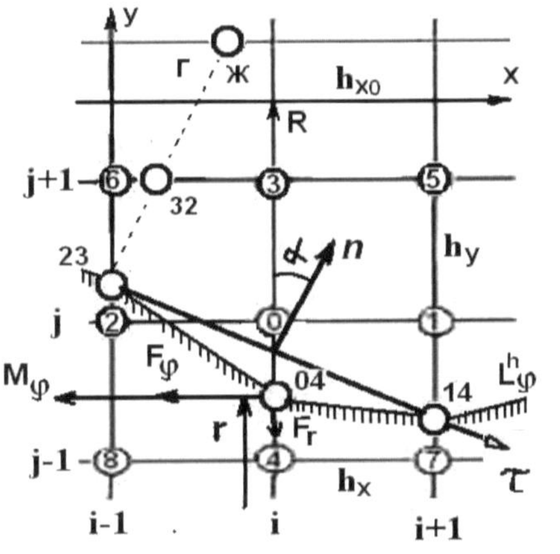

Figure 1. The determination in a discrete form the dynamic values (radial and tangential force, moment of force) from one computational cell.

To determine the active forces moment $M\varphi$ on the cylinders axis from the surfaces of contacting with ink, let us calculate the integral values of the following form with regard to the main liquid stress and the conditions of incompressibility and adhesion of viscous liquid on solid boundaries (at constant cylinder radius $r = R$):

$$F_r = \iint_s \left[-(P - P_0) - 2\nu\rho\left(\frac{\partial U_r}{\partial r}\right)\right] R d\varphi dl \qquad (1)$$

where r, φ, l are the surface cylindrical coordinates; S, P are the pressure on the cylinder and the liquid surfaces; P_0 is the external atmospheric pressure.

For determining the acting moment of forces M_φ relatively to the cylinders axes it is necessary to integrate the tangent stresses $\nu\rho\left(\frac{\partial U_\varphi}{\partial r} r\right)$ over the surface of the liquid contact with the cylinder while considering the positive moments of forces anticlockwise:

$$M_\varphi = \iint \left[-\nu\rho\left(\frac{\partial U_\varphi}{\partial r}\right)\right] R^2 d\varphi dl \qquad (2)$$

To find out the total radial force and the moment of force for the given let us do the summation over the calculated cells of the deformed surfaces, which generates force stress, interacting with the contacting liquid, with which a power generated voltage.

Thus, if the computational point «0», nearest by index i to the boundary point «04» of the cylinder, belongs to the liquid region, then in this boundary point we calculate discrete values F_{ri}, $F_{\varphi i}$, $M_{\varphi i}$ at radius value $r_{i,j(04)} = R + y_{i,j(04)}$ and angle $\varphi_i = x_i / R$.

Normal and tangent directions n, τ with angle α to the radius-vector $\boldsymbol{r}$ can be determined here by the neighbor points coordinates $(x_{i-1}, y_{i-1(23)})$, $(x_{i+1}, y_{i+1(14)})$ and by angle β, being a discrete step along φ in a polar coordinate system and defined on a uniform grid in the form of $\beta = h_{x0} / R$.

Thus, let us define this angle α in the following form:

$$\alpha = arctg\left(\frac{(r_{i,j(14)} - r_{i,j(23)}) \cdot \cos\beta}{(r_{i,j(14)} + r_{i,j(23)}) \cdot \sin\beta}\right) \qquad (3)$$

For approximate definition of the values *$(P - P_0)$* when calculating F_r, M_φ in place of P let us use $P_{i,j(04)}$, defined either by the pressure value in the nearest computational point $P_{i,j(0)}$ with the precision $O(h_y)$, or $O(h_y^2)$ by recomputation from the expression, obtained on the basis of the equation of motion by component *y* at $h_x = \beta \cdot R, h_x' = x_{i(0),j(0)} - x_{i(02),j(02)}$, $h_y = r_{i,j(3)} - r_{i,j(0)}$, $h_y' = r_{i(0),j(0)} - r_{i(04)}$ *n* the following form:

$$
\begin{aligned}
P_{i,j(04)} = {} & P_{i,j(0)} \frac{h_y^2 - h_y'^2}{h_y^2} - P_{i,j(3)} \frac{h_y'^2}{h_y^2} + \rho \frac{h_y'(h_y + h_y')}{h_y} \times \\
& \times \left[\begin{array}{l} -\left(\frac{\partial U_r}{\partial t} + U_r \frac{\partial U_r}{\partial r} + \frac{U_\varphi R}{r} \frac{\partial U_r}{\partial \varphi R} - \frac{(U_\varphi + \omega r)^2}{r} \right)_{i,j} + \\ + \nu \left(\nabla^2 U_r - \frac{U_r}{r^2} - \frac{2R}{r^2} \frac{\partial U_\varphi}{\partial \varphi R} \right)_{i,j} \end{array} \right] + \\
& + O(h_y^2)
\end{aligned}
\qquad (4)
$$

where $\nabla^2 U_r = \frac{1}{r}\frac{\partial U_r}{\partial r} + \frac{\partial^2 U_r}{\partial r^2} + \frac{R^2}{r^2}\frac{\partial^2 U_r}{(\partial \varphi R)^2}$, ω – angular rotation speed.

For calculating $P_{i,j(04)}$ it is necessary to replace the first and the second partial derivatives of U_r, U_φ to finite-difference.

Here, it is necessary to use all additional points at the boundary intersection with the nodal lines (Figure 2).

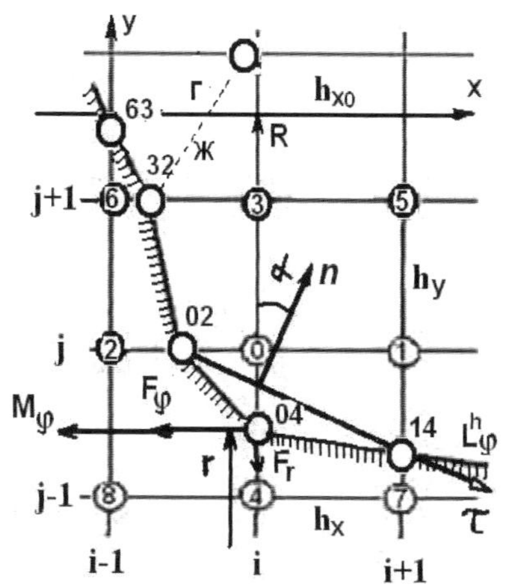

Figure 2. The definition of the dynamic forces (radial and transversal forces, moment of force) in the discrete form from one computational cell at additional grid points on the grid lines of various directions under very curved boundaries.

Besides, for approximate definition of the derivatives in the expression $\frac{\partial U_\varphi}{\partial r} Cos(\alpha) + \frac{\partial U_r}{r \partial \varphi} Sin(\alpha)$, used in the stress projections, let us use finite-difference approximations $\Lambda_r^h U_\varphi$ and $\Lambda_\varphi^h U_r$ for $\frac{\partial U_\varphi}{\partial r}$

and $\frac{\partial U_r}{\partial \varphi R}$ (on the non-uniformity grid points with steps h and h' let us use notations $\Lambda_r^{h'} U_\varphi$ and $\Lambda_\varphi^{h'} U_r$ for them):

$$\frac{\partial U_\varphi}{\partial r} = \frac{1}{\left(h_y + h_y^{'}\right)} \left(\frac{h_y}{h_y^{'}} U_{\varphi i,j(04)} - \frac{\left(h_y^2 - h_y^{'2}\right)}{h_y h_y^{'}} U_{\varphi i,j(0)} - \frac{h_y^{'}}{h_y} U_{\varphi i,j(3)} \right) + \\ + O\left(h_y^2\right) = \Lambda_r^{h'} U_\varphi + O\left(h_y^2\right) \tag{5}$$

$$\frac{\partial U_r}{\partial \varphi R} = \frac{1}{\left(h_x + h_x^{'}\right)} \left(\frac{h_x^{'}}{h_x} U_{ri,j(1)} + \frac{\left(h_x^2 - h_x^{'2}\right)}{h_x h_x^{'}} U_{ri,j(0)} - \frac{h_x}{h_x^{'}} U_{ri,j(02)} \right) + \\ + O\left(h_x^2\right) = \Lambda_\varphi^{h'} U_r + O\left(h_x^2\right) \tag{6}$$

Based on the expressions of types (5) – (6) for calculating the pressure on the computational region boundary $P_{i,j(04)}$ instead of (4) we obtain the following approximate formula:

$$P_{i,j(04)} = P_{i,j(0)} \frac{h_y^2 - h_y^{'2}}{h_y^2} - P_{i,j(3)} \frac{h_y^{'2}}{h_y^2} + \rho \frac{h_y^{'}\left(h_y + h_y^{'}\right)}{h_y} \times \\ \times \left[\begin{array}{l} -\left(\Lambda_t^{\tau} U_r + U_r \Lambda_r^{h'} U_r + \frac{U_\varphi R}{r} \Lambda_\varphi^{h'} U_r - \frac{\left(U_\varphi + \omega r\right)^2}{r} \right)_{i,j} + \\ + \nu \left(\frac{1}{r} \Lambda_r^{h'} U_r + \Lambda_{rr}^{h'} U_r + \frac{R^2}{r^2} \Lambda_{\varphi\varphi}^{h'} U_r - \frac{U_r}{r^2} - \frac{2R}{r^2} \Lambda_\varphi^{h'} U_\varphi \right)_{i,j} \end{array} \right] \tag{7}$$

Besides, for an approximate definition of the derivatives in the expression $\frac{\partial U_\varphi}{\partial r} Cos(\alpha) + \frac{\partial U_r}{r \partial \varphi} Sin(\alpha)$, used in (5)-(6), let us take finite-difference approximations $\Lambda_r^h U_\varphi$ and $\Lambda_\varphi^h U_r$ for $\frac{\partial U_\varphi}{\partial r}$ and $\frac{\partial U_r}{\partial \varphi R}$ (at the non-uniformly grid points with steps h and h' let us denote them $\Lambda_r^{h'} U_\varphi$ and $\Lambda_\varphi^{h'} U_r$).

From here, values F_{ri} , $F_{\varphi i}$, $M_{\varphi i}$ near the boundary of the considered cylinder with computational node *(i,j)* could be obtained from the following expressions (if there is no the cylinder boundary contact with the fluid, these values for the ***i***-th radial direction are set to zero):

$$F_{ri} = L_z \left(h_x r_{i,j(04)} / R\right) / Cos(\alpha) [-\left(P_{i,j(04)} - P_0\right) \cdot Cos(\alpha) - \\ - \nu\rho \left(\Lambda_r^{h'} U_\varphi Cos(\alpha) + \left(R / r_{i,j(0)}\right) \Lambda_\varphi^{h'} U_r Sin(\alpha)\right) \cdot Sin(\alpha)] \tag{8}$$

$$F_{\varphi i} = L_z \left(h_x r_{i,j(04)} / R\right) / Cos(\alpha) [-\left(P_{i,j(04)} - P_0\right) \cdot Sin(\alpha) - \\ - \nu\rho \left(\Lambda_r^{h'} U_{\varphi i,j(0)} Cos(\alpha) + \left(R / r_{i,j(0)}\right) \Lambda_\varphi^{h'} U_{ri,j(0)} Sin(\alpha)\right) \cdot Cos(\alpha)] \tag{9}$$

$$M_{\varphi i} = r_{i,j(04)} L_z \left(h_x r_{i,j(04)} / R\right) / Cos(\alpha) [-\left(P_{i,j(04)} - P_0\right) \cdot Sin(\alpha) - \\ - \nu\rho \left(\Lambda_r^{h'} U_{\varphi i,j(0)} Cos(\alpha) + \left(R / r_{i,j(0)}\right) \Lambda_\varphi^{h'} U_{ri,j(0)} Sin(\alpha)\right) \cdot Cos(\alpha)] \tag{10}$$

Total values F_r, M_φ over the all cylinder surface boundary points, contacting with the liquid, can be obtained in the following form: (8) и (10)

$$F_r = \sum_{i=1}^{N_1} F_{ri} e_{ri} , \; M_\varphi = \sum_{i=1}^{N_1} M_{\varphi i} \tag{11}$$

where N_1 is the number of calculated points over coordinate x (or φ); e_{ri} is the unit vector for the ***i***-th direction of the vector radius near the boundary *(i,j)* - th point of the computational grid.

5. Results and discussion

The computed values of radial forces on the axis and of the moments of forces for cylinders during the interaction with the flowing fluid including the moment of the fluid division between the cylinders surfaces are shown. Figure 3 presents the change in radial forces for the case with no cylinders surface deformation; while Figure 4 shows the detailed calculation for the case of the interaction with the deforming cylinder surfaces and the substrate.

It can also be noted, that when there is no boundaries deformation, in the middle of the period of interaction between the cylinders and the liquid, the peak of power loads occurs when the liquid is compressed to the maximum.

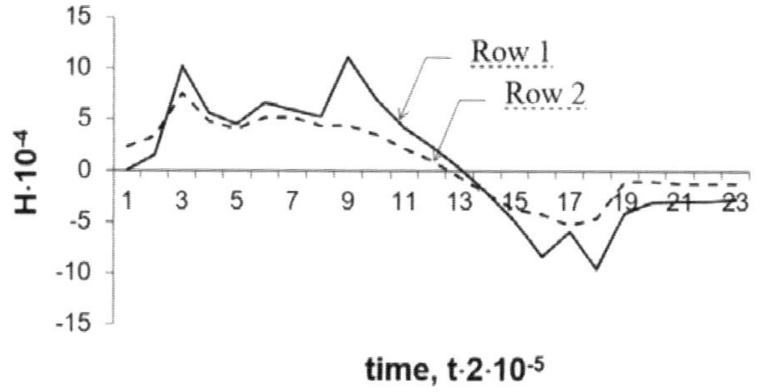

Figure 3. Radial forces of the pressure on cylinders 1 and 2 without deformation during filtration on cylinder 2 (row 2).

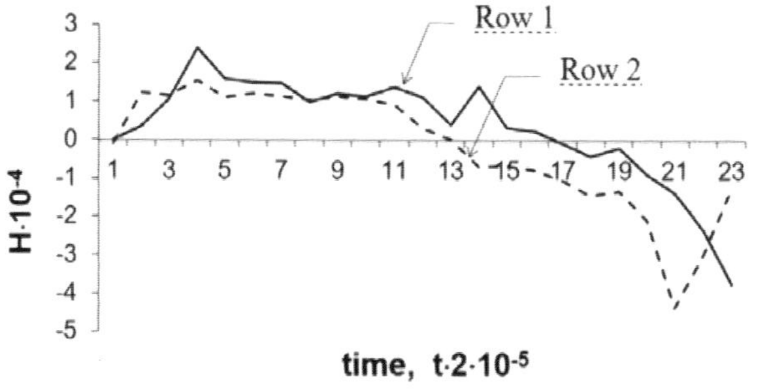

Figure 4. Radial forces of the pressure on cylinders 1 and 2 during deformation and filtration on cylinder 2 (row 2).

It may be noted that in the absence of boundaries deformation power loads peak occurs in the middle of the period of the cylinders interaction with fluid, when its maximum compression occurs. The values of the radial forces are several times greater than for the case of deformation. During the deformation of the cylinders surfaces relative maxima are observed at the beginning of the cylinders interaction during the collision of cylinder 2 with the fluid, and finally during the division of fluid along the two cylinders when compression forces are replaced by the tensile forces that occurs with some delay in contrast to the reaction without deformation. Such values behavior can be also observed for those moments of forces, for which the forces are calculated in transverse directions.

6. Conclusions

The implementation of the approach suggested for solving the task of defining the acting (operating) moments of forces:

- taking account of boundaries deformation reduces the computational dynamic quantities by 3-5 times and the ink spreading around the boundaries of the printed area;

- maximum values occur both at the initial moment of the hydrodynamic shock in the collision of the printing cylinder with liquid and at the end of the separation (gap) of the liquid in the cylinders and its transfer to the substrate;
- when there is no boundary deformation, maximum strain at the cylinders can occur at the moment of the maximum fluid compression. Taking into consideration the deformation, the forces change the direction into the reverse (opposite) one with some delay after the maximum compression of the fluid, when the deformed boundaries recover their original shape.

References

[1] ISO 12647-2:2004 2012 Graphic technology, Process control for the production of half-tone colour separations, proof and production prints, Part 2: Offset lithographic processes p 19
[2] Ou Y-R 1992 Decision and Control relative to a rotating cylinder *Decision and Control*, IEEE Conf. Publications vol 4 pp 3399–3404
[3] Yu J and Wei P 2011 Simulation the axial oil distribution of offset printing *Electrical and Control Engineering (ICECE)*, IEEE Conf. Publications pp 4968–4971
[4] Huang W-X, Lee S-H, Sung H J , Lee T-M and Kim D-S 2008 Simulation of liquid transfer between separating walls for modeling micro-gravure-offset printing *Int. J. of Heat and Fluid Flow* **29(5)** 1436–1446
[5] Reis N C Jr, Griffiths R and Santos J M 2004 Numerical simulation of the impact of liquid droplets on porous surfaces *J. Comput. Phys.* **198** 747–770
[6] Panichkin A V and Varepo L G 2014 The Numerical Calculation of a viscous Incompressible Fluid transfer onto Poros Surface between Rotating Cylinders *Int. Multidisciplinary Microscopy Congress, Springer Proc. in Physics* vol 154 pp 79–83
[7] Panichkin A V and Varepo L G 2014 The simulation of the viscous fluid splitting at the outlet of the engagement zone between cylinder surfaces and substrate *Dynamics of Systems, Mechanisms and Machines,* IEEE Conf. Publications pp1–4
[8] Tai C H, Zhao Y and Liew K M 2005 Parallel computation of unsteady incompressible viscous flows around moving rigid bodies using an immersed object method with overlapping grids, J. of Computational Phys. **207(1)** 151–172
[9] Yu J and Lin J 2012 The ink preset algorithm based on the model optimized by chaotic bee colony *5th Int. Congress on Image and Signal Processing* pp 547–551

IOP Conf. Series: Journal of Physics: Conf. Series **858** (2017) 012026 doi:10.1088/1742-6596/858/1/012026

Dynamics of rotationally reciprocating stirred tank with planetary actuator

A A Prikhodko[1], **A I Smelyagin**[1]

[1]Kuban State Technological University, 2, Moskovskaya st.,Krasnodar, 350072, Russia

E-mail: sannic92@gmail.com

Abstract. The article investigates the dynamics of rotationally reciprocating stirred tank (RRST), whose actuator is the original planetary mechanism with elliptical gears. The dynamic model is constructed by reduction of driving forces, masses and moments to the reduction link (the input shaft of the actuator). The study of the resulting dynamic model was carried out by energy-mass method. As a result of the dynamic analysis we determined the necessary moment of driven forces and found the reduction link law of motion. The flywheel has been designed to ensure the required coefficient of rotation irregularity. Resulting dynamic model can be used for development and research of rotationally reciprocating stirred tanks.

1. Introduction

Mechanically stirred vessels are widely used in machinery, petrochemical, chemical, food and many other industries [1, 2]. Currently, the most common and investigated stirred tanks are agitators with a rotational movement of impellers, because they are highly reliable, easy to manufacture and use.

Angular velocity of impeller is constant in such vessels, so over time stirred liquid velocity and impeller velocity are equalized, which leads to low intensity of heat and mass transfer. For better performance of stirred tanks it is necessary to change velocity and direction of impeller rotation [3–5].

In [6, 7] RRST is investigated, and rotationally reciprocating motion of its impeller is provided by reversing motion of the stepping motor. The developed machine is simple, due to the wide range of the rotation angle it allows to carry out various experiments in the small volume reactors. However, the use of such device is impractical on an industrial scale, since the stepping motor has a low efficiency and can not be used at high oscillation frequency of impeller. Therefore, [8, 9] proposed a planetary converter of rotational motion into reciprocating rotational motion, which can be used as an actuator of the stirred tank. It is a double row planetary gear with two external gearing, in which one pair of cylindrical gears is replaced by elliptical ones. Angle of the output shaft rotationally reciprocating motion is defined by eccentricity of elliptical gears. Mechanism with two pairs of elliptical gears will increase the output shaft rotation angle at the same size (Fig. 1).

Planetary gear (Fig. 1) consists of a rack 0, the input shaft 1, the carrier 2, the output shaft 3, the sun elliptic gear 4, elliptic gear 5, satellite elliptical gears 6 and 7, a shaft 8, which connects the satellite gears. Reciprocating rotational motion is provided by a rational sizing of actuator links.

IOP Conf. Series: Journal of Physics: Conf. Series **858** (2017) 012026 doi:10.1088/1742-6596/858/1/012026

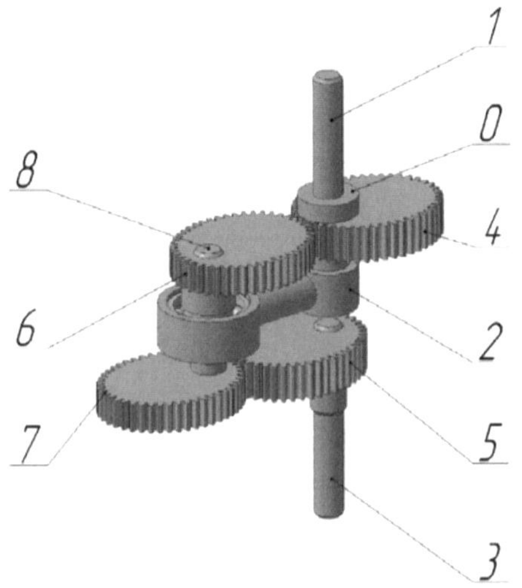

Figure 1. Planetary mechanism with elliptical gears.

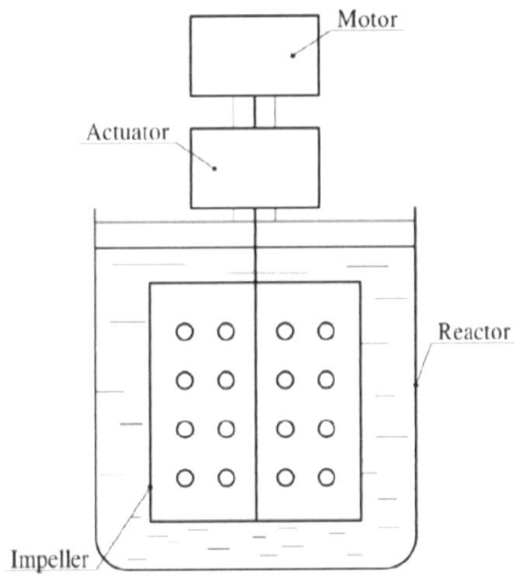

Figure 2. Rotationally reciprocating stirred tank (RRST).

Connecting the input shaft of investigated mechanism with a motor and output shaft with impeller, we obtain rotationally reciprocating stirred tank (RRST) (Fig. 2).
The considered vessel achieves high gradient of stirred liquid velocities, leading to increase in the heat and mass transfer intensity. Selection of the optimal amplitude and frequency of vibrations will reduce the flow time of many processes by 1.5-2 times, and operating costs by 1.2-1.8 times [10].
One of the most important steps in the design of new machines is the study of dynamic processes.To investigate the rotationally reciprocating stirred tank it is necessary to build dynamic model and conduct its analysis.

2. Dynamic model of the actuator

Since the planetary actuator has one degree of freedom [11] and its links are rigidly interconnected, for the rational solution it is advisable to take the input shaft 1 as the reduction link (Fig. 1). Then single-mass dynamic model of RRST (Fig. 2) takes the following form (Fig. 3).

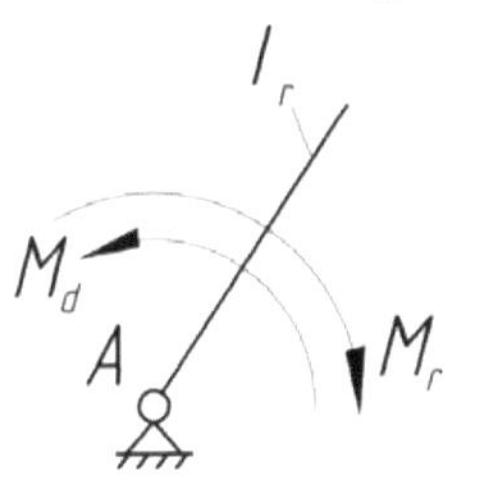

Figure 3. Single-mass dynamic model of stirred tank.

To find the law of reduction link motion, it is necessary to define the following parameters of the dynamic model: reduced moment of inertia I_r, and the reduced moment of resistance forces M_r. The required drive moment M_d is calculated during the analysis of the dynamic model.

2.1. Determination of the reduced moment of inertia

According to [12-14] reduced moment of inertia given by:

$$I_r = \sum_{i=1}^{n} m_i S_i'^2 + \sum_{i=1}^{n} I_{s_i} \varphi_i'^2, \tag{1}$$

where n is the number of movable links, which masses and moments of inertia are known; $m_i$is the mass of the i-th link; I_{Si} is the moment of inertia of the i-th link about an axis passing through the center of mass; $S_i' = \frac{dS_i}{d\varphi_1}$ is the velocity analogue of the i-th link center of mass; $\varphi_i' = \frac{d\varphi_i}{d\varphi_1}$ is the angular velocity analogue of the i-th link.
For the investigated mechanism the equation (1) takes the following form:

$$I_r = (I_d + I_1 + I_2)\cdot\varphi_1'^2 + m_6\cdot S_6'^2 + m_7\cdot S_7'^2 + m_8\cdot S_8'^2 + (I_6 + I_7 + I_8)\cdot\varphi_8'^2 + (I_3 + I_5 + I_{im})\cdot\varphi_3'^2, \quad (2)$$

where $I_d$is the motor moment of inertia, $I_{im}$is the impeller moment of inertia. The moments of inertia and velocity analogues of the actuator links identified in accordance with Figure 1.
Differentiating (2) according to the generalized coordinate, we get:

$$\frac{dI_r}{d\varphi_1} = 2[m_6\cdot S_6'\cdot S_6'' + m_7\cdot S_7'\cdot S_7'' + m_8\cdot S_8'\cdot S_8'' + (I_6 + I_7 + I_8)\cdot\varphi_8'\cdot\varphi_8'' + (I_3 + I_5 + I_{im})\cdot\varphi_3'\cdot\varphi_3''].\quad(3)$$

As seen from equations (2), (3), to find the reduced moment of inertia and its derivative it is necessary to find kinematic characteristics of planetary actuator – angular velocity and angular acceleration analogues of mechanism links, as well as linear velocity and linear acceleration analogues of link's centers of mass.

2.2. Kinematic analysis of planetary actuator

To carry out the kinematic analysis we shall represent the kinematic scheme of the mechanism in one of the positions and construct a plan of linear velocities (Fig. 4).

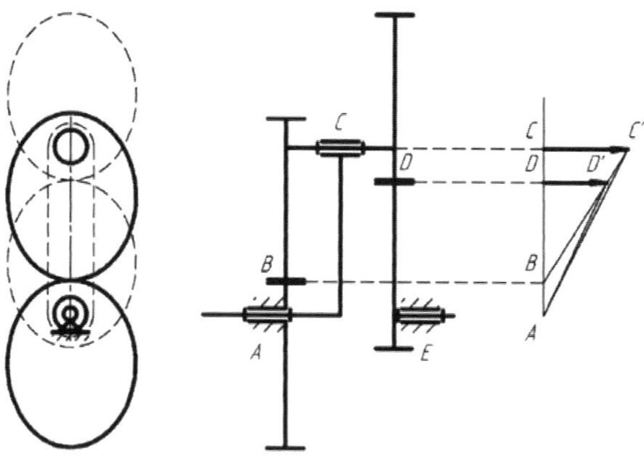

Figure 4. The plan of the mechanism velocities.

Angular velocity analogue of the output shaft, in accordance with Figure 4, is determined as:

$$\varphi_3' = \frac{\omega_3}{\omega_1} = \frac{\upsilon_D\cdot AC}{\upsilon_C\cdot DE} = \frac{BD\cdot AC}{BC\cdot DE}. \quad (4)$$

To determine the distances included in the equation (4), let's consider the equation of the ellipse in polar coordinates. The focus of the driving ellipse will be taken as the pole, and the major axis will be taken as the polar axis (Fig. 5), then we will get the following ellipse equation [15]:

$$\rho(\varphi) = \frac{p}{1 - e\cos\varphi}; \quad (5)$$

$$p=a(1-e^2), \quad (6)$$

where φ is a drive gear rotation angle; p is an ellipse focal parameter; e is eccentricity of the ellipse; a is a semi-major axis of the ellipse.

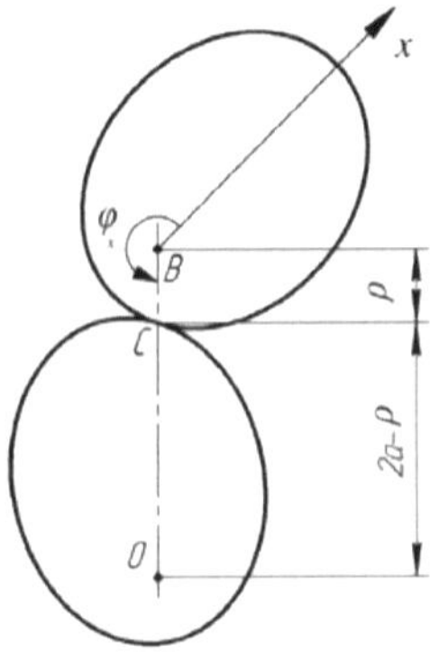

Figure 5. Elliptical gear.

According to Figure 6, taking into account (6), the equation (5) for the ellipses 4 and 7 will have the following form:

$$\rho_4(\varphi_4) = \frac{a(1-e_1^2)}{1-e_1\cos\varphi_4};$$
$$\rho_7(\varphi_7) = \frac{a(1-e_2^2)}{1-e_2\cos\varphi_7}, \qquad (7)$$

where φ_4, φ_7 are angles of ellipses rotation; e_1 is eccentricity of the first ellipse gear pair (4, 6); e_2 is eccentricity of the second ellipse gear pair (5, 7).

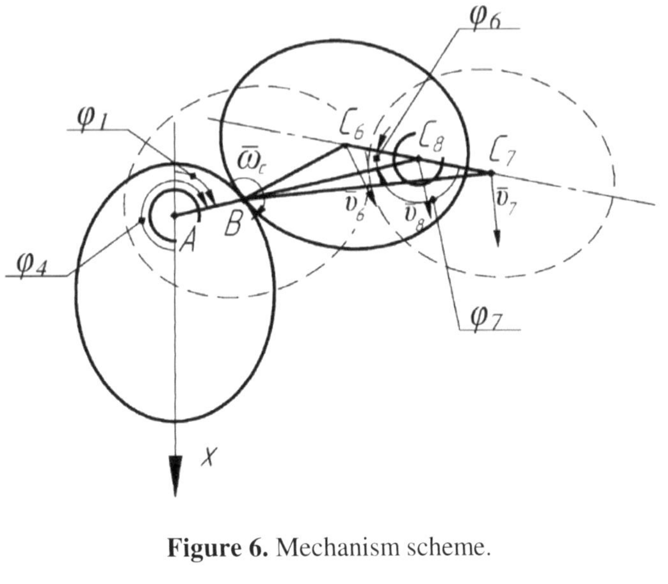

Figure 6. Mechanism scheme.

According to [16] transmission function of elliptical gear can be written as:

$$u_{21} = \frac{1-e^2}{1+e^2-2e\cos\varphi}. \qquad (8)$$

Taking into account (8) we established the relation between the rotation angles of elliptical gearwheels (Fig. 6):

$$\varphi_4 = \varphi_1 + \pi;$$
$$\varphi_6 = \int u_{64} d\varphi_4 = \int \frac{1-e_1^2}{1+e_1^2 - 2e_1\cos(\varphi_1+\pi)} d(\varphi_1+\pi); \tag{9}$$
$$\varphi_7 = \pi - \varphi_6 = \pi - \left(\int \frac{1-e_1^2}{1+e_1^2 - 2e_1\cos(\varphi_1+\pi)} d(\varphi_1+\pi) \right).$$

It follows from Figure 6 that the distances in the equation (4) are defined as:

$$BD = 2a - \rho_4 - \rho_7; \tag{10}$$

$$BC = 2a - \rho_4; \tag{11}$$

$$AC = 2a; \tag{12}$$

$$DE = 2a - \rho_7. \tag{13}$$

Substituting (10)...(13) into equation (4) we find the angular velocity analogue of the planetary mechanism output shaft 3:

$$\varphi_3' = \frac{(2a - \rho_4 - \rho_7)\cdot 2a}{(2a-\rho_4)\cdot(2a-\rho_7)}. \tag{14}$$

Figure 6 indicates that the satellite 8 and elliptical gearwheels 6 and 7 make the plane-parallel motion, and point B for satellite will be instantaneous center of velocity. Figure 6 also shows that the satellite shaft center of mass (C_8) velocity υ_8 and the angular velocity ω_8 of the satellite are determined as:

$$\upsilon_8 = \omega_1 \cdot AC_8 = \omega_1 \cdot 2a; \tag{15}$$

$$\omega_8 = \frac{\upsilon_8}{BC_8} = \frac{\omega_1 \cdot 2a}{2a - \rho_4}. \tag{16}$$

Taking into account (15) and (16) velocity analogues φ_8' and S_8' will take the following form:

$$\varphi_8' = \frac{2a}{2a-\rho_4}; \tag{17}$$

$$S_8' = 2a. \tag{18}$$

Velocities of elliptical gearwheels mass centers C_6 and C_7 are determined as:

$$\upsilon_6 = \omega_8 \cdot BC_6; \tag{19}$$

$$\upsilon_7 = \omega_8 \cdot BC_7. \tag{20}$$

Distances BC_6 and BC_7(Fig. 5) are determined by the law of cosines from BC_8C_6 and BC_8C_7 triangles:

$$BC_6 = \sqrt{C_6C_8^2 + BC_8^2 - 2C_6C_8 \cdot BC_8 \cdot \cos\varphi_6}; \tag{21}$$

$$BC_7 = \sqrt{C_7C_8^2 + BC_8^2 - 2C_7C_8 \cdot BC_8 \cdot \cos\varphi_7}, \tag{22}$$

where $C_6C_8 = a \cdot e_1$, $C_7C_8 = a \cdot e_2$ are the focal distances of elliptical gears, BC_8 is determined by (11), angles φ_6 and φ_7 are determined from equation (9).

Then, considering (16), (19)...(22), the velocity analogues S_6' and S_7', are determined as:

$$S_6' = \frac{2a\cdot\sqrt{(a\cdot e_1)^2+(2a-\rho_4)^2-2ae_1(2a-\rho_4)\cdot\cos\varphi_6}}{(2a-\rho_4)}; \tag{23}$$

$$S_7' = \frac{2a\cdot\sqrt{(a\cdot e_2)^2+(2a-\rho_4)^2-2ae_2(2a-\rho_4)\cdot\cos\varphi_7}}{(2a-\rho_4)}. \tag{24}$$

Substituting (14) and (17), (18), (23), (24) to (2), (3), we obtain the reduced moment of inertia and its derivative at any rotation angle of reduction link.

2.3. Determination of the fluid resistance moment

A useful fluid resistance acts on stirred tank impeller, which determines motion laws of the mechanism links. To construct a dynamic model it is necessary to find a moment of force, which arises at interaction between fluid and impeller, and reduces it to the input shaft of the planetary actuator.

The stirred product is a liquid, so in general, the viscous resistance force will act on the impeller, and it can be either linear or quadratic function of the velocity:

$$F_1 = k_1\cdot\upsilon; \tag{25}$$

$$F_2 = k_2\cdot\upsilon^2, \tag{26}$$

where k_1, $k_2$are coefficients of resistance, υ is the impeller linear velocity.

The linear velocity of different impeller areas is variable; therefore, the variable force of fluid resistance will act on the RRST impeller (Fig. 7).

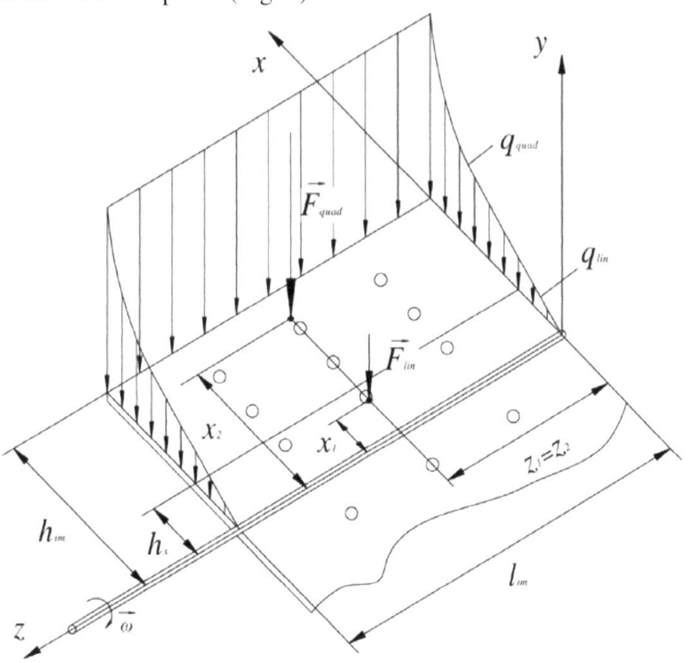

Figure 7. Force factors acting on the impeller.

In [17] there is obtained the following equation for determining the resistance moment of impeller that perform rotationally reciprocating motion:

$$M = \frac{k}{3} B'_{lin} \cdot \omega \cdot l_{im} \cdot h_x^3 + \frac{k}{4} B'_{quad} \cdot \omega^2 \cdot l_{im} (h_{im}^4 - h_x^4) \cdot sign(\omega), \tag{27}$$

where B'_{lin}, B'_{quad} are reduced coefficients of linear and quadratic resistance, ω is the angular velocity of impeller, l_{im} is an impeller length, h_{im} is the blade width, h_x is the distance from the rotation axis to the boundary between the laminar and turbulent regimes, k is the number of impeller blades (in investigated stirred tank $k = 2$).
For most technological processes there we can observe turbulent mode [1], so (27) takes the form:

$$M = \frac{1}{2} B'_{quad} \cdot \omega^2 \cdot l_{im} \cdot h_{im}^4 \cdot sign(\omega). \tag{28}$$

Denote $B = 0.5 B'_{quad} \cdot l_{im} \cdot h_{im}^4$, as a result we get:

$$M = B \cdot \omega^2 \cdot sign(\omega). \tag{29}$$

According to [10], reduced resistance moment is generally defined as follows:

$$M_r = \sum_{i=1}^{n} \left(\sum_{i=1}^{m} F_i l'_i + \sum_{i=1}^{q} M_i \varphi'_i \right), \tag{30}$$

where n is total number of mobile links; m is number of forces F, acting on the i-th link; l'_i is velocity analogue of force application point; q is the number of moments M, acting on the i-th link.
Taking into account that only the impeller resistance moment acts in the stirred tank, then (30) to determine the reduced resistance moment takes the form:

$$M_r = M \cdot \varphi'_3. \tag{31}$$

Given that $\omega_3 = \omega_1 \cdot \varphi'_3$, we substitute (29) to (31) and obtain:

$$M_r = B \cdot \omega_1^2 \cdot \varphi_3'^3 \cdot sign(\varphi'_3). \tag{32}$$

Equation (32) allows to find resistance moment, reduced to the input shaft 1.

3. Investigation of the dynamic model

As an example, we consider a stirring device with the following parameters (link numbers correspond to Fig. 1): I_m = 100 g·cm^2 (motor); I_1 = 9.8 g·cm^2; I_2 = 1233 g·cm^2; I_3 = 22.4 g·cm^2; I_5 = 627 g·cm^2; I_6 = 564 g·cm^2, m_6 = 0.09 kg; I_7 = 564 g·cm^2, m_7 = 0.09 kg; I_8 = 19.2 g·cm^2, m_8 = 0.04 kg; I_{im} = 15 g·cm^2, h_{im} = 0.045 m, l_{im} = 0.12 m (impeller); B=6·10^{-7}; allowable coefficient of rotation irregularity [δ] = 0.05. The eccentricities of the first and second pairs of elliptical wheels are respectively e_1=0.3 and e_2=0.4, semi-major axes of elliptical wheels a=25mm. Input shaft of actuator driven by a motor, which rotating speed $\omega_1 = 52 rad/s$ (n_1=500 rpm).

A study of the dynamic model is performed using the energy-mass method [12, 13], which is widely used in the dynamic analysis of machines. In accordance with the selected method we will find the increment of the kinetic energy ΔT :

$$\Delta T = A_d - A_r, \tag{33}$$

where A_d is work of driving forces, A_r is work of resistant forces.
Works in (33) are determined as:

$$A_d = M_d \cdot \varphi_1; \tag{34}$$

IOP Conf. Series: Journal of Physics: Conf. Series **858** (2017) 012026 doi:10.1088/1742-6596/858/1/012026

$$M_d = \frac{\int_0^{2\pi} M_r d\varphi_1}{2\pi}; \tag{35}$$

$$A_r = \int_0^{\varphi_1} M_r d\varphi_1. \tag{36}$$

For investigated RRST we construct graphs of A_r, A_d, ΔT (Fig. 8).

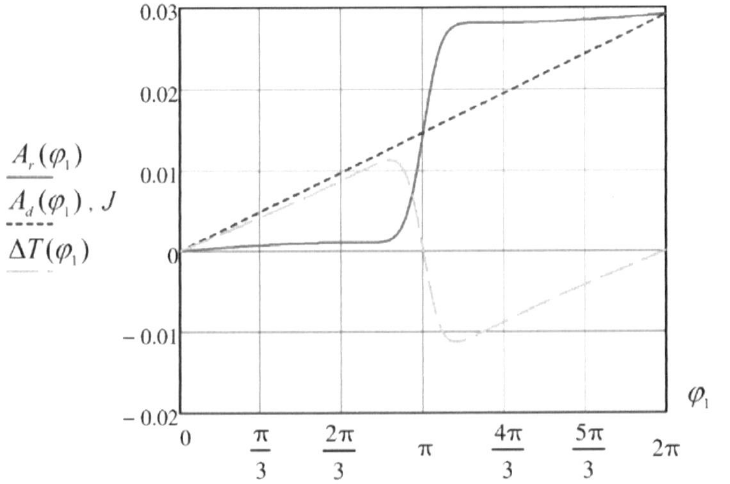

Figure 8. Graphs of functions A_r, A_d, ΔT .

Angular velocity of reduction link is determined as [13]:

$$\omega_1 = \sqrt{\frac{2\Delta T - C_{\max} - C_{\min}}{I_r}}, \tag{37}$$

where:

$$C_{\max} = \Delta T - \frac{1}{2} I_r \omega_{av}^2 (1 + [\delta]);$$

$$C_{\min} = \Delta T - \frac{1}{2} I_r \omega_{av}^2 (1 - [\delta]).$$

Using (37) and the calculation results (Fig. 8), we construct a graph $\omega_1(t)$ (Fig. 9).

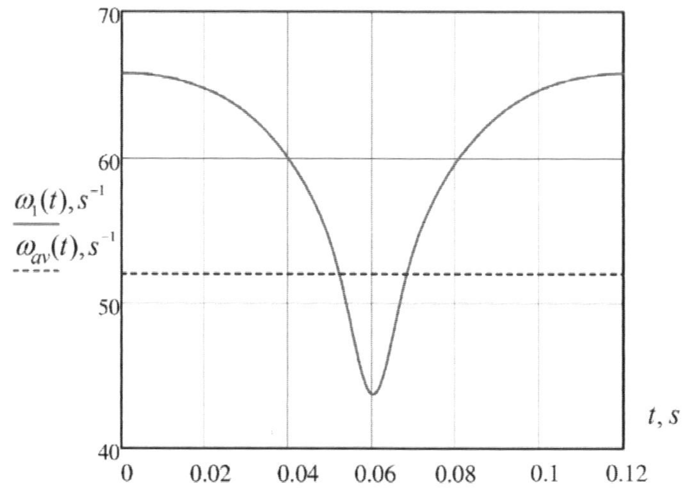

Figure 9. Graph of function $\omega_1(t)$.

The graph shows that the reduction link angular velocity is not constant and varies around the average value. Velocity oscillations are determined by the intracyclic changes in gear ratio of mechanism with elliptical gearwheels and force changes on impeller. Since the angular velocity of reduction link is variable, then we define coefficient of rotation irregularity δ [13]:

$$\delta = \frac{\omega_{\max} - \omega_{\min}}{\omega_{av}}. \tag{38}$$

It is seen from Figure 9 that $\omega_{\max} = 65.8 rad/s$, $\omega_{\min} = 43.7 rad/s$, $\omega_{av} = 52 rad/s$. Then coefficient of irregularity $\delta = 0.42$. It can be seen that the found coefficient does not satisfy the previously allowable coefficient $[\delta] = 0.05$. Therefore, it is necessary to install the flywheel in the investigated RRST.

According to [13], flywheel moment of inertia I_f will be defined as:

$$I_f = \frac{C_{\max} - C_{\min}}{[\delta] \cdot \omega_{av}^2}. \tag{39}$$

Substituting in (39) C_{max}, C_{min}, $[\delta]$ and ω_{av} we find the required flywheel moment of inertia. Knowing the flywheel, we find the angular velocity of reduction link [13]:

$$\omega_1 = \sqrt{\frac{I_f \cdot \omega_{av}^2 (1 + [\delta]) - 2(C_{\max} - \Delta T)}{I_f + I_r}}. \tag{40}$$

Angular velocity of reduction link for RRST is shown in Figure 10.

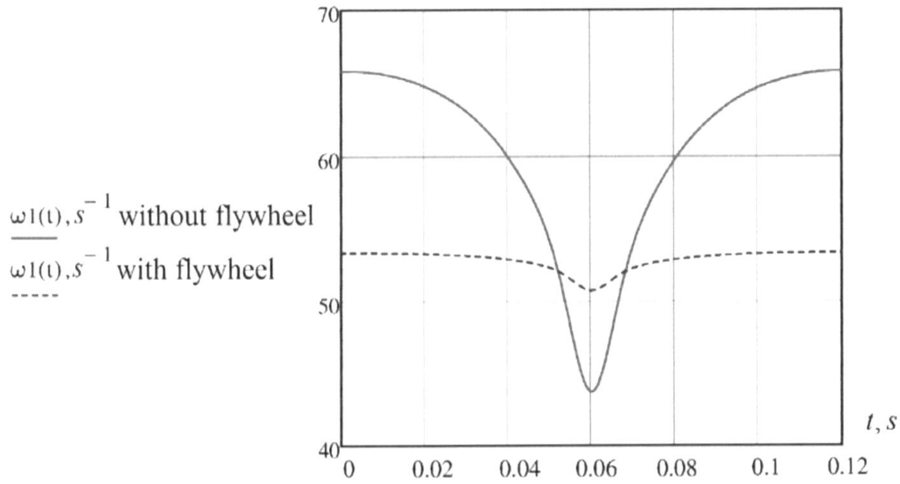

Figure 10. Graphs of functions $\omega_1(t)$ with and without flywheel.

As can be seen from the graphs, the installation of a flywheel reduced rotation irregularity of reduction link. Coefficient of rotation irregularity decreased to allowable value $[\delta]=0.05$.

4. Conclusion

In this paper we construct and investigate dynamic model of the stirred tank with a rotationally reciprocating motion of impellers. As the actuator of such machines, the new planetary mechanism for converting rotational motion into reciprocating rotational with elliptic gears was proposed. The investigations produced the following results:

- the reduced link (input shaft of the mechanism) law of motion was found;
- we calculated the value of the drive moment, which is required for the stirred tank operation;
- we determined flywheel moment of inertia, which is necessary to reduce coefficient of rotation irregularity to allowable value.

The resulting mathematical model can be used in the calculations, design and investigation of stirred tanks with rotationally reciprocating motion of impellers.

References

[1] Cullen P J2009*Food mixing: Principles and applications*(New York: Wiley-Blackwell)
[2] Hemrajani R R, Tatterson G B2004*Mechanically stirred vessels. Handbook of industrial mixing: science and practice* (Hoboken: Wiley-Interscience) pp 345-390
[3] Kato Y, Tada Y, Ban M, Nagatsu Y, Iwata S, Yanagimoto K 2005*Journal of chemical engineering of Japan***38 (9)**688–691
[4] Senda S, Komoda Y, Hirata Y, Takeda H, Suzuki H, Hidema R 2014*Journal of chemical engineering of Japan***47 (2)**pp151–158
[5] Prikhodko A A, Smelyagin A I 2016*Vibroengineering Procedia***8** pp 102–107
[6] Senda S, Yamagami N,Komoda Y, Hirata Y, Suzuki H, Hidema R 2015*Journal of chemical engineering of Japan***48 (11)**pp885–890
[7] Senda S, Komoda Y, Hirata Y, Takeda H, Suzuki H, Hidema R 2016*Journal of chemical engineering of Japan***49 (4)**pp341–349
[8] Smelyagin A I, Prikhod'ko A A 2016*Journal of Machinery Manufacture and Reliability* **45(6)** pp 500–505
[9] Prikhodko A A, Smelyagin A I 2015*Procedia Engineering* **129** pp87–92

IOP Conf. Series: Journal of Physics: Conf. Series **858** (2017) 012026 doi:10.1088/1742-6596/858/1/012026

[10] Smelyagin A I1986Theory, synthesis and investigation of machines and mechanisms with electromagnetic drive *Doctoral thesis*
[11] Prikhodko A A, Smelyagin A I 2015*Vestnik Donskogo gosudarstvennogo tekhnicheskogo universiteta***15(4)**pp69–75
[12] Wittenbauer F1923Graphische Dynamik: ein Lehrbuch für Studierende und Ingenieure mit zahlreichen Anwendungen und Aufgaben(Berlin: Julius Springer)
[13] Smelyagin A I2014Mechanism and machine theory(Moscow: Infra-M)
[14] Prikhodko A A, Smelyagin A I2015*Proc. Int. Conf. on Mechanical Engineering, Automation and Control Systems*(Tomsk: IEEE) pp 1-5
[15] Coxeter H S M 1969*Introduction to Geometry* (New York:Wiley)
[16] Litvin F L, Fuentes A 2004*Gear geometry and applied theory*(Cambridge: Cambridge University Press)
[17] Prikhodko A A, Smelyagin A I2015*Proc. Int. Conf. on Innovations in Mechanical Engineering* (Kemerovo, KuzSTU) pp 516-519

Acknowledgments
The research is done using scientific equipment of the Center of Shared Usage "The research center of engineering, materials, construction, and transportation systems" in Kuban State Technological University.

IOP Conf. Series: Journal of Physics: Conf. Series **858** (2017) 012027 doi:10.1088/1742-6596/858/1/012027

Creation of graphic database specifying android arm mechanism work envelope taking into account forbidden zones position

F N Pritykin[1] and V I Nebritov [1]

[1] Omsk State Technical University, 11, Mira ave., Omsk, 644050, Russia

E-mail: pritykin@mail.ru

Abstract. The structure of graphic database specifying the shape and the work envelope projection position of an android arm mechanism with various positions of the known in advance forbidden zones is proposed. The technique of analytical assignment of the work envelope based on the methods of analytical geometry and theory of sets is represented. The conducted studies can be applied in creation of knowledge bases for intellectual systems of android control functioning independently in the sophisticated environment.

1. Introduction

Creating universal algorithms for examination of manipulator mechanism functionality is still a topical problem [1-4]. In this respect 3D development of the virtual environment to model experiments to analyze the behavior of independently functioning androids in organized environments is one of the important problems of robotics [5-7]. The solution to the specified problem is inter-related with work envelope construction of the robot arm mechanism taking into account the known in advance environment. An analytical assignment of the work envelope allows analyzing goals position of the synthesized trajectory for the output link and defining their attainability at the initial stage of arm motion realization [8-9]. The most frequent motional task for an android is putting a working object on or taking it off the tool rack. Therefore let us examine the influence of the forbidden zones on the work envelope of android arm mechanism, these zones being the tool racks of different height.

2. Problem statement

In Figure 1 the images of android AR-600E arm mechanism kinematic diagram and forbidden zones P_1 and P_2 are shown. These forbidden zones specify the position of tool-racks P_1 and P_2. The figure illustrates the virtual safe zone of l_δ = 50mm in thickness around obstacles. In the course of experiments this zone proves to be forbidden, the intersection of arm configuration and this zone as a result of a calculation error not resulting in the damage of mechanism or working objects and tool-racks. In Figure 1 points O_1, O_2 and so on define the origin of coordinates used for assignment of a kinematic chain model. The values of generalized coordinates assigned by the vector $\boldsymbol{q}$ $(q_1, \ldots, q_5)$ (Fig. 1) are varied within the specified range and defined by the inequalities:

$$q_i^{min} \leq q_i \leq q_i^{max} \quad (1)$$

where q_i^{min}, q_i^{max} are the innermost and outermost limits of generalized coordinates. In the considered example the specified values are equal to q_i^{min} (-10°, -100°, -90°, -120°, -120°) and q_i^{max} (10°, 120°, 120°, 120°, 120°) respectively. The parameters x_t and x_{11} in Figure 1 then define the assigned safe distance between the body and the tool-racks and minimum permissible value of the coordinate x_{11} specifying the center O_{11} of the output link (OL). The coordinates x_{B1} and z_{B1} specify the position of the reference point B_1 of the forbidden zone P_1.

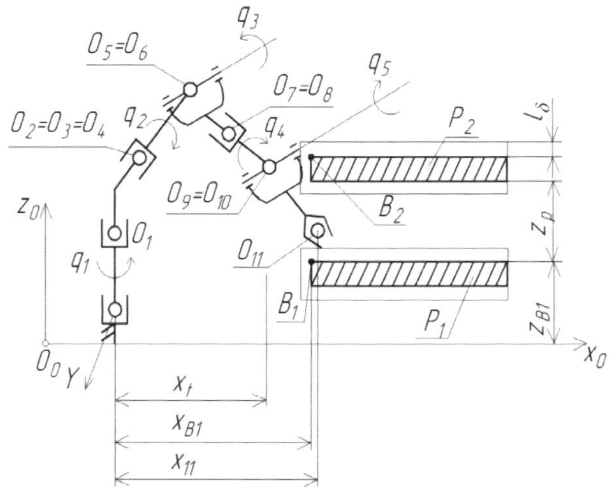

Figure 1. Relative position of the android arm mechanism and tool-racks.

Let us construct the set of gripping centers positions (possible positions of point O_{11}) defining the work envelope of an android arm mechanism provided the intersection of the arm configuration and forbidden zones is absent. Let the alteration in pitch of generalized coordinates Δq_i be equal to 5° in construction of the work envelope.

3. Computational results for the work envelop projection of android arm mechanism with forbidden zones taken into account

The block diagram in Figure 2 represents the construction of OL center (point O_{11}) positions set of the work envelope.

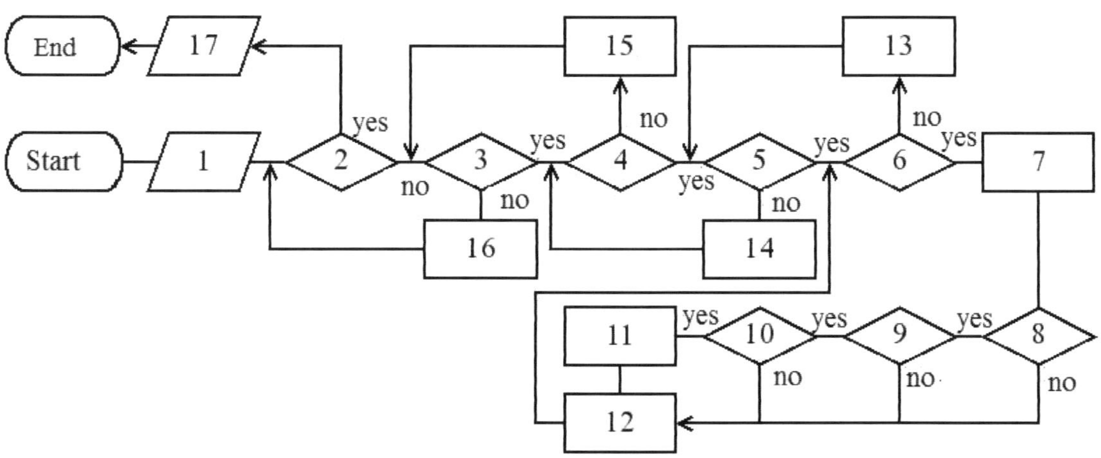

Figure2. The algorithm of points construction of android arm mechanism work envelope taking the position of forbidden zones P_1 and P_2 into account.

The following notation is introduced in Figure 2: 1 – data input, n, q_i^{min}, q_i^{max}, Δq_i, x_{11}^{min}, x_{B1}, z_{B1}, x_{B2}, z_{B2} and z_P; 2 – condition test $q_1 < q_1^{max}$; 3 – $q_2 < q_2^{max}$; 4 – $q_3 < q_3^{max}$; 5 – $q_4 < q_4^{max}$; 6 – $q_5 < q_5^{max}$; 7 – calculation $p_1 = 0$ if $q_i > q_i^{max}$, if not $p_1 = 1$, calculation $p_2 = 0$ in case the intersection of the arm configuration and the obstacle is absent, if not $p_2 = 1$; 8 – condition test $p_1 = 0$ (if values of q_i satisfy the limit values of generalized coordinates); 9 – condition test $p_2 = 0$ (the condition of the configuration and the obstacle intersection); 10 – $x_{11} > x_{11}^{min}$ (where x_{11}^{min} — assigned minimum permissible values of the coordinate x_{11}, in the considered example x_{11}^{min} = 400 mm); 11 – Point belongs to the section of the working region. Imaging the given point; 12 – $q_5 = q_5 + \Delta q_i$; 13 – $q_4 = q_4 + \Delta q_i$, $q_5 = q_5^{min}$; 14 – $q_3 = q_3 + \Delta q_i$, $q_4 = q_4^{min}$, $q_5 = q_5^{min}$; 15 – $q_2 = q_2 + \Delta q_i$, $q_3 = q_3^{min}$, $q_4 = q_4^{min}$, $q_5 = q_5^{min}$; 16 – $q_1 = q_1 + \Delta q_i$, $q_2 = q_2^{min}$, $q_3 = q_3^{min}$, $q_4 = q_4^{min}$, $q_5 = q_5^{min}$; 17 – array extraction of point data belonging to the section of the work envelope.

Figure 3abc shows an image of the work envelope for accordingly specified values z_{B1} = 350 mm, z_{B1} = 200 mm and z_{B1} = 500 mm at x_{B1} = 400 mm. The parameter z_p is set to be 100 mm. Figure 3d presents the top view of the work envelope being the same for all three represented versions of the work envelope in Figure 3abc. The work envelope is designed for the configurations of the arm not intersecting the forbidden zones P_1 and P_2 and being located at the safe distance of $l_\delta < 50$ mm.

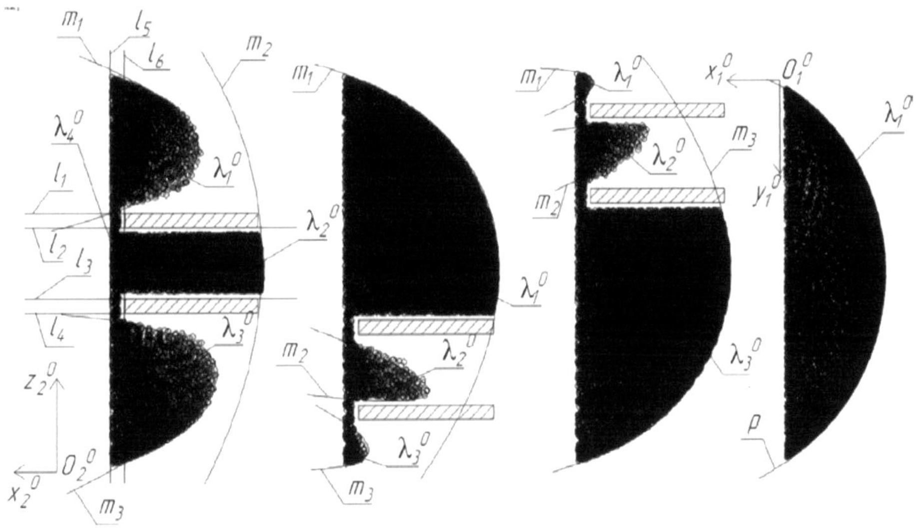

Figure 3. Representation of the work envelope with the tool-racks position on the elevation and top views being taken into account when x_{B1} = 400 mm: (a) – z_{B1} = 1200 mm; (b) – z_{B1} = 1350 mm; (c) – z_{B1} = 1500 mm; (d) – representation of the work envelope on the top view.

As the figure indicates the specified graphic images of top views are defined with six lines l_i and three second-order curves that is parabolas m_i. The parabolas are most accurately fit the contours of the work envelop. The work envelop contour on the top view is assigned by the radial arc p.

4. Analytical assignment of the work envelop projection contours

Mathematical models of graphic images λ^0_i for the work envelop projection regions are defined based on theory of sets [9]. Lines l_i position data and parabolas m_i position and shape are prescribed by algebraic equations coefficients for these graphic objects in the inertial system with the graphic data base depending on the values of z_B and z_P.

To define the goals belonging to the work envelop let us assign this envelop with the theory of sets, these goals prescribing the end position of the OL center on synthesized trajectories [9]. The work envelop is defined as the intersection of regions $\Delta_1, \Delta_2, \ldots \Delta_9$ on the elevation view.

The regions $\Delta_{1\text{-}6}$ (assigned by the lines $l_1 - l_6$ on the elevation view) are defined by inequalities of the form (Fig. 3abc):

$$-x^0\left(z_{i-1}^0-z_i^0\right)+z^0\left(x_{i-1}^0-x_i^0\right)-x_{i-1}^0\cdot z_i^0+x_i^0\cdot z_{i-1}^0\geq 0\,, \tag{2}$$

or

$$A_i\,x^o+B_i\,y^o+C_i\geq 0, \tag{3}$$

where x_i^0, z_i^0 and x^0_{i-1}, z^0_{i-1} are the point coordinates wherethrough lines l_1-l_6 are drawn, x^0, z^0 are the coordinates of the current point line respectively.
The three regions on the elevation view Δ_7, Δ_8 and Δ_9 are defined by the parabolic curves m_1, m_2 and m_3 (Fig. 3abc). The points assigning inequalities within the regions $\Delta_{7\text{-}9}$ appear as follows:

$$\begin{aligned}
&\left(x_0\cos\varphi^{\Delta 7}+z_0\sin\varphi^{\Delta 7}+m_x^{\Delta 7}\right)^2-2p^{\Delta 7}\left(x_0\sin\varphi^{\Delta 7}+z_0\cos\varphi^{\Delta 7}+m_z^{\Delta 7}\right)\geq 0,\\
&\left(x_0\cos\varphi^{\Delta 8}+z_0\sin\varphi^{\Delta 8}+m_x^{\Delta 8}\right)^2-2p^{\Delta 8}\left(x_0\sin\varphi^{\Delta 8}+z_0\cos\varphi^{\Delta 8}+m_z^{\Delta 8}\right)\geq 0,\\
&\left(x_0\cos\varphi^{\Delta 9}+z_0\sin\varphi^{\Delta 9}+m_x^{\Delta 9}\right)^2-2p^{\Delta 9}\left(x_0\sin\varphi^{\Delta 9}+z_0\cos\varphi^{\Delta 9}+m_z^{\Delta 9}\right)\geq 0,
\end{aligned} \tag{4}$$

where $m_x^{\Delta 7}$, $m_y^{\Delta 7}$, $p^{\Delta 7}$, $\varphi^{\Delta 7}$ and so on define the form and position parameters of the curves m_1, m_2 and m_3 determining points within the regions $\Delta_{7\text{-}9}$ (Fig. 4ab). The parameters $m_x^{\Delta 8}$, $m_y^{\Delta 8}$ in Figure 4 define the center of the coordinate system $O'x'z'$ connected with the parabola. The parabola point has the coordinate $z'=0$ (Fig. 4ab). The parameter $\varphi^{\Delta 8}$ assigns the inclination angle of the axis x' determining the parabola inclination to the axis x^o. The value $p^{\Delta 8}$ prescribes a focus point of parabola. The inequalities (4) are obtained by transforms of coordinates assigning the transition from the system $O'x'z'$ to the system $O^0x^0z^0$ (Fig. 4ab). Figure 4ab shows the position of the three lines l_2, l_3, l_6, the curve m_2 and the region λ_2^0 resulted from the intersection of regions Δ_2, Δ_3, Δ_6 and Δ_8 with parameters $z_{B1}=500$ mm, $z_{B1}=350$ mm and $x_{B1}=400$ mm.

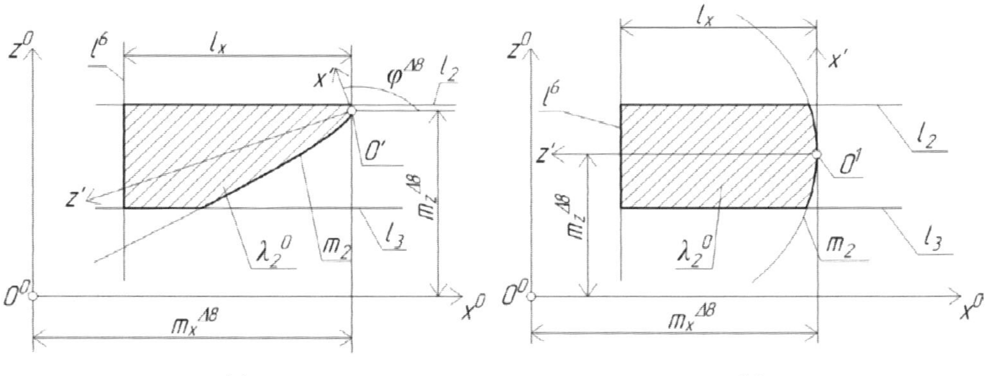

Figure 4. Different versions of analytical assignment of the region λ_2^0 of the android arm mechanism work envelop fragment on the elevation view at: a) – $z_{B1}=500$ mm, $x_{B1}=400$ mm; (б) – $z_{B1}=350$ mm, $x_{B1}=400$ mm.

In Figure 4 the value of the parameter l_x defines the outermost point of region λ_2^0 of the work envelop fragment from the line l_6. The intersection of regions is performed with the operations of Boolean algebra [9]:

IOP Conf. Series: Journal of Physics: Conf. Series **858** (2017) 012027 doi:10.1088/1742-6596/858/1/012027

$$\begin{aligned}
\lambda_1^0 &= \Delta_5 \cap \Delta_7, \\
\lambda_2^0 &= ((\Delta_2 \cap \Delta_3) \cap \Delta_6) \cap \Delta_8, \\
\lambda_3^0 &= \Delta_5 \cap \Delta_9, \\
\lambda_4^0 &= ((\Delta_5 \cap \Delta_6) \cap \Delta_1) \cap \Delta_4.
\end{aligned} \tag{5}$$

The set of R-functions reflecting the logic of work envelops projection graphic imaging (5) is also predetermined in the array, based on regions Δ_i these envelops are defined by the lines l_1 - l_6, parabolas m_1 - m_3 and curve p.. The graphic database structure fragment is shown in Tables 1 and 2.

Table 1. Coefficients values of lines $l_1 \div l_6$ equations.

z_p	z_B		l_1			l_2			l_3	
mm	*mm*	A_1	B_1	C_1	A_2	B_2	C_2	A_3	B_3	C_3
100	1200	0	-210	69300	0	-210	60900	0	-210	48300
	1500	0	-210	132300	0	-210	123900	0	-210	111300
z_p	z_B		l_4			l_5			l_6	
mm	*mm*	A_4	B_4	C_4	A_5	B_5	C_5	A_6	B_6	C_6
100	1200	0	-210	39900	-80	0	28000	-80	0	31200
	1500	0	-210	102900	-80	0	28000	-80	0	31200

Table 2. Coefficients values of parabolas $m_1 \div m_3$ equations.

z_p	z_B		m_1				m_2				m_3		
mm	*mm*	$m_x^{\Delta 7}$	$m_y^{\Delta 7}$	$\varphi^{\Delta 7}$	$p^{\Delta 7}$	$m_x^{\Delta 8}$	$m_y^{\Delta 8}$	$\varphi^{\Delta 8}$	$p^{\Delta 8}$	$m_x^{\Delta 9}$	$m_y^{\Delta 9}$	$\varphi^{\Delta 9}$	$p^{\Delta 9}$
100	1200	622	1535	108	249.4	510	1212	73	2.64	432	1136	77	3.61
	1500	413	1755	111	2.69	494	1572	63	2.55	627	1524	73	251.1

5. Conclusion

In Figure 5 the results of movements synthesis along the velocity vector [8] using the graphic database with a focus on working object transfer from the point A^H to the point A^K are presented. The points A^H and A^K in the figure are specified by the projections A_1^H , A_2^H, A_1^K and A_2^K. At the initial stage of android motion virtual modeling the algorithm defines the goal A^K membership in the region λ_2^0 (Fig. 3abc). If a specified point is in the region λ_2^0, motion synthesis is performed in specified trajectory sections along the velocity vector.

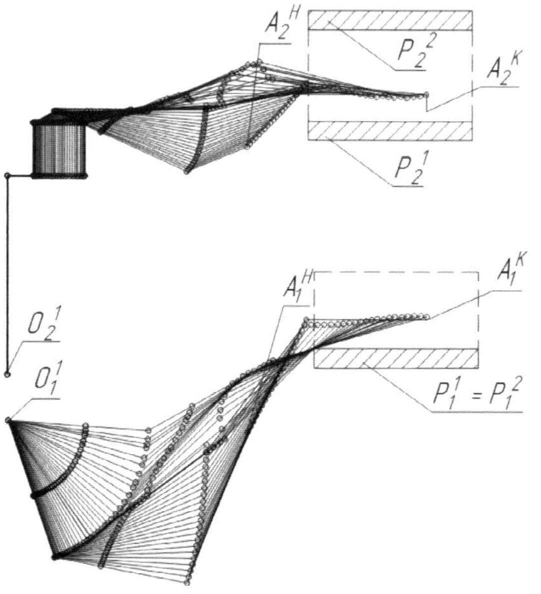

Figure 5. The result of motion synthesis to the end point of the synthesized trajectory.

The results of computational experiments demonstrate the calculation time reduction for tests related to putting a working object on or taking it off the tool rack with the use of the developed graphic databases. The presented research results can be applied in virtual modeling of an android motion in the known in advance sophisticated environment.

References

[1] Pratt J, Dilworth P and Pratt G 1997 Virtual model control of a bipedal walking robot *Proceedings of Int. Conf. on Robotics and Automation* 1 pp 193–198

[2] You B, Zou Y, Xiao W and Wang J 2010 Telerobot control system based on dual-virtual model and virtual force *Int. Forum on Strategic Technology* pp 246–250

[3] Hrr J, Pratt J, Chew C-M, Herr H and Pratt G 1998 Adaptive virtual model control of a bipedal walking robot *Proceedings of Int. Joint Symp. on Intelligence and Systems* pp 245–251

[4] Tsukamoto H K, Takubo T, Ohara K, Mae Y and Arai T 2010 Virtual impedance model for obstacle avoidance in a limb mechanism robot *Int. Conf. on Information and Automation* pp 729-734

[5] Hasegawa T, Suehiro T and Takase K 1992 A model-based manipulation system with skill-based execution *IEEE Transactions on Robotics and Automation* **8** pp 535–544

[6] Hrr J, Pratt J, Chew C, Herr H and Pratt G 1998 Adaptive virtual model control of a bipedal walking robot *IEEE Int. Simp. Intellegence and Systems* pp 245–251

[7] Lopatin P 2016 Investigation of a Target Reachability by a Manipulator in an Unknown Environment *IEEE International Conference on Mechatronics and Automation* pp 37–42

[8] Pritykin F N and Tevlin A M 1987 Procedure for construction of manipulator motions from a given local grip path in the presence of obstacles *Soviet machine science* **4** pp 30–33

[9] Rvachev V L 1974 Methods of Logic Algebra in Mathematical Physics (Kiev) 256

IOP Conf. Series: Journal of Physics: Conf. Series **858** (2017) 012028 doi:10.1088/1742-6596/858/1/012028

Control of influence of a thread on a bending of screws

N E Proskuriakov[1], **I V Lopa**[1], **E V Trapeznikov**[2]

[1] Tula State University, 92, Lenina ave., Tula, 300012, Russia
[2] Omsk State Technical University, 11, Mira ave., Omsk, 644050, Russia

E-mail: vippne@mail.ru

Abstract. The influence of the threads and the bending of screw on their moments of inertia of the cross section considered. This problem is actual since existing methods exclude from calculations the influence of supporting the thread, using as the basic geometrical parameter such as the internal diameter of the thread (diameter of cavities). Fundamental difference of a bend of the screw from a bend of a smooth rod consists that moment of inertia of the screw is a variable. It is shown that the change in cross-section moment of inertia along the length of the screw are essential and have periodic character. Analytical interrelation of the bending of the screw and the decreasing of moment of inertia of its cross section is established and equation describing this phenomenon is suggested. The greatest decrease of the moment of inertia occurs in the middle of the screw length, and the lowest - at its ends. Function and approximate coefficients for the main types of thread are proposed, which take into account this change.

1. Introduction

The pipeline fittings equip many installations and aggregates in chemical, oil-extracting and oil-processing industry, in metallurgy and power industry.

One of the most important classes of pipeline fittings are the shutoff valves — the devices used to periodic or one-time turning on, or shutdown of the pipeline or valves.

The screw-nut transfer is widely applied in gates of pipelines. Theoretical and experimental researches on the causes of failures or inefficiencies of mechanisms of gates under dynamic loading have shown that one of the weaknesses are the screws of mechanism of pipeline gates.

During the operation, the screws of pipeline gates are subjected to considerable forces and moments. If the screw length exceeds the diameter of more than eight times, then calculation of the longitudinal stability is necessary for these screws. When determining the constructional characteristics of the screws there is a problem to define of the influence of thread on their strength and rigidity parameters [1].

The existing methods exclude from calculations effect of support of a thread as use in calculations as the critical geometrical parameter - diameter of an internal thread of screw or use empirical relationships so-called resulting moment of inertia of the screw [2, 3].

2. Formulation of problem

Figure 1 shows three arbitrary cross sections of the screw. It can be seen that this section has a complex configuration, which cannot be described by elementary functions.

IOP Conf. Series: Journal of Physics: Conf. Series **858** (2017) 012028 doi:10.1088/1742-6596/858/1/012028

Therefore, to determine the moments of inertia of the cross section of the screw we have conducted a numerical experiment using the combined methods of calculations based on the software "SolidWorks2011" (Table 1).

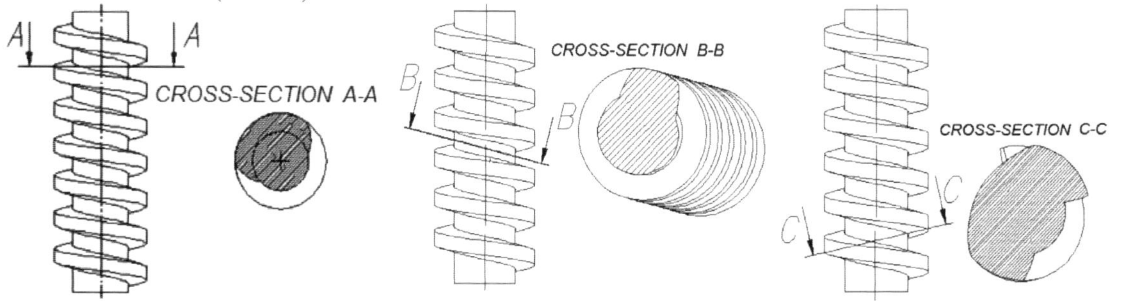

a b c

Figure 1. Different cross-sections of the screw: a) perpendicular; b) on the line of the helix of the thread; c) arbitrary.

3. Theory

Table I shows that the function describing the change of the moments of inertia along the length of the screw are clearly the periodic nature and can be approximated by type $J(z) = J_0 + a \cdot \sin(\omega \cdot z + \varphi)$, which for the definition of the coefficients can be expressed as:

$$J_x(z) = J_0 + a \cdot \sin\left(\frac{\pi \cdot z}{b} + \varphi\right) \tag{1}$$

where a, φ, ω and J_0, approximation coefficients, $b = \pi / \omega$. Function coefficients (1) are presented in Table 2 and 3.

Table 2 shows when you increase the thread pitch, amplitude and initial phase of harmonic functions (1) are increased, and the average value of the moment of inertia decreases and the angular frequency is the same for different types of thread.

Table 1. Changes the moments of inertia along the length of the screw, mm^4

Thread type		CROSS-SECTION, mm									
		0.3	0.6	0.9	1.2	1.5	1.8	2.1	2.4	2.7	3.0
M10	J_x	250.9	227.6	235.7	264.6	273.8	250.9	227.6	235.7	264.6	273.8
	J_y	250.1	273. 6	265.1	236.6	227.0	250.1	273.6	265.1	236.6	227.0
M20	J_x	5481.9	5341.1	5308.9	5428.7	5536.2	5481.9	5341.1	5308.9	5428.7	5536.3
	J_y	5357.5	5497.4	5530.1	5410.6	5302.6	5357.5	5497.4	5530.1	5410.6	530.6
TR10	J_x	290.8	288.4	296.6	303.6	299.6	290.8	288.4	296.6	303.6	299.6
	J_y	301.2	303.1	294.9	288.3	291.4	301.2	303.1	294.9	288.3	291.4
TR20	J_x	5912.9	5962.4	5940.9	5973.6	5954.4	5912.8	5902.0	5940.8	5973.6	5954.5
	J_y	5902.0	5970.8	5932.3	5901.6	5916.3	5962.4	5970.8	5932.0	5901.3	5916.3
U10	J_x	313.5	328.7	323.4	304.2	297.7	313.5	328.7	323.4	304.2	297.7
	J_y	314.4	297.7	303.8	323.1	328.1	314.4	297.7	303.8	323.1	328.1
U20	J_x	6081.5	6287.4	6236.0	5988.6	5891.6	6081.5	6287.4	6236.0	5988.6	5891.6
	J_y	6121.5	5899.7	5960.8	6208.8	6292.3	6121.5	5899.7	5960.8	6208.8	6292.3

Table 2. Approximation coefficients for different types of thread with a diameter of 20 mm

Coefficients	Metrical			Trapezoidal			Buttress		
	thread pitch, h, mm								
	1	2	3	1	2	3	1	2	3
a, mm^4	25	58	125	5	15	38	94	160	225
J_0, mm^4	6918	6113	5425	7120	6485	5936	7175	6600	6100
φ	0.06	0.11	0.17	0.18	0.4	0.68	0.04	0.06	0.08
π/ω, mm	0.23	0.45	0.77	0.23	0.45	0.77	0.23	0.45	0.77

Table 3. Coefficients for different threads with the same thread pitch of 3 mm

Coefficients	Metrical			Trapezoidal			Buttress		
	diameter of thread, d, mm								
	10	20	30	10	20	30	10	20	30
a, mm^4	25	125	570	8	38	90	17	225	900
J_0, mm^4	250	5425	31340	296	5936	32830	314	6100	33400
φ	0.17	0.17	0.17	0.68	0.68	0.68	0.05	0.05	0.05
π/ω, mm	0.77	0.77	0.77	0.77	0.77	0.77	0.77	0.77	0.77

Table 3 shows that the angular frequency of the harmonic function (1) does not depend on the diameter and thread type.

The initial phase of function (1) is constantly for the screw thread with the same type and the same thread pitch of and is not changed when the diameter changed.

Figure 2 shows the variation of the axial moment of inertia of the screw with thread TR20 and a thread pitch of 3 mm, built by the formula (1) (solid line) and tabular values of table 1 (dashed line).

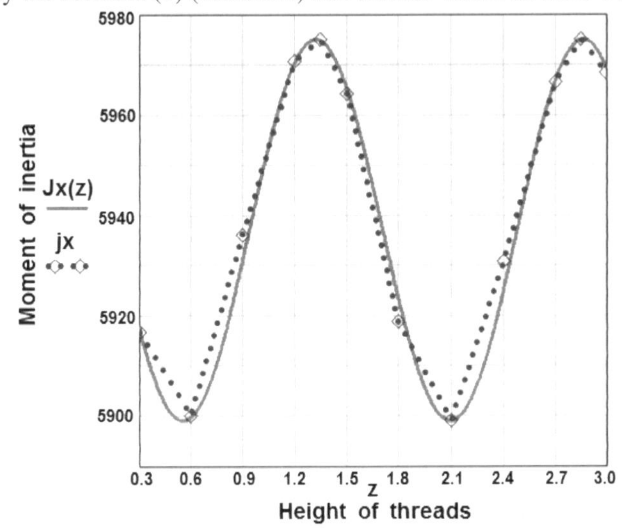

Figure 2. The variation of the axial moment of inertia of the screw TR20×3: the formula (1) - solid line; the table values - dotted line.

The analysis of the Figure 2 allows making a conclusion about high accuracy of approximating the actual table values by formula (1).

Further, we study the influence of bending on the geometrical characteristics of the cross-section of the screw.

IOP Conf. Series: Journal of Physics: Conf. Series **858** (2017) 012028 doi:10.1088/1742-6596/858/1/012028

The longitudinal compressive forces in power screw jacks, screw presses, and the spindles of pipeline gates can achieve big values, when the screws may lose longitudinal stability, as a result, the bending deflections reach high values, and the screw material is experiencing of plastic deformations [4, 5].
As a result of numerical experiment, using a 3D-modeling package SolidWorks-2011, it was found that if the screw is bent, then its moment of inertia in all axial cross sections decreases proportional to the bending deflection (Figure 3).
The results of computer researches of a bending screw TR20×3 and l = 1000 *mm* are given in Table 4.

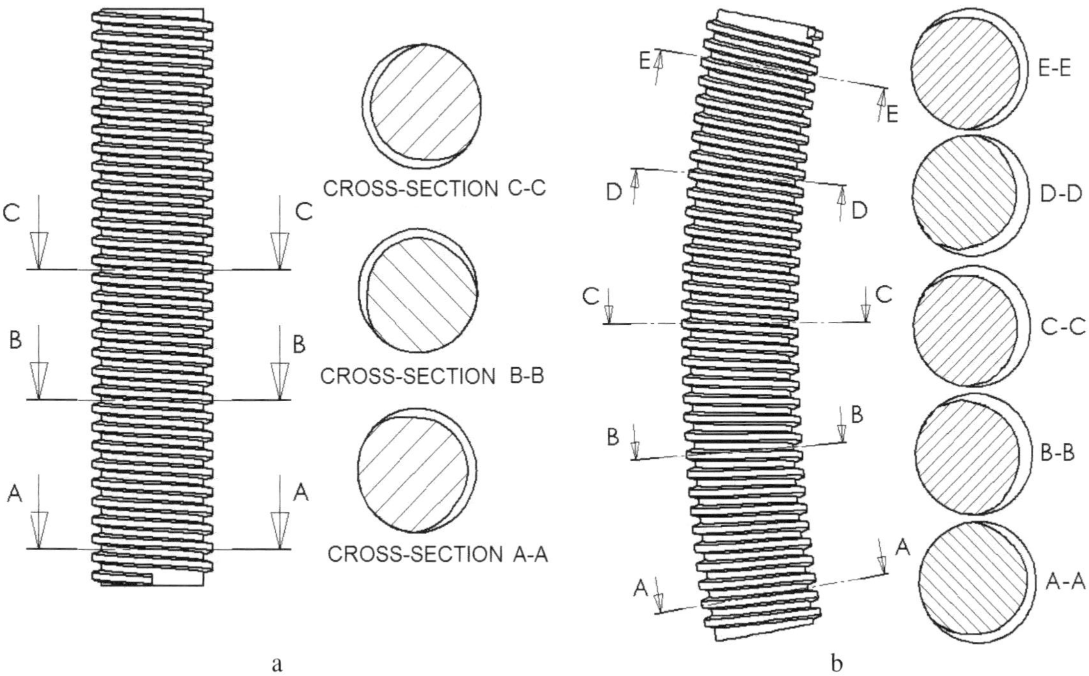

Figure 3. The cross-sections of the screw cavity of the threads: a) straight screw; b) bending screw.

Table 4. Change of the moment of inertia, mm^4, in different sections of screw with the thread TR20×3, length L = 1000 mm, depending on the deflection Y

0	5954.45	5954.45	5954.45
1	5928.12	5930.06	5948.75
2	5896.54	5910.48	5936.36
3	5860.37	5892.23	5920.68
4	5840.43	5878.29	5910.22
5	5808.26	5858.59	5901.31
6	5770.12	5835.17	5891.05
7	5745.45	5815.86	5886.16
8	5710.26	5798.36	5879.24
9	5688.38	5779.62	5869.45
10	5659.24	5758.35	5860.45

4. Experimental results

Analysis of the data showed that the fundamental difference between the bending of the screw from bending smooth rod is that the cross-sectional moment of inertia of the screw is variable. It is shown that the change of the cross-section moments of inertia along the length of the screw is significant and

IOP Conf. Series: Journal of Physics: Conf. Series **858** (2017) 012028 doi:10.1088/1742-6596/858/1/012028

has the periodic in nature. An analytical relationship of the bending of the screw and reducing the cross-section moment of inertia and proposed the equation describing this phenomenon.
The greatest decrease of the cross-section moment of inertia occurs in the middle of the screw and least at its ends.
The adequacy of the developed models is confirmed by the results of experimental researches conducted in limited company "Livgidromash" under the leadership of Professor V. M. Ryazantsev [6].
This phenomenon is due to the fact that during bending of the screw in the stretched region of the threads of fibers are diverged. Moreover, if in the direct screw of the cross-section gets most of the thread, and then in a distorted screw in cross-section gets a smaller part of the thread, thereby decreasing the cross-section moment of inertia of the screw.
The numerical experiment and processing of the obtained results allowed us to propose a convenient approximation of the dependence of changes in cross-section moments of inertia of the screw along its length. The use according to formula (1) more accurately determines the various parameters of screw mechanisms, such as the deflection of screws under transverse bending or characteristics to ensure their stable position of equilibrium at buckling.

5. Discussion of results

Analysis of the data leads to the conclusion that although the experimental and calculated values are close enough; however, the approximation coefficients «*a*» for various cross-sections differ significantly.
Comparison of experimental and calculated data of changes the moment of inertia in different cross-sections is shown in Figure 4.
It is proposed to use in the calculations the average value, for example, at $l = 300$ mm (see Figure 4, middle line). It can be concluded that the screw is less resistance to bending than a smooth rod the same size and, as a function describing the change of the moment of inertia of half the length of the screw, we propose to use (2):

$$J_0(y, z) = J_0 - a \cdot z \cdot y \tag{2}$$

where a, mm^2, J_0. mm^4 - approximation coefficients, y – deflection, mm.

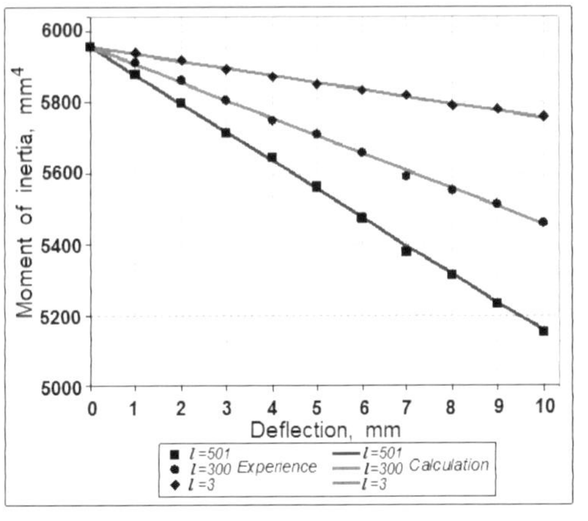

Figure 4. The results of the researches of screw with thread TR20×3.

6. Conclusions

Thus, the functions describing the change of the cross-section moments of inertia of the along the length of the screw are proposed, it is shown that they are periodic in nature. The analytical relationship of the bending of the screw and reducing the cross-section moment of inertia installed and the equation (2) is proposed to account for this phenomenon. Developed method may be useful for a more accurate calculation of the power screw mechanisms.

IOP Conf. Series: Journal of Physics: Conf. Series **858** (2017) 012028 doi:10.1088/1742-6596/858/1/012028

References

[1] Panic D, Beier Th H and Vormwald M 2014 Damage assessment of threaded connections based on an advanced material model and local concepts *Procedia Engineering* **74** 119 – 128
[2] Kanga H, Yanga J and Menga X 2015 Tests and analysis of mechanical behaviours of rock bolt components for China's coal mine roadways *J. of Rock Mechanics and Geotechnical Engineering* **7(1)** 14–26
[3] Noda N-A, Chen X, Sano Y, Wahab M A, Maruyama H, Fujisawa R and Takase Y 2016 Effect of pitch difference between the bolt–nut connections upon the anti-loosening performance and fatigue life *Materials & Design* **96** 476-489
[4] Yu H, Niu Z, Kong L, Hao C and Cao P 2015 Mechanism and technology study of collaborative support with long and short bolts in large-deformation roadways *Int. J. of Mining Science and Technology* **25(4)** 587-593
[5] Proshiriakov N E, Lopa I V 2014 Effect of thread on transverse bending screw *Izvestiya of the Tula State University. Technical Sciences* **11(2)** 472-476
[6] Ryazantsev V M 2005 Screw pumps with cycloidal toothing *Mashinostroenie* p 346

Two dimensional measuring methods of geometrical adjectives of cylindrical pins

V V Pshenichnikova[1], D B Martemyanov[1] and V A Penner[1]

[1] Omsk State Technical University, 11, Mira ave., Omsk, 644050, Russia

E-mail: pshenichnikova@omgtu.ru

Abstract. Both instrumental errors and method errors of measuring tools, depending on reference values deviation, should be taken into account in measurement accuracy tolerance calculation. Uncorrected form deviations of measuring surfaces originate from dimensions procedural errors. A geometrical representation of a smooth cylindrical pin specifying primary errors, coordinate systems and bases is developed. The measuring methods of cylindrical pin form deviation in longitudinal and cross sections are developed based on this model. According to the measurement data realistic pin surfaces are constructed. The form deviation measurement rules are proposed. The mentioned measurement methods reveal procedural errors of measurement that depend on the kind of form deviation and error values.

1. Introduction

Operating standards specify many geometrical adjectives reference systems such as spacing and form deviation, linear and angular dimensions, so that geometry precision of components and goods cannot be normalized in complex. This is resulted in quality deterioration of machine and instrument-making production still at the stage of the design process. Both Russian and foreign specialists conduct researches of smooth cylindrical surface mating behavior. Little attention is paid to geometrical representations and measurement methods development of pin geometrical adjectives.

Geometry precision of components and goods in the manufacturing process should be controlled by the measurement technique (measurement methods) defined in Standards being the principle metrological document. A measurement method is specific and developed for a particular measuring object. It should include specific bases, systems of reference, geometrical adjectives composition of every component, a measuring object model, measurement design and certificate conforming measurement data authenticity based on measurement error estimation of each value. Hence, the method of measurement is to be indicated in an operational procedure.

2. Problem statement

Paper [1] describes the technique of linear and angular elemental and coordinates dimensions, component parts spacing and form deviation formed by plane and cylindrical surfaces. The technique is represented by the specific examples of pin connections. The deviations of spacing and form are shown to affect the fit class and assemblability of compound assembly units based on geometrical models of pin connections. The exclusion methods of locating redundancy with parts mating relative to one another (cover and shell joining) using two pins are proposed.

IOP Conf. Series: Journal of Physics: Conf. Series **858** (2017) 012029 doi:10.1088/1742-6596/858/1/012029

The precision of components and goods is formed and revealed at all stages of a product life cycle: design – manufacture – control – operation.
The metrology provision of components geometrical adjectives measurement should be coupled with the metrology provision of a design process and manufacturing technology. It should be supported by proper regulation of value tolerances in coordinate systems generated by set of bases taking into account their informativeness based on static location definability.
Thus, a special emphasis should be made on the measurement error estimation as the primary component of a procedural error, which estimates specifically for a particular measurement and is not included into the structure of measurement permissible errors.

3. Theory

A measurement method should correspond to the standard term of deviation and be developed based on the construction of the measuring object geometrical model. The measuring object geometrical model should include all dimensional and geometric initial errors of the measuring object and two sets of bases. They are a generalized coordinate system of a component and an auxiliary coordinate system where an examined element fulfills its operational function. Based on the pin geometrical model (Figure 1), the measurement methods of a smooth cylindrical pin are developed [2].

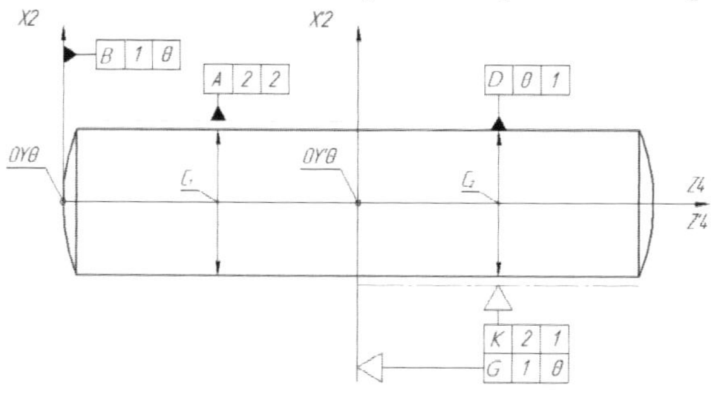

Figure 1. Smooth cylindrical pin: bases and coordinate system.

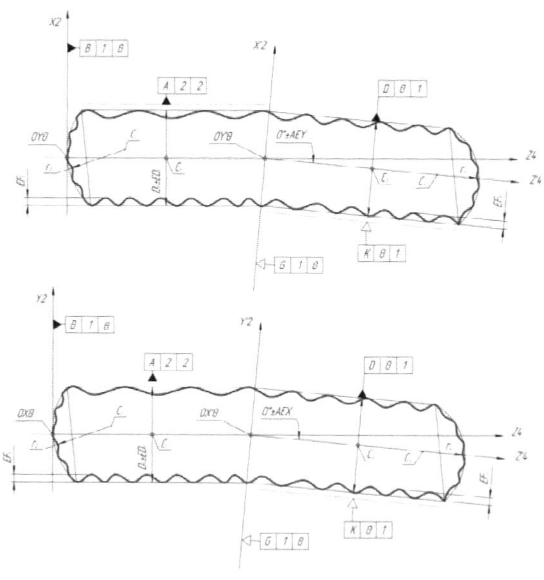

Figure 2. Smooth cylindrical pin: geometrical model.

IOP Conf. Series: Journal of Physics: Conf. Series **858** (2017) 012029 doi:10.1088/1742-6596/858/1/012029

All actual surfaces have form deviations, these surfaces having proper dimensions and tolerances of AUSS (GOST) P 53442-2009 (IOS 1101:2004). Form deviations are the deviations from a proper geometrical form of a surface in longitudinal and cross sections. More accurate measurement can be made in measuring statically definable components due to error of locating exclusion.

The primary errors of pin elements should be taken into account for a component geometrical model to be developed and for all deviations to be recorded. Measurement data derived through cylindrical pin measurements are required for the provision of components and goods quality. The measuring surfaces themselves are the measuring form deviations measurement bases. The deviation readout takes place on adjacent (actual) surfaces. Surface undulation and irregularity should be excluded by selecting the radius of a gaging tip in measuring form deviation. The above said allows the procedural errors of measurement to be excluded or significantly reduced.

The pin form deviation measurement schemes include the roundness measuring diagram (Figure 3), cross section measuring diagram (cylindricity deviation) (Figure 4) and pin diameter measuring diagram (pin axis straightness) for diameters minimum (L) and maximum (M) materials to be found (Figure 5).

4. Experimental results

The crossing profile form deviation measuring diagram (cylindricity deviation) is shown in Figure 3. The nature of the crossing profile form deviation is determined by selecting the measuring scheme. The two-point measuring scheme can detect out-of-roundness, and faceting can be detected exclusively by three-point measuring scheme.

Figure 3 shows the three-point measuring diagram. A narrow prism gives two points of contact with the pin surface. A plunger of the dial gage gives the third contact point with the pin surface. All three points are located in one cross section plane, which is perpendicular to the pin base axis (base A, Figure 1).

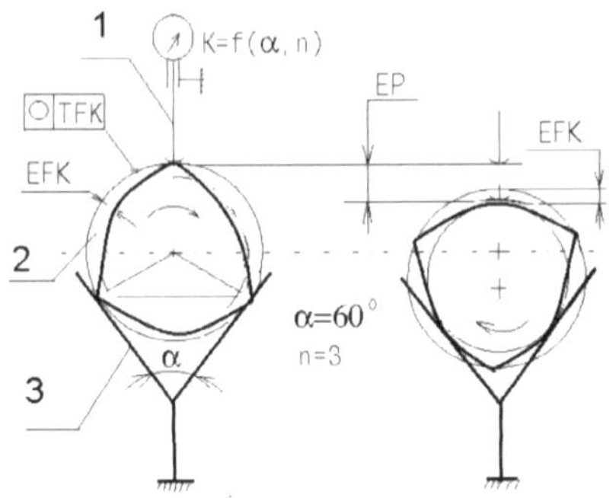

Figure 3. Measuring diagram of pin crossing profile deviation (cylindricity deviation): 1 – measuring transducer; 2 – unit under test (pin); 3 – prism.

The prism angle depends on the number of pin surface faces (n). As a rule the measurement line coincides with the prism bisector. Depending on the prism angle and the number of faces, the calibration constant of diagram K is selected and the form deviation value is calculated as the difference between the biggest and the least readout in one section in a single pin revolution: EFfacet=(Pmax-Pmin)/K=EP/K.

The procedure of the pin crossing profile form deviation measurement (cylindricity deviation):

- a pin is set on the two narrow prisms with the angle α;
- a digital indicator is mounted on a rack and the instrument is tuned for measuring (the tuning reference is a cylindrical measure);
- measurements on six or seven points at three cross sections are carried out, two of them being boundary and one being central, within the mating surface;

IOP Conf. Series: Journal of Physics: Conf. Series **858** (2017) 012029 doi:10.1088/1742-6596/858/1/012029

The boundary values of the pin diameter and the faceting with n number of faces are determined by the results obtained.

- the results of measurement are entered into the report.

The five point scheme is required for the cross section pin form deviation to be measured (Figure 4). Five points are measured, the four are formed with the two narrow prisms and the fifth is formed with the plunger of the dial gage. The prisms are spaced at the extreme pin sections as far as the mating length is the length of the pin. All three points of the measuring scheme should be spaced in the cross section plane and be contacted with the one generator cylindrical surface for the cylindrical deviation (cylindrical part axis straightness) to be measured. The cylindrical deviation values are calculated by the difference method as the semidifference of the biggest and the least values: EFcyl=0.5EFPP. The measurement treatment can be simplified if the measuring transducer is tuned for zero on one of the sections, although the measurement time increases in this case.

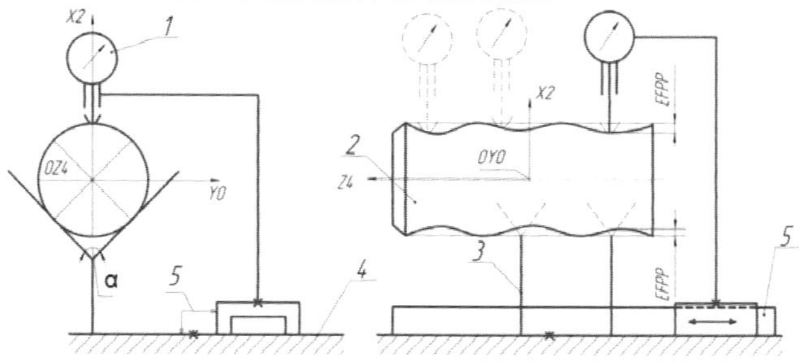

Figure 4. Pin crossing profile measuring diagram (straightness deviation): 1– indicator; 2 – unit under test (pin); 3 – prism; 4 – testing plate; 5 – toolmaker's straightedge.

The procedure of longitudinal section pin form deviation measurement:

- a pin is set in the prism on the angle α;
- a measuring transducer is mounted on a rack and the instrument is tuned for measuring;
- measurements on three points, at three cross sections are carried out, measurement is carried out by a workpiece rotation;
- the results of measurement are entered into the report.

In such schemes the measuring surface of testing plate is the degree of rectilinearity, its strain making force measurement error contribution affected by the workpiece and rack gravity, and that should be taken into account.
A pin is a workpiece with an external cylindrical base having informativeness which equals to four. It is based on two narrow prisms reifying four base points for rotation.
The diameter measuring procedure of minimum (L) and maximum (M) pin materials is described schematically (Figure 5):

- a pin is set in the prism on the angle 90°;
- a measuring transducer is mounted on a rack and the instrument is tuned for measuring;
- measurements on three points, at three cross sections are carried out;
- the results of measurement are entered into the report.

The biggest and the least cylindrical pin diameters are located on the two regular cylinders encircling actual pin surface. In this case the base axis is the axis of an external (maximum d M) cylinder which participates in components mating (with an aperture in a plane workpiece) with its exposed points.

IOP Conf. Series: Journal of Physics: Conf. Series **858** (2017) 012029 doi:10.1088/1742-6596/858/1/012029

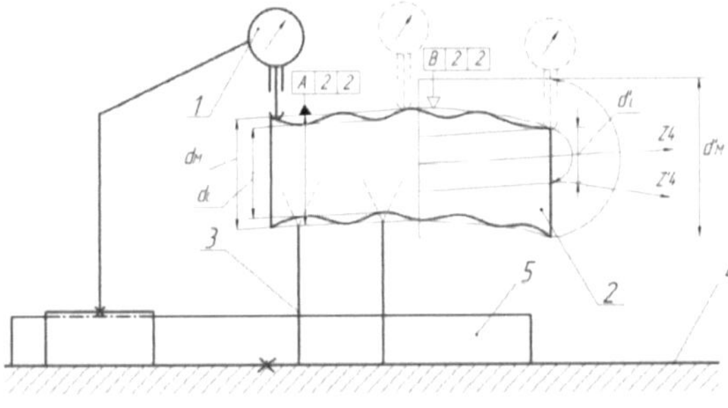

Figure 5. Pin actual diameter measuring diagram. 1– measuring transducer; 2 – unit under test (pin); 3 – prism; 4 – plate.

Rectangular coordinates are exclusively used for defining geometrical adjectives of a smooth cylindrical pin. Measurements of cross section prolife and rounding-off radii in two sections are used for defining smooth cylindrical pin geometric adjectives.

The pin longitudinal section profile coordinates and graphical representation of realistic longitudinal section profile and rounding-off radii according to the measurement results obtained, are shown in Figure 6.

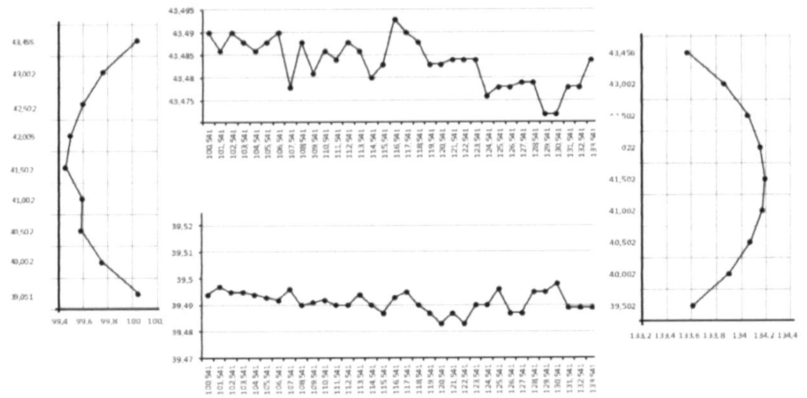

Figure 6. Graphs of longitudinal section actual profile and rounding-off radii of the left and the right pin radius.

Similar measurements were performed after turning the pin through 90° about axis Z, the results are shown in Figure 7.

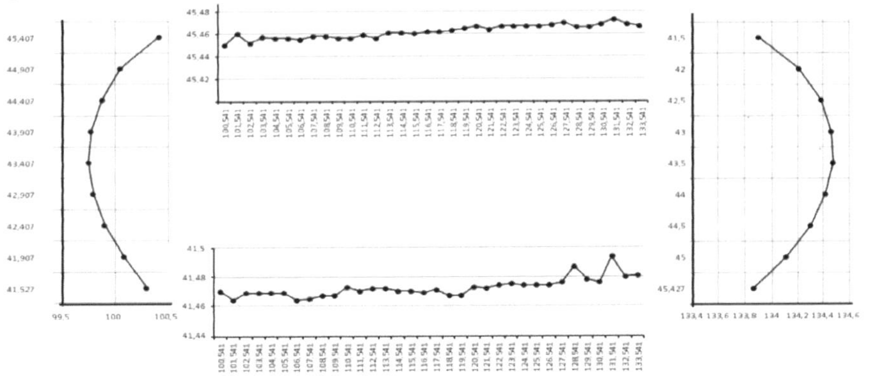

Figure 7. Graphs of longitudinal section actual profile and rounding-off radii of the left and the right pin radius after turning through 90 °.

5. Experimental result discussion

In the view of the foregoing the rules of the form deviation measurement are brought up for discussion:

- the schemes of form deviation measurement should strictly correspond to standardized definitions;
- the form deviation measurement means finding its maximum value within the limits of a normalized range, a mating length;
- the measurement schemes are developed based on the geometrical model of the measuring object;
- the main assembly bases should provide orientation of the measuring object in space of the measuring instrument the same way as the workpiece is used in operation;
- measuring bases are measuring surfaces and readout deviation bases are the adjacent surfaces;
- surface undulation and irregularity should be excluded by selecting the radius of a gaging tip.

6. Summary and conclusion

A realistic pin model differs from its nominal prototype by the dimensions deviation, the element surface form deviation from the proper geometrical form and these elements deviation relative to workpiece bases due to inevitable manufacturing process errors. Measuring forms of deviation using different measuring schemes will produce different results. Thus, measuring schemes with standardized instruments allow two actual dimensions to be found – the biggest and the least pin diameter, but they produce measuring procedural errors that depend on the forms deviation kind and their values. Hence, in searching for cross section pin smooth cylindrical surface deviation we should take the number of faceting faces into account and in terms of this one should select the adjusting prism angle and the calibration constant of the measuring method. The mating length and measuring error force component should be taken into account in measuring cross section pin form deviation. A measuring scheme error arises if the geometrical model of the measuring object is not considered, incorrect measuring scheme (to use two point circuit instead of the three point, and not to take into account the mating surface of a workpiece) is used, as well as when the sensing element form of the instrument is incorrectly selected.
Following certain rules in measuring cylindrical surfaces form deviations allows measuring procedural errors to be excluded or substantially reduced. A procedural error can be excluded by introducing a correction, the pin surface form deviation having been preliminary measured.

References

[1] Pshenichnikova V V 2014 Geometrical product specifications: Basing precision of pin connection *Dynamics of Systems, Mechanisms and Machines (Dynamics),* IEEE Conf. Publications pp 1–4
[2] Glukhov V I 2014 Geometrical product specifications: Alternative standardization principles, coordinate systems, models, classification and verification *Dynamics of Systems, Mechanisms and Machines (Dynamics),* IEEE Conf. Publications pp 1–9
[3] Ped S E and Darienko E V 2011 Error appraisal of coordinate metrology of cylindrical surfaces section parameters *Measuring equipment* **8** 17–19
[4] Qiaoa L, Wua J, Zhua Z and Cuia Y 2016 Approach to the deviation representation of non-ideal cylindrical surfaces based on the curvilinear coordinate system. *14th CIRP Conf. on Computer Aided Tolerancing (CAT)* vol 43 pp 17-22

[5] Schleich B, Anwer N, Mathieu L and Wartzack S 2015 Contact and mobilitysimulation for mechanical assemblies based on skin model shapes *J. of Computing & Information Science in Engineering* **15(2)** 021009-021009-7

[6] Zhang X D, Zhang C, Wang B and Feng S C 2005 Unified functional tolerancing approach for precision cylindrical components *Int. J. of Production Research* **43(1)** 25-47

[7] ISO 841:2001 Industrial automation systems and integration. Numerical control of machines. Coordinate system and motion nomenclature

[8] ISO 286-1:2010 Geometrical product specifications (GPS) – ISO code system for tolerances on linear sizes – Part 1: Basis of tolerances, deviations and fits

IOP Conf. Series: Journal of Physics: Conf. Series **858** (2017) 012030 doi:10.1088/1742-6596/858/1/012030

Analytical study of the cord path character in rubber-cord flexible elements

V AShepetkov[1], A OZvonov[1] and AGYanishevskaya[2]

[1] FSUE "RPE "Progress", Omsk 644018, Russia

[2]Omsk State Technical University, Omsk 644050, Russia

E-mail: anna-yanish@mail.ru

Abstract. The mathematical model of the cord path character in rubber-cord flexible elements is investigated. The applicability boundaries of known classical expressions for the determining of the cord path character are analysed. It is shown that the nominal initial cord angle on the assembly drum exists and is uniquely determined for the given assembly drum radius, the element carcass radius as well as the ply inclination to the meridian at the element surface equator. Sinus multiplication formulas which relate the initial cord angle on the drum, fiber tilt to the manufactured element meridian and the fiber tilt at an arbitrary point of manufactured flexible element carcass are obtained. The dependence of the technological constant on the initial ply angle is analysed. The coefficient of dependence between the ply angle and the effective radius in manufactured flexible element are obtained.

1. Introduction

Rubber-cord flexible elements (RCE) are widely used due to their elastic properties, which allow considerable mechanisms movement under the influence of external loads [1].

The RCE carcass consists of several layers or plies of rubberized cord forming a fiber network where the plies are superimposed at different angles. Usually, the carcass is a surface of revolution, working under the distributed loads and loads, caused by the axial, radial and other forces in its embedding.

This model case is well described in terms of pneumatic tires theory with its basic equations [2–5] which can be used to:

–determine the fiber tilt angle β at any point of the manufactured element if the angle $\beta_k$at the element surface equator (see fig. 1) is known:

$$\sin\beta = \frac{R}{R_\kappa}\sin\beta_\kappa \tag{1}$$

– determine the slope angle of the normal at an arbitrary point of the element surface

$$\sin\varphi = \frac{(R - R_o^2)\cos\beta}{(R_k^2 - R_o^2)\cos\beta_k} \tag{2}$$

In the expressions (1) and (2), R_0 is the effective radius; R_k is the radius of the element equator; R is the radius value at an arbitrary point of the element surface

Expressions (1–2) are obtained by using the formula

$$\sin\beta_\kappa = \frac{\sin\alpha}{R_{SB}(1+\delta)}R \tag{3}$$

In expression (3)

$$\chi = \frac{\sin\alpha}{R_{SB}(1+\delta)} \tag{4}$$

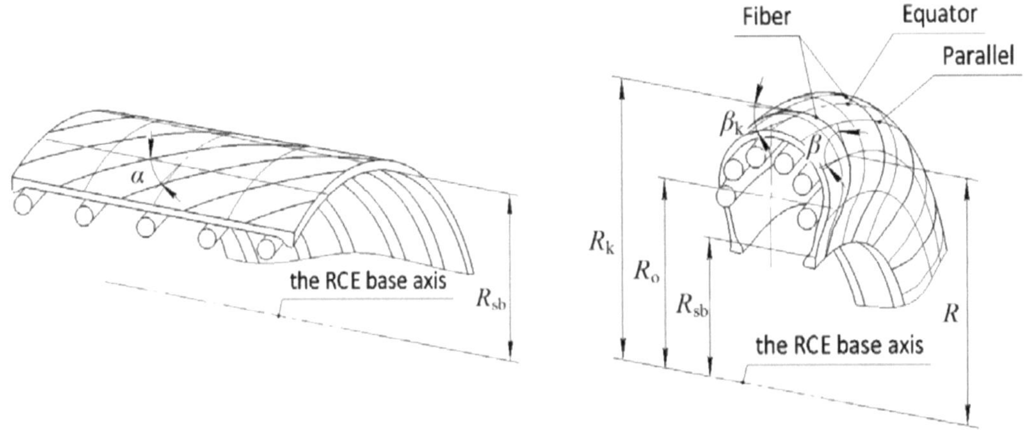

Figure 1. The RCE carcass geometry: on the assembly drum (a) and in manufactured RCE (b)

– technological constant (TC), determined for the fixed inclination α to the meridian of the assembly drum with R_{sb} radius and the cord stretch δ.

Taking into account (4), the law of the fibers arrangement in the fiber network (on the surface of the element) is obtained:

$$\sin\beta = \chi R. \tag{5}$$

The above equations have their own applicability boundaries, which are primarily related to the element and the assembly drum dimensions and the fiber inclination on the assembly drum.

2. The application range of the tire geometry basic equations

The first step is to determine the pneumatic tires theory basic assumptions which are limiting its usage [6, 7].

Let us consider the RCE condition on the assembly drum. The conditional cord stretch δ is taken equal to zero. We will also assume that the element manufacturing process is not causing the fibers elongation.

The change in technological constant χ is due to a change in its argument α, determined on a finite interval $[0, \pi/2]$:

$$\chi(\alpha) = \frac{\sin\alpha}{R_{SB}} \tag{6}$$

If R_{sb} is measured in meters, the base unit of χ will be [1/m]. The assembly drum radius R_{sb} is constant, therefore, the minimum and maximum values of $\chi(\alpha)$ are attainable at the end points of the α variation interval. The minimum value is reached when α_o=0, in this case $\chi(0) = 0$. The maximum value is reached for $\alpha = \pi / 2$:

$$\sin\frac{\pi}{2} = 1. \tag{7}$$

Wherein $\chi(\alpha)$ is the assembly drum radius reverse value:

$$\chi\left(\frac{\pi}{2}\right) = \frac{1}{R_{SB}}. \tag{8}$$

IOP Conf. Series: Journal of Physics: Conf. Series **858** (2017) 012030 doi:10.1088/1742-6596/858/1/012030

Thus, the graph of $\chi\ (\alpha)$ monotonically increasing from zero to the assembly drum radius R_{sb} reverse value:

$$0 \le \chi(\alpha) \le \frac{1}{R_{SB}} < 1 \tag{9}$$

Reverse values of $\chi\ (\alpha)$ will have the base unit [m] and can be represented as

$$R_p(\alpha) = \frac{1}{\chi(\alpha)} = \frac{R_{SB}}{\sin\alpha}, \tag{10}$$

where $R_p(\alpha)$ is the calculated reverse value of the $\chi\ (\alpha)$ from expression (10) . It should be noted that

$$R_p(\alpha)\chi(\alpha) = 1. \tag{11}$$

Maximum of $R_p(\alpha)$ grows without bound while α approaches zero; minimum value is achievable when

$$\alpha = \frac{\pi}{2},\ \sin\alpha = 1. \tag{12}$$

Thus, the $R_p(\alpha)$ reaches a minimum value equal to the assembly drum radius R_{sb}:

$$[R_p(\alpha) \to \infty\ when\ \alpha \to 0] \ge R_p(\alpha) \ge \left[R_p\left(\frac{\pi}{2}\right) = R_{SB}\right]. \tag{13}$$

Under given inclination to the meridian at the element surface equator β_k there is the following dependence between the calculated value R_p and the RCE equator radius R_k:

$$R_\kappa(\alpha) = R_p(\alpha)\sin\beta_\kappa. \tag{14}$$

3. Nominal angle of the cut on the assembly drum

Under given angle β_k, there is an initial fiber inclination on the assembly drum α_H for which the following relation comprising the $\sin\beta_k$, R_k and R_{sb} radius exists:

$$\sin\beta_\kappa = \frac{R(\alpha_н)}{R_{SB}}\sin\alpha_н. \tag{15}$$

Angle value α_H satisfying the equation (15) for given R_{sb}, R_k and $\sin\beta_k$ will be called *the nominal initial fiber inclination* on the assembly drum or simply *the initial angle*.

If put the equator radius $R_k(\alpha_H)$ (corresponding to theinitial angle α_H) in correspondence to the following value of calculated radius $R_p(\alpha_d)$:

$$R_p(\alpha_d) = R_K(\alpha_н), \tag{16}$$

then, with an account to (10), the following equation will take a place:

$$R_p(\alpha_d) = \frac{R_{SB}}{\sin\alpha_d}. \tag{17}$$

By equating (16) to (17), we obtain

$$\sin\alpha_d = \frac{R_{SB}}{R_K(\alpha_н)}, \tag{18}$$

$$R_{SB} = R_K(\alpha_н)\sin\alpha_d \tag{18.1}$$

and finally

$$\alpha_d = \arcsin\frac{R_{SB}}{R_K(\alpha_н)}. \tag{18.2}$$

The obtained equation (18.2) allows the calculation of the fiber inclination α_d under given values of the assembly drum radius R_{sb} (or diameter D_{sb}) and radius R_K (or diameter D_K) of the RCE carcass.

4. The interconnection between the angles α_H, α_D and β_K

With the obtained expression (18.2), the RCE carcass geometry can be fully described by its fiber angles. Converting (15), using expression (18), we obtain

IOP Conf. Series: Journal of Physics: Conf. Series **858** (2017) 012030 doi:10.1088/1742-6596/858/1/012030

$$\sin\beta_\kappa = \frac{\sin\alpha_\textit{и}}{\sin\alpha_\mathrm{d}} \tag{19}$$

or

$$\sin\alpha_\textit{и} = \sin\alpha_d \sin\beta_\kappa . \tag{20}$$

The expression (21) can be represented as following:

$$\alpha_\textit{и} = \arcsin(\sin\alpha_d \sin\beta_\kappa). \tag{21}$$

In expression (21), the fiber inclination to the meridian on the assembly drum is less than 1. Consequently, the angle exists and is uniquely determined for a given radius of the assembly drum R_{sb}, the radius of the equator R_k, and the value β_k of the fiber inclination to the meridian of the manufactured RCE.

The expression (20) will be called **the first sinus multiplication** formula which determines the sine of the fiber nominal angle on the assembly drum.

All obtained expressions were considered for a constant value of fiber inclination to the manufactured RCE meridian and different values of fiber inclination on the assembly drum. It also allows to formulate a number of new equations-consequences, which are given below.

Implication 1. Similarly to formula (20), expression

$$\sin\alpha_m = \sin\alpha_d \sin\alpha_\textit{и} \tag{22}$$

will be **the second sinus multiplication** formula. It determines the smallest value α_m of the fiber inclination on the assembly drum.

The expression (22) can be represented as following:

$$\alpha_m = \arcsin(\sin\alpha_d \sin\alpha_\textit{и}) . \tag{23}$$

Converting (22), using expression (20), we obtain

$$\sin\alpha_m = \sin^2\alpha_d \sin\beta_\kappa . \tag{24}$$

General case is the formula

$$\sin\alpha_{m_j} = \sin^{j+1}\alpha_d \sin\beta_\kappa \tag{25}$$

where j is integer.

Implication 2. The third sinus multiplication formula has the following forms:

$$\sin\alpha_g = \sin\alpha_\text{н} \sin\beta_\text{к}, \tag{26}$$

$$\sin\alpha_g = \sin\alpha_\text{d} \sin^2\beta_\text{к}. \tag{27}$$

The position of the considered point α_g with respect to the point α_m on the interval [0, π / 2] is defined by the following equations:

$$((\alpha = \beta_\kappa) > \alpha_\text{d}) \to (\alpha_g > \alpha_\text{m}), \tag{28}$$

$$((\alpha = \beta_\kappa) < \alpha_\text{d}) \to (\alpha_g < \alpha_\text{m}). \tag{29}$$

Implication 3. The interval [0, π / 2] contains a point where

$$\alpha = \beta_\kappa, \tag{30}$$

and the following relation occurs:

$$\frac{\sin\beta_\kappa}{R_\kappa(\alpha)} = \left(\frac{\sin\alpha}{R_{SB}} = \frac{\sin\beta_\kappa}{R_{SB}}\right). \tag{31}$$

Taking into account (30), theequation (31) can be represented as following:

$$R_{SB} = R_K(\alpha) = R_\kappa(\beta_\kappa). \tag{32}$$

5. Change of technological constant

Consideration of the technological constant as the angle α function leads to a visual table 1 of the geometric parameters at the considered points corresponding to certain values of the angle.

IOP Conf. Series: Journal of Physics: Conf. Series **858** (2017) 012030 doi:10.1088/1742-6596/858/1/012030

In the table 1 the geometric values of the RCE equator radius and assembly drum radius in $R_k(\alpha)$ function correspond to the specified initial conditions, while the fiber inclination to the RCE surface equator and initial fiber inclination on the assembly drum are interchanged relatively to named radiuses.

Table 1. The RCEgeometric parameters under $\alpha_\beta > \alpha_d$

α	0	α_m	α_g	α_H	α_d	α_β	$\pi/2$
$\sin\alpha$	0	$\sin\alpha_m$	$\sin\alpha_g$	$\sin\alpha_H$	$\sin\alpha_d$	$\sin\alpha_\beta$	1
$\chi(\alpha)=\dfrac{\sin\alpha}{R_{SB}}$	0	$\chi(\alpha_m)$	$\chi(\alpha_g)$	$\chi(\alpha_H)$	$\chi(\alpha_d)$	$\chi(\alpha_\beta)$	$1/R_{sb}$
$R_K(\alpha)=\dfrac{\sin\beta_K}{\chi(\alpha)}$	∞	$\dfrac{R_K(\alpha_H)}{\sin\alpha_d}$	$\dfrac{R_{SB}}{\sin\alpha_H}$	$R_k(\alpha_H)$	$R_k(\alpha_H)\sin\beta_k$	R_{sb}	$R_{sb}\sin\beta_k = R_k(\alpha_H)\sin\alpha_H$

To match the angles β on the RCE surface with their radiuses, the following transformation can be applied:

$$R_r(a) = R_{SB}\frac{\sin\alpha}{\sin\alpha_m}, \tag{33}$$

Where $R_r(\alpha)$ gives the radius values on the RCE surface, arranged in ascending order and corresponding to the fiber inclination values on the assembly drum.

Matched angle β is determined by the expression

$$\beta = \alpha\sin(\chi(\alpha_H)\mathrm{R}_r(\alpha)). \tag{34}$$

6. The RCE effective radius

For further consideration of the RCE geometric parameters, three specific radiuses determining the RCE shape as shown in fig. 2 can be pointed out.

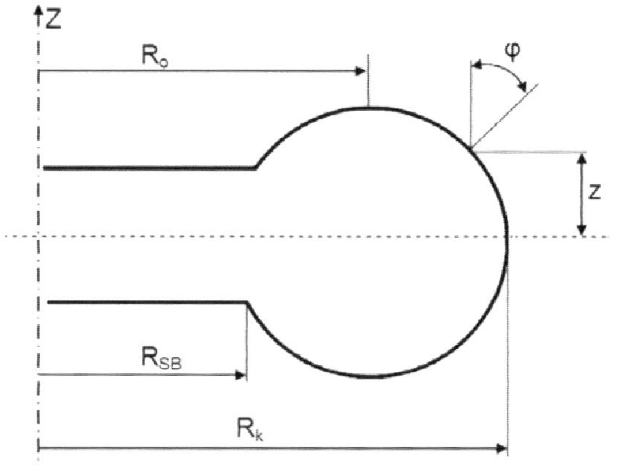

Figure 2. The RCE specific radiuses:
R_o – effective radius;
R_{sb} – assembly drum radius;
R_k – RCE equator radius.

By transforming equation (2) with respect to the effective radius R_o, under $R(\alpha_d) = R_d$,we obtain that

$$R_o^2 = \frac{R_d^2\cos\beta_K + R_{SB}^2\cos\alpha}{\cos\beta_K + \cos\alpha}. \tag{35}$$

The next designations will be useful:

$$C_1 = \frac{\cos\alpha}{\cos\alpha + \cos\beta_K}; \tag{36}$$

$$C_2 = \frac{\cos\beta_K}{\cos\beta_K + \cos\alpha}. \tag{37}$$

It should be noted that

$$C_1 + C_2 \equiv 1. \tag{38}$$

By using (36)-(38), the equation (35) can be formed as

$$R_o^2 = (R_{SB}^2 - R_d^2)C_1 + R_d^2, \tag{39}$$

or for C_2:

$$R_o^2 = (R_d^2 - R_{SB}^2)C_2 + R_{SB}^2. \tag{40}$$

The obtained formulas are equivalent and allow estimation of the effective radius R_0 changes caused by changing of angle α from zero to α_d.
Consideration of the α changing interval starting point obtains: $\alpha = 0$, $\beta = 0$ while $C_1 = C_2 = 0{,}5$.
For both formulas (33) - (34) the following expression is actual:

$$R_o = \sqrt{\frac{R_{SB}^2 + R_d^2}{2}} = \frac{R_{SB}}{\sqrt{2}}\left(1 + \frac{1}{\sin^2\alpha_d}\right) \tag{41}$$

Consideration of the α changing interval end point obtains: $\alpha = \alpha_d$, $\beta = \pi/2$ while $R_r = R_H$, $C_1 = 1$, $C_2 = 0$ and $R_o = R_{sb}$.
Thus, when the angle α changes from zero to α_d, the effective radius decreases with a coefficient

$$K_{R_o} = 0{,}7071\left(1 + \frac{1}{\sin^2\alpha_d}\right). \tag{42}$$

7. Conclusion

The obtained equations can be used during the RCE design process for the selection of the optimal fiber initial angle under given angles in manufactured RCE. Analysis of the given expressions applicability boundaries shows the character of RCE carcass working conditions changing depending on the initial angles laid during the design. The angles exit beyond the permissible values leads to loss of the RCE structural stability under the influence of external loads [8].

8. References

[1] Tribel'skii I A and Zubarev A V 2008 Strength and working parameters of covers with rubber-cord casings of different designs *Russian Engineering Research* **28(7)** pp632–635. DOI:10.3103/S1068798X08070022.
[2] Paris A, Costello G 1999 Bending of Cord Composite Cylindrical Shells *J. Appl. Mech* **67(1)** 117–127 DOI: 10.1115/1.321156.
[3] Clark S, Dodge R 1990 Nonlinear Cord-rubber Composites *Tire Sci. and Tech* **18(3)** pp 191–200. DOI: 10.2346/1.2141699.
[4] Entwistle K 1981The behaviour of braided hydraulic hose reinforced with steel wires *Int. J. of Mech. Sci.* **23(4)**pp 229–241. DOI: 10.1016/0020-7403(81)90048-5.
[5] Zorowski C, Dunn S and Shah N 1971 The Influence of Geometric and Cord-Load Design Parameters on the Profile Shape and Cord Tensions in Inflated Pneumatic Tires, *Textile Res. J.* **41(6)**pp 501-509 DOI: 10.1177/004051757104100605.
[6] Besselink I, Schmeitz A and Pacejka H 2010 An improved Magic Formula/Swift tyre model that can handle inflation pressure changes *Vehicle Sys. Dyn.* **48(1)**pp 337-352 DOI: 10.1080/00423111003748088.
[7] Tang Z, Weng X, Zhu S and Lou J 2011Analysis of the cord layer's influence on the air spring's lateral stiffness *Remote Sensing, Environment and Transportation Engineering (RSETE)*pp 2908-2911 DOI: 10.1109/RSETE.2011.5964922.
[8] Rao S, Daniel I and Gdoutos E 2004 Mechanical properties and failure behaviour of cord/rubber composites *Applied Composite Mat.,* **11(6)**pp 353–375 DOI: 10.1023/B:ACMA.0000045312.61921.1f.

IOP Conf. Series: Journal of Physics: Conf. Series **858** (2017) 012031 doi:10.1088/1742-6596/858/1/012031

Accounting the mechanical resistance of testing device for straight-flow branch pipe in the transverse direction

D V Sitnikov, S V Klishin[2], A V Zubarev[2], A A Shvarts[2]

[1]Omsk State Technical University, 11, Mira ave., Omsk, 644050, Russia
[2]FSUE «FRPC «Progress», 4, 5-th Kordnaya, Omsk, 644018, Russia

E-mail: d.sitnikov@list.ru

Abstract. The flexible inserts as a rubber-cord branch pipe are used to reduce the vibrations distributed from units of a hydraulic system through the pipelines. The device with completely rigid elements is used for the experimental determination of the stiffness characteristics of the branch pipes that determine their anti-vibration properties. In the paper the study of the mechanical resistances of an experimental device construction is carried out by method of the harmonic analysis in a frequency range of 1 Hz to 1000 Hz. Then these mechanical resistances are compared with the mechanical resistance of the branch pipe with diameter 100 mm obtained experimentally. It is found as a result it is necessary to take into account the value of the device mechanical resistance when the stiffness characteristics of the rubber-cord branch pipes with diameter 100 mm in the transverse direction at frequencies above 735 Hz are determined experimentally.

1. Introduction

During the tests to determine the mechanical resistance of the branch pipes by existing methods the construction of testing device is considered to be completely rigid. Before test the mechanical resistances of the stand elements are determined experimentally (Fig. 1). According to the existing requirements they should be not less 10 times higher than the mechanical resistances of tested branch pipe in the predetermined frequency range (typically from 1 Hz to 1000 Hz). The disadvantage of this method is the reduction of accuracy of the vibration isolation properties determination of the branch pipes and in the some cases – a restriction of the frequency range of the experimental researches. The effect of these disadvantages is enhanced by increasing of the branch pipe stiffness (with increasing the pipe diameter and an operating pressure).

In consideration of the foregoing the study of the input and transfer resistance values of the device at transverse oscillations in a frequency range from 1 Hz to 1000 Hz by the harmonic analysis is carried out in this paper. The module Harmonic Response of software package ANSYS was used for calculation. The same study of the device mechanical resistances for longitudinal oscillations was carried out in the paper [1].

2. Task description

The construction of the device for experimental determination of the mechanical resistances of the straight-flow branch pipes in the transverse direction is shown in the Fig. 1 [2]. The Fig. 1 illustrates: 1

IOP Conf. Series: Journal of Physics: Conf. Series **858** (2017) 012031 doi:10.1088/1742-6596/858/1/012031

– branch pipe; 2 – angle section; 3 – plane; 4 – transmission unit with force sensor; 5 – vibrator; 6 – columns; 7 – elastic suspension; 8 – support construction. The angle sections 2 and the columns 6 are mounted on a heavy plate (it is not shown on the fig.) which is attached to the basis.

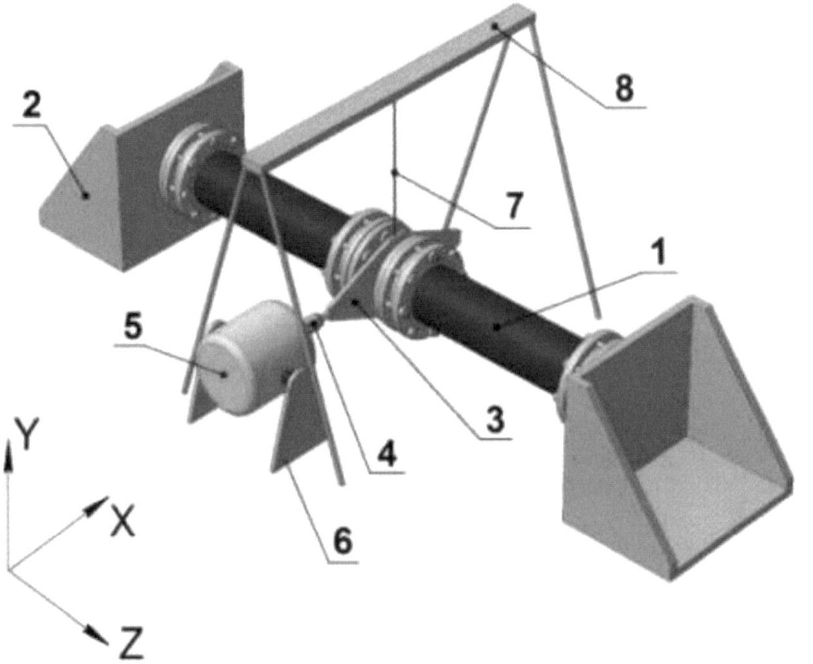

Figure 1. Device scheme for determination of the mechanical resistances of the branch pipes in the transverse direction

In the tests on the transverse oscillations two identical branch pipes are mounted on the device to provide the boundary conditions. The inlet flanges of the branch pipes 1 are fixed with the plate 3 and are driven by the vibrator 5 which affects on the plate through the force sensor 4. The outlet flanges are fixed on the angle sections 2. The accelerometers mounted on the inlet and outlet flanges of the branch-pipes are intended to determinate the transverse oscillations rate of the inlet and outlet flanges of the branch pipe by indirect method.

The mechanical resistances of the branch pipe are determined experimentally as a ratio of force to rate [2]. The input mechanical resistance of the branch pipe

$$Z_{22} = Q_{in} / V_{in} = Q_2 / V_2.$$

The transfer mechanical resistance of the branch pipe

$$Z_{29} = Q_{in} / V_{out} = Q_2 / V_9,$$

where $Q_{in} = Q_2$ – input force,

$V_{in} = V_2$, $V_{out} = V_9$ – the rates of the inlet and outlet flanges of the branch pipe.

Indexing corresponds to a mechanical resistances matrix of the straight-flow branch pipe, and it is given, for example, in [[1]].

Except the branch pipe the oscillations from the vibrator to the angle sections distribute through the device construction (the columns – the plate – the angle sections), which have a finite mechanical resistance. To estimate the value of this resistance in the predetermined frequency range a calculation model was made.

IOP Conf. Series: Journal of Physics: Conf. Series **858** (2017) 012031 doi:10.1088/1742-6596/858/1/012031

3. Calculation model

The calculation of the device mechanical resistance was carried out by harmonic analysis in the software package ANSYS using the module HarmonicResponse [8]. The calculation model of the device with a finite-element mesh is shown on the Fig. 2. The boundary conditions: the lower surface of the plate – an anchorage. The connections: the contact surface of the angle sections and the columns are «adhered» on the plate surface (bolted connection is used practically). Loading: harmonic force with amplitude Q_{in} = 1000 N acts in X-direction and is applied to the holes in the column (uniformly is distributed over cylindrical surfaces of the holes). The number of mesh cells is 104202.

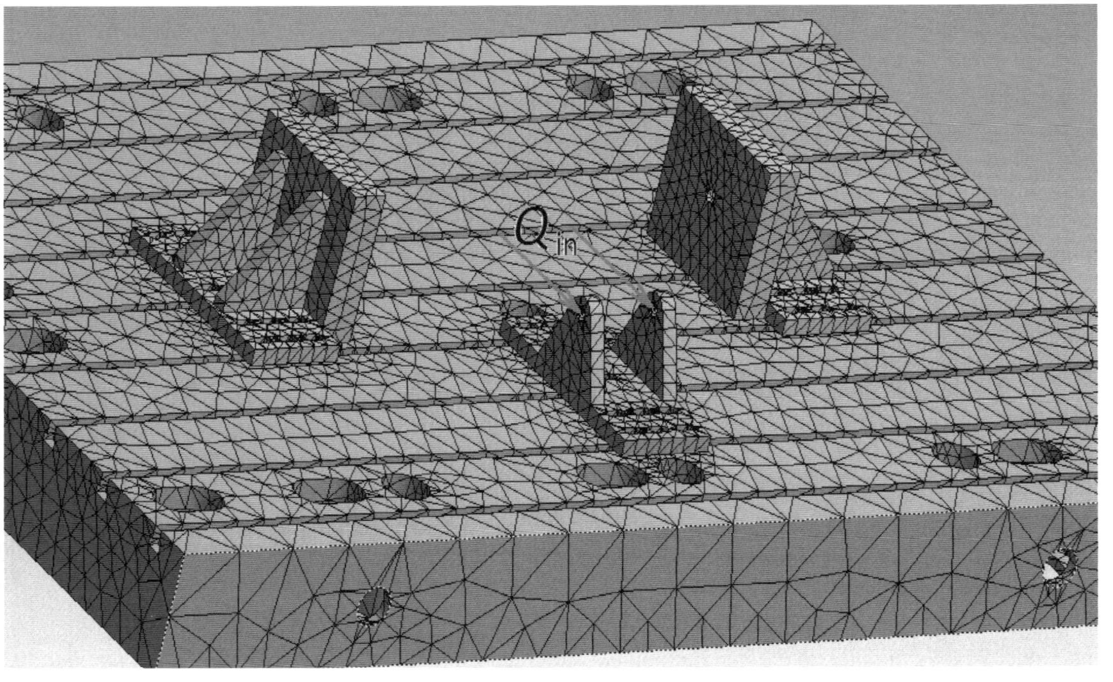

Figure 2. The calculation model of the device

The frequency range of the harmonic analysis is from 1 Hz to 1000 Hz with step 1 Hz.

During the calculating the oscillation rates on the column, in the force point ($V_{a.in}$) and in the central part of the angle section vertical surface ($V_{a.out}$) are defined. The input and transfer mechanical resistance of the device were calculated by formulas:

$$Z_{in} = Q_{in} / V_{a.in},$$
$$Z_{trans} = Q_{in} / V_{a.out}.$$

4. Calculation results

The frequency characteristics of the input and transfer mechanical resistance of the device obtained as a result are shown on the Fig. 3, 4 together with the mechanical resistances of the rubber-cord branch pipe with diameter 100 mm which were obtained experimentally.

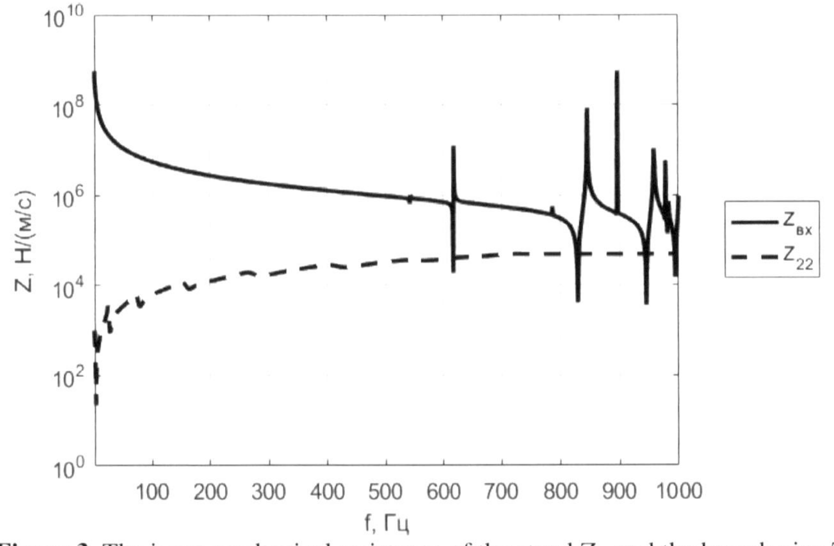

Figure 3. The input mechanical resistance of the stand Z_{in} and the branch pipe Z_{22}

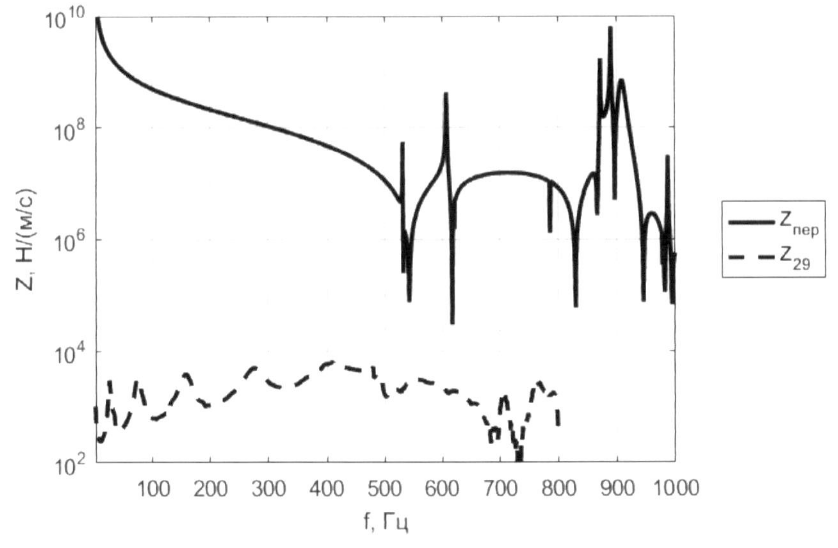

Figure 4. The transfer mechanical resistance of the stand Z_{trans} and the branch pipe Z_{29}

5. Result discussion

On the Fig. 5 the ratio of the input mechanical resistance of the device and the branch pipe Z_{in} / Z_{22} is shown. When the frequency is more than 735 Hz this ratio is less than 10. To obtain the reliable result for determination of the branch pipe characteristics it is should include accounting the mechanical resistance of the device in the experimental technique. Also a resonance of one device elements at the frequency 617 Hz takes place resulting in unreliable measurements without excluding a pliability of the stand near this frequency.

The transfer resistance of the stand in entire frequency range exceeds the transfer resistance of the branch pipe more than 10 times (Fig. 4). In the resonance points 542 and 617 Hz the ratio Z_{out} / Z_{92} is respectively 25,4 and 15,5. In this case the requirements of the device stiffness are satisfied. However

IOP Conf. Series: Journal of Physics: Conf. Series **858** (2017) 012031 doi:10.1088/1742-6596/858/1/012031

accounting the mechanical resistance of the device will increase the determination accuracy of the branch pipe mechanical resistances.

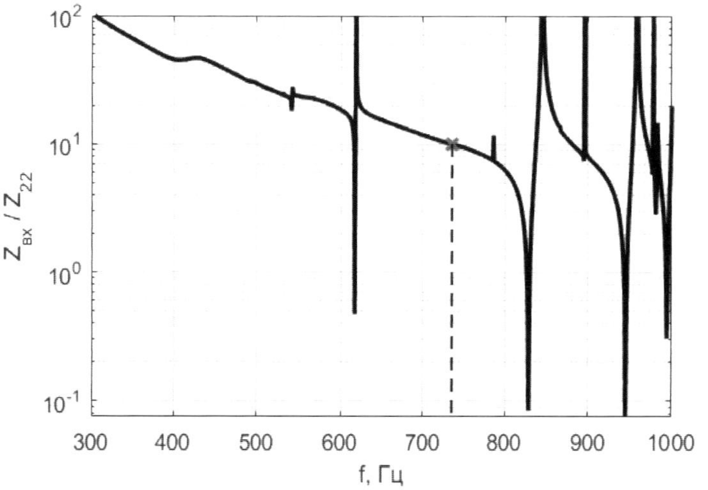

Figure 5. The ratio of the input mechanical resistances Z_{in} / Z_{22}

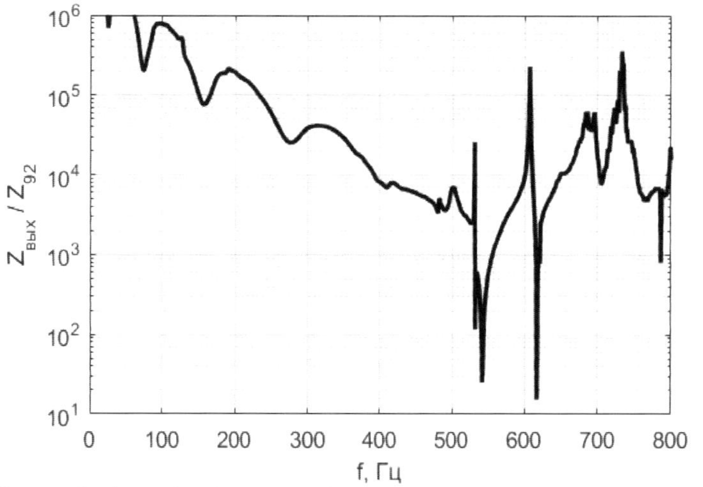

Figure 6. The ratio of the transfer mechanical resistances Z_{out} / Z_{92}

Near the resonance the frequency step should be decreased to increase the accuracy of determination of the device resonance frequencies during the experimental study.

6. Summary and conclusion

In the process of the experimental determination of the stiffness characteristics of the rubber-cord branch pipes with diameter 100 mm at the frequencies 735 Hz it is necessary to take into account the values of the device mechanical resistances, which should be determined experimentally and a numeric calculation by the harmonic analysis is used as an evaluation (for example, in designing or modifying elements of the test device).

References

[1] Sitnikov D V, Klishin S V, Ugrenev M V, Bohan V V, Zubarev A V 2015 Evaluation of mechanical resistances of the stand for dynamic tests of straight-flow branch pipe *Omsk Scientific Bulletin* **3(143)** pp 95–99

[2] Popkov V I, Popkov S V 2009 Vibrations of mechanisms and designs (St. Petersburg: Sudarynya) 490

[3] Ugrenev M V, Shvarts A A, Bohan V V, Grakov S A and Zubarev A V 2015 Influence of the design parameters of rubber-cord branch pipe on their mechanical resistances *Problems of tires, rubber articles and elastomeric composite, Moscow, 12-16 October 2015* pp 429–435.

[4] Ivanov V.S., Petrova V.V. 2010 Mechanical resistance of longitudinally oriented basis in the shell of submarine hull *Materials of Krylov State Research Centre* **52** pp 97–118.

[5] Ivanov V S and Petrova V V 2011 Mechanical resistance of the orthotropic cylindrical shell for concentrated forces and moments acting on this shell *Materials of Krylov State Research Centre* **61** pp 13–26

[6] Korneev S A, Korneev V S, Pen'kov I A and Tribelsky M.I. 2014 Technique and results of the static test of rubber-cord branch pipe for connection pipelines *Omsk Scientific Bulletin* **2(130)** pp 139–145.

[7] Aver'yanov G S, Bel'kov V N, Bur'yan Y A, Korchagin A B, Balashov V S 2014 Vibrational protection of te pumping stations in water-supply systems *Russian Engineering Research* **34(4)** pp 199–204

IOP Conf. Series: Journal of Physics: Conf. Series **858** (2017) 012032 doi:10.1088/1742-6596/858/1/012032

Boiling process modelling peculiarities analysis of the vacuum boiler

E N Slobodina[1], **A G Mikhailov**[1]
[1]Omsk State Technical University, 11 Mira ave., Omsk 644050, Russia

E-mail: slobodina_e@mail.ru

Abstract. The analysis of the low and medium powered boiler equipment development was carried out, boiler units possible development directions with the purpose of energy efficiency improvement were identified. Engineering studies for the vacuum boilers applying are represented. Vacuum boiler heat-exchange processes where boiling water is the working body are considered. Heat-exchange intensification method under boiling at the maximum heat-transfer coefficient is examined. As a result of the conducted calculation studies, heat-transfer coefficients variation curves depending on the pressure, calculated through the analytical and numerical methodologies were obtained. The conclusion about the possibility of numerical computing method application through RPI ANSYS CFX for the boiling process description in boiler vacuum volume was given.

1. Introduction

One can identify several areas of low and medium powered boiler technology development:

- energy efficiency improvement by heat losses reducing and maximum use of the fuel energy potential;
- boiler unit dimensions reduction by means of the fuel combustion and heat-exchange intensification in the furnace and on the heating surfaces;
- hazardous (polluting) emissions reduction;
- boiler unit operational reliability improvement.

The analysis of low and medium powered boiler equipment by domestic and foreign manufacturers showed that horizontal fire-tube boilers for vapour and hot water producing have become widely used nowadays [1-3].

The boilers of the given type require considerable steps to install chemical water treatment system of make-up water. Considerably sophisticated installations requiring expensive components for its operation are used for make-up water softening and desalting. Furthermore, when in operation, penalty payments for the discharge of salt concentrate poisoning the environment are possible [1]. The given issues can be avoided when using the vacuum hot water boiler.

The boiler operation peculiarity is that purified water is not discharged from the boiler while in operation. There is no need in boiler water regular chemical treatment, moreover, waterquality is the essential condition of boiler unit reliable and continuous operation [1].

Vacuum hot water boiler application as a heat source can solve a number of problems arising during typical boiler unit operation.

Vacuum boiler efficient operation is connected with intensive boiling and condensing processes.

Additionally, by means of the given boiler working pressure reducing taking into account boiling temperature and as a result of maximum heat-transfer coefficients achieving at boiling and vapour condensation, significant decrease of boiler unit mass and size characteristics is possible. However, at subatmospheric pressures, heat-exchange process is known to get worse as a consequence of boiling and condensing processes intensity change. The given paper is devoted to the boiling intensification issues.

High values of heat-transfer coefficients at subatmospheric pressures are possible to achieve only by means of intensification methods application.

The article deals with the boiling process peculiarities, the possible method of boiling process intensification at subatmospheric pressures is proposed.

2. Materials and methods

Boiling is a complex process, its conditions are determined by various thermal and physical factors. Boiling intensity depends on the number of thermodynamic parameters and liquid thermal and physical properties. The parameter having the maximum impact on boiling process is pressure.

Pressure reduction and vapour bubble critical radius increase result in considerable difficulties of vapour phase formation on the heating surface that causes a significant change of boiling process internal characteristics such as growth and departure diameter rates, vapour bubbles departure frequency. Moreover, boiling transition characterized by vaporization process instability and heat-exchange non-stationarity is extended [1].

Therefore, method taking into account every change happening at the low pressure and decreasing unstable boiling risk is necessary to be used for the boiling process intensification. Finning is possible to be applied as an intensification method (Fig. 1).

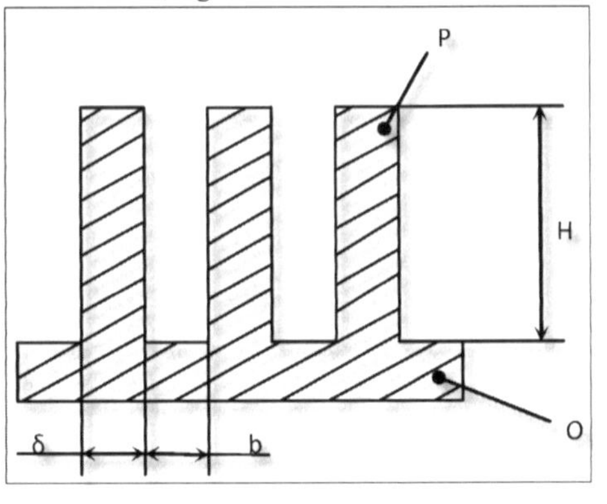

Figure 1. The finned surface computational scheme

Fin geometrical characteristics (height (H), length (l), thickness (δ) as well as adjoining fins spacing (b) should be accurately calculated to get the maximum effect, besides, the determining factor is the pressure. For instance, fins spacing should not exceed boiling liquid capillary constant value to avoid boiling crisis. The given paper concerns the research of the fin geometrical parameters to get maximum effect at the liquid boiling process intensification in the boiler vacuum volume.

3. The study of the boiling process modelling peculiarities of the vacuum boiler

The boiler heat calculation was performed by the zone method, the boiler can be conditionally divided into three zones: the furnace chamber volume (1), the convective heat exchange area (2), the vacuum volume (3) (Fig. 2).

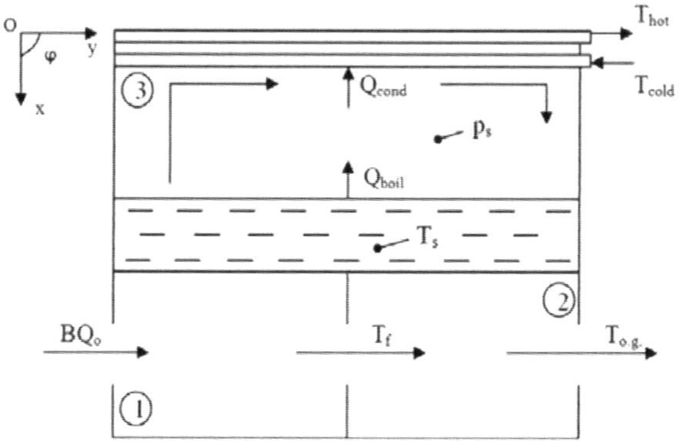

Figure 2. The calculation model of the vacuum hot-water boiler

Fig. 2 presents the following: T_T is the furnace outlet absolute temperature; $T_{o.g.}$is the outlet gas temperature; p_s, T_s are the saturation pressure and temperature; BQ_o, $Q_{boil.}$, $Q_{cond.}$ is the heat quantity generated under the fuel combustion, boiling and condensation correspondingly; T_{hot}, T_{cold} are the outlet and inlet heat carrier temperatures.

The heat exchange calculation in the hot-water boiler furnace is based on the similarity theory application to the burning processes. On the basis of this theory the heatingcalculation normative method of boiler units was developed [4].

There are different techniques for heat-transfer coefficients (a) calculation at liquid boiling. Many different parameters influence the heat transfer coefficient, such as the reduced pressure p_s, the heat flux q, the roughness and material of the heating surface and the thermophysical properties of the fluid. The heat transfer coefficient itself is defined as the quotient of the heat flux and the wall superheat:

$$a = \frac{q}{\Delta T} = \frac{q}{T_W - T_s} \tag{1}$$

Liquid heat-transfer coefficient depends only on the operating parameters (q, p), empirical dependences are applied for practical calculations [1, 5]. The calculations showed that at different techniques application the result does not change. There is a tendency of heat-transfer coefficients decreasing at pressure reducing [6]. Generally, heat-transfer coefficients decreasing demonstrates heat-exchange process intensity reducing. The given fact is unacceptable, therefore, the decision to intensify the process with the application of finning in the boiling zone was made.

Under the boiling process organization in the system of capillary slotted channels to obtain the maximum effect, the boiling liquid capillary constant variable value calculated by the presented below formula is taken into account [7]:

$$l_o = \sqrt{\frac{\sigma}{g(\rho_l - \rho_v)}} \tag{2}$$

where σ is the surface tension coefficient of the boiling liquid; g is the gravity acceleration; p_l, p_v are the liquid and vapour densities.

Simultaneously, the vapour bubble critical radius varying in proportion to pressure changing is taken into account. The calculation formula is given below:

$$R_{c.r.} = \frac{2\sigma T_s}{r\rho_v (T_w - T_s)} \tag{3}$$

where σ is the surface tension coefficient of the boiling liquid; p_v is the vapour density; r is the

IOP Conf. Series: Journal of Physics: Conf. Series **858** (2017) 012032 doi:10.1088/1742-6596/858/1/012032

vapourization heat; T_w is the wall temperature; Ts is the saturation temperature [1].

Based on the above mentioned characteristics values it is possible to conclude that the slot gap is chosen taking into consideration bubble critical radius and should not exceed boiling liquid capillary constant value [1, 7].

Regarding the fin height, the optimal value is possible to be chosen taking into account fin effectiveness, the calculation formula is presented below:

$$E = t\left(\frac{H}{\delta}\right)\sqrt{2Bi} \Big/ \frac{H}{\delta}\sqrt{2Bi} \tag{4}$$

In paper [7] the calculation equation for heat-transfer coefficient determining at liquid boiling on the finned surface in capillary channels under two-dimensional position is represented. The equation under discussion makes possible to evaluate the obtained effect of finning.

However, analytical solution of the heat distribution problem in the fin and liquid volume under multidimensional position is connected with considerable difficulties. It is not always possible to take into account fin end surface heat-exchange.

For the most reliable results obtaining, two methods (analytical and numerical) combination is necessary to be applied.

Different subcooled nucleate boiling models based on local parameters (temperature, void fraction or for example turbulence intensity) could be used without any difficulty in CFD. One of them was presented by Kurul and Podowki and it is the model applied in this solver [8].

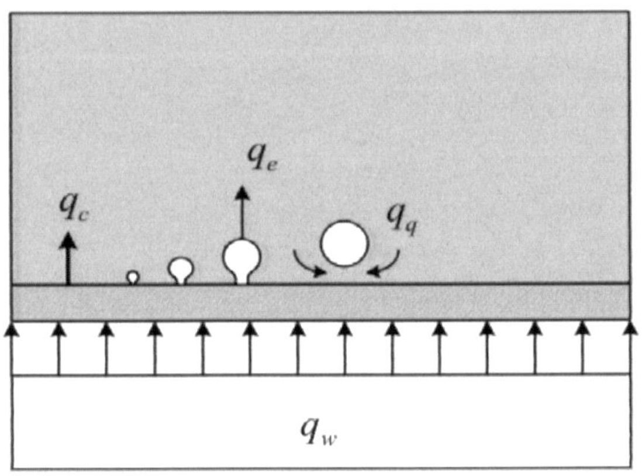

Figure 3. Heat flux partitioning according

The basic idea in the model is that the heat transfer originates from three different mechanisms between the heated wall and the liquid phase:

$$q_w = q_c + q_e + q_q \tag{5}$$

whereq_c, q_e and q_q denote respectively single-phase convective heat flux density, evaporation heat flux density and quenching heat flux density (Fig. 3).

At the heated wall, bubbles are formed at the nucleation sites due to evaporation of liquid at the wall and one therefore assumes that part of the wall heat ux density is directly used to transform the liquid into vapor. Bubbles grow and reach a critical size at which they detach from the wall. Once a bubble leaves the wall, the volume previously occupied by it is filled with cooler liquid, which will receive heat from the wall. The heat transfer to this cooler liquid is called quenching. The ratio of the wall influenced by quenching Aq is thus closely depending on the evaporation and the bubble lift-off size. Finally, the rest of the wall 1 – Aq is governed by single-phase convection.

Boiling model RPI is possible to be used as a numerical method. In the given model, boiling

IOP Conf. Series: Journal of Physics: Conf. Series **858** (2017) 012032 doi:10.1088/1742-6596/858/1/012032

process followed by convection is considered in the channels system.

The presented mathematical models contains 8 independent equations:

- 2 mass conservation equations for the liquid and gas phases;
- 2 momentum conservation equations for the liquid and gas phases;
- 1 energy conservation equation for the liquid phase;
- 1 interfacial area concentration transport equation;
- 2 additional partial derivative equations for the turbulence of the liquid phase.

Using various closures laws, all the parameters appearing in these equations have been expressed as a function of 8 independent variables:

- the void fraction;
- the gas velocity;
- the liquid velocity;
- the specific enthalpy of the liquid phase;
- the interfacial are concentration;
- the turbulent kinetic energy of the liquid phase;
- the turbulent dissipation rate of the liquid phase;
- the pressure.

Furthermore, the initial and boundary conditions are defined [9].

As a result of the conducted calculation studies, heat-transfer coefficients variation curves calculated by using analytical and numerical methods depending on pressure were obtained.

Calculation and empirical data comparison curves by the following authors: V.P. Isachenko, V.A. Osipov, A.S. Sukomel, Ju.M. Lipov, Ju.M. Tretjakov [6], using correlation equations, are represented in Fig. 4. The comparison was carried out with calculations under similar conditions obtained in ASYSCFX software. Calculation ratio a_r/a_s, where $a_r$is heat-transfer coefficient on the finned surface, a_s is heat-transfer coefficient on the smooth surface, the variable is pressure in vacuum volume, is represented in Fig. 5. Calculations were conducted taking into account subatmospheric pressures.

Having fixed pressure to the heat-transfer coefficient maximum value (60 kPa), the fin geometry impact analysis was conducted. At the fixed height value, fin length changed, variation range amounted from 0.05 to 0.1 m, and slot gap value varied from 0.001 to 0.01 m. Obtained computational data made possible to choose geometrical characteristics with maximum effect of finning.

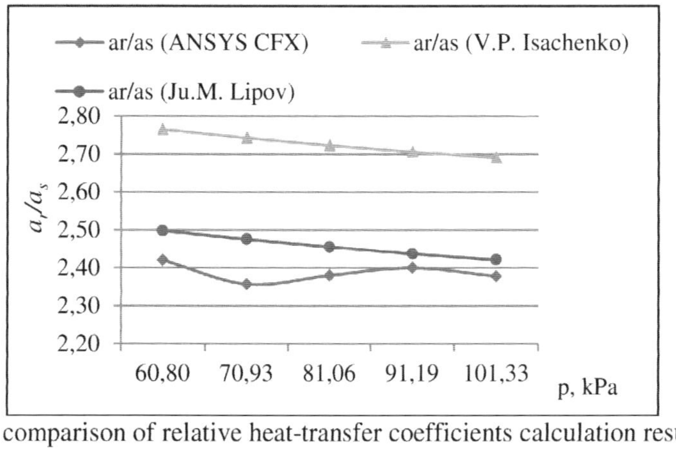

Figure 4. The comparison of relative heat-transfer coefficients calculation results at boiling according to different approaches

Reduced heat-transfer coefficient dependence on the finned surface geometrical characteristics is represented in Fig. 5.

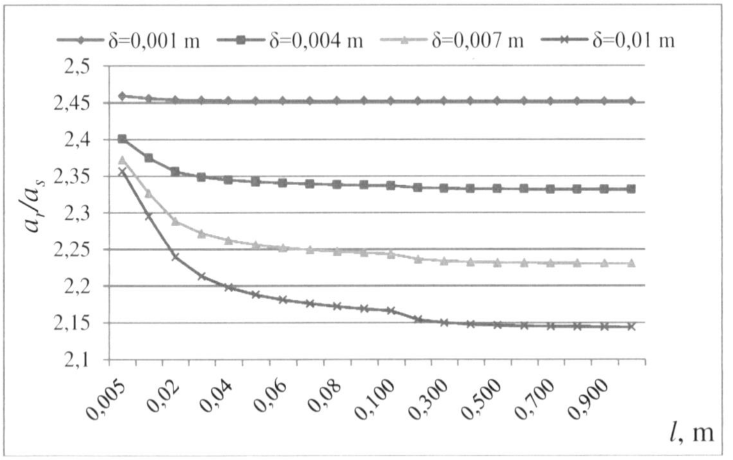

Figure 5. Reduced heat-transfer coefficient dependence on the finned surface geometrical characteristics

Heat-transfer coefficients values mean difference (Fig. 4) obtained at heat-exchange process calculated modelling, under boiling, varies from 2 to 12 % at the pre-determined pressure from 60 to 101 kPa.

Maximum difference between numerical and analytical values calculated according to Ju.M. Lipov, Ju.M. Tretjakov are minimal at specified initial and boundary conditions as a result of uneven temperature distribution impact on the heated surface and convective phenomena.

According to the computational data, it may be concluded that fin length (l) shortening (Fig. 5) results in heat-exchange specific influence increasing on the side surface provided that fin end surface area value tends to the minimal values. Thus, one could affirm that rectangular-cutout tenons fins appliance taking into account slot gap not exceeding capillary constant is the most effective option of finning for liquid boiling process intensification. Under the given conditions, reduced heat-transfer coefficient increases.

4. Conclusion

The studies have shown that at heat-exchange process modelling according to the boiling approach, RPI compared with analytical approaches ranges from 2 to 12 % under predetermined pressure from 60 to 101 kPa. There are minimal differences when comparing with Ju.M. Lipov, Ju.M. Tretjakov approach at specified initial and boundary conditions by taking into account subatmospheric pressures. Therefore, model RPI ANSYS CFX calculation numerical method is possible to be used for boiling process describing in the boiler vacuum volume.

The calculation showed that the most effective finning option for the liquid boiling process intensification is the rectangular-cutout tenons fins appliance taking into account the slot gap not exceeding capillary constant. The given technical solution resulting in reduced boiler unit dimensions is possible for applying in the boiler vacuum volume to intensify boiling process.

Research made in the framework of the research work №17087B Omsk State Technical University.

References

[1] Slobodina E N 2016 Oil and Gas Engineering *The peculiarities of the boiling process at subatmospheric pressure* pp 395–399

[2] Annaratone D 2008 *Steam generators* (Springer-Verlag Berlin Heidelberg) p 434

[3] Mikhailov A G, Batrakov P A, Terebilov S V, E. N. Slobodina 2014 Dynamics of Systems, Mechanisms and Machines *Gas-tube boilers: issues of classification and thermal calculation* pp 1–4

[4] Slobodina E N, Mikhailov A G 2016 Oil and Gas Engineering *The analysis of the condensation process impact on the vacuum boiler operating efficiency* pp 395–399

IOP Conf. Series: Journal of Physics: Conf. Series **858** (2017) 012032 doi:10.1088/1742-6596/858/1/012032

[5] Feldmann H and Luke A 2008 International Refrigeration and Air Conditioning Conference at Purdue *Nucleate boiling in water for different pressures* pp 1–8
[6] Slobodina E N 2015 Omskij nauchnyj vestnik *Vlijanie processa kipenija v razrjazhennoj polosti na jeffektivnost' raboty vakuumnogo kotla* (Ser. Pribory, mashiny i tehnologi №3 (143)) pp 185–187
[7] Korneev S D 2006 Izvestija MGIU 1 (2) *Vlijanie teplofizicheskih svojstv materiala orebrennoj poverhnosti na vybor ee optimalnyh geometricheskih harakteristik* pp 56–61
[8] Kurul N and Podowski M Z 1991 Proceedings of the 27th National Heat Transfer Conference *On the modeling of multidimensional effects in boiling channels* pp 21–26
[9] Inc. ANSYS. ANSYSCFX-Solver Theory Guide, 12.1 edn.(November 2009) p 372

IOP Conf. Series: Journal of Physics: Conf. Series **858** (2017) 012033 doi:10.1088/1742-6596/858/1/012033

The Specific Features of designing and technological preparation of production in territorially remote branch of the industrial enterprise

V V Sosedko[1] **and A G Yanishevskaya**[1]

[1] Omsk state technical university, 11, Mira ave., 644050, Omsk, Russia

Abstract. Output of products at the industrial enterprise means - the debugged working mechanisms at each stage of product's life cycle: from initial design documentation to product and finishing it with utilization.

The topic of the article is development of mathematical model of system of designing and technological preparation of production in branch of the industrial enterprise, and also statistical processing of estimated implementation result of the developed mathematical model in branch, and demonstration of its advantages at application at this enterprise.

During the creation of model of a data flow about driving of information, orders, details and modules in branch of the enterprise groups of divisions were classified. Proceeding from the analysis of activity of divisions, a data flow, details and documents the state graph of system of designing and technological preparation of production was constructed, transitions were described and coefficients are appropriated. To each condition of system of the constructed state graph the corresponding limiting state probabilities were defined, and also Kolmogorov's equations are worked out. When integration of sets of equations of Kolmogorov the probability of conditions of activity of the specified divisions and production as function of time in each instant is defined.

On the basis of the developed mathematical model of uniform system of designing and technological preparation of production and manufacture, and a state graph by authors statistical processing of results of application of mathematical model was carried out, and also its advantage at application at this enterprise is shown. Researches on studying of probability of loading of services of branch and third-party contractors (the orders received from branch within a month) were conducted.

The developed mathematical model of system of designing and technological preparation of production and manufacture can be applied to definition of probability of conditions of activity of divisions and manufacture as function of time in each instant that will allow to keep account of loading of performance of work in branches of the enterprise.

1. Introduction

Production output at the industrial enterprise means the debugged working mechanisms at each stage of the product's life cycle: from initial design documentation for the product (depending on a sort of the work – the preliminary design, the engineering design, development of working design documentation, further DD) [1], and finishing it with utilization. In the part of development and preparation of production of products this refers to uniform system of designing and technological preparation of production (DTPP). It's considered processes of design, design-preparation, engineering studies, preparation of documentation (design, technological, program) for production and directly production.

2. Problem definition

The work of territorially remote branch which is in other city is carried out by means of connection to informational resources and databases of head plant through the protected Internet channel. At the same time licenses for operating system, office programs, and CAD-systems, take also via the Internet.

IOP Conf. Series: Journal of Physics: Conf. Series **858** (2017) 012033 doi:10.1088/1742-6596/858/1/012033

The branch carries out development of DD and manufacture of a prototype of a product with the subsequent small-scale and serial production. In this regard technological preparation of production passes on the second plan, however in the long term at expansion of private production base these works will be performed (proceeding from technological equipment of branch and optimization of production).
Development of DD is conducted in CAD-system the Kompas. For elements, necessary to application (reference products, materials) requests for adding in databases and reference books are made out, and business processes in the PDM-system Lotsman are started.
Generally the scheme of processes of design and technological preparation of production has the appearance presented in fig. 1. In branch of the enterprise in the conditions of original lack of equipment, the mastered technology and licenses for CAD/CAM/CAE systems [2] the scheme of processes of design preparation is submitted in fig. 2.

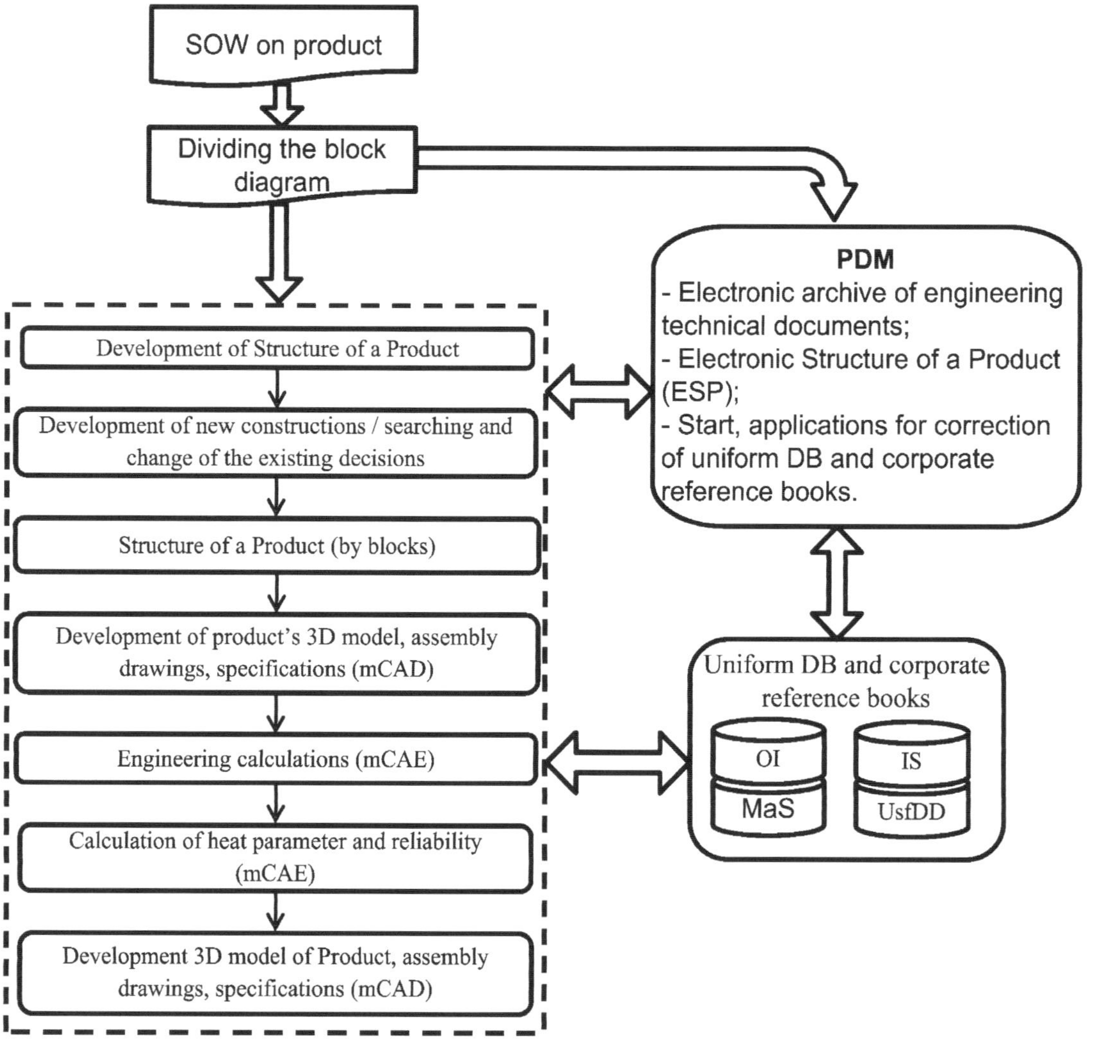

Figure 1. Scheme of processes of design-technology preparation of production.

he principle of action according to the scheme of processes consists in work in PDM-system with an array of the corporate reference books (directory system, reference products, materials and ranges) and the ESKD qualifier).

The main theme of the article is development of mathematical model of system of design and technological preparation and production in branch of the industrial enterprise, and also statistical processing of estimated results of application of the developed mathematical model in branch, and demonstration of its advantages at application at this plant.

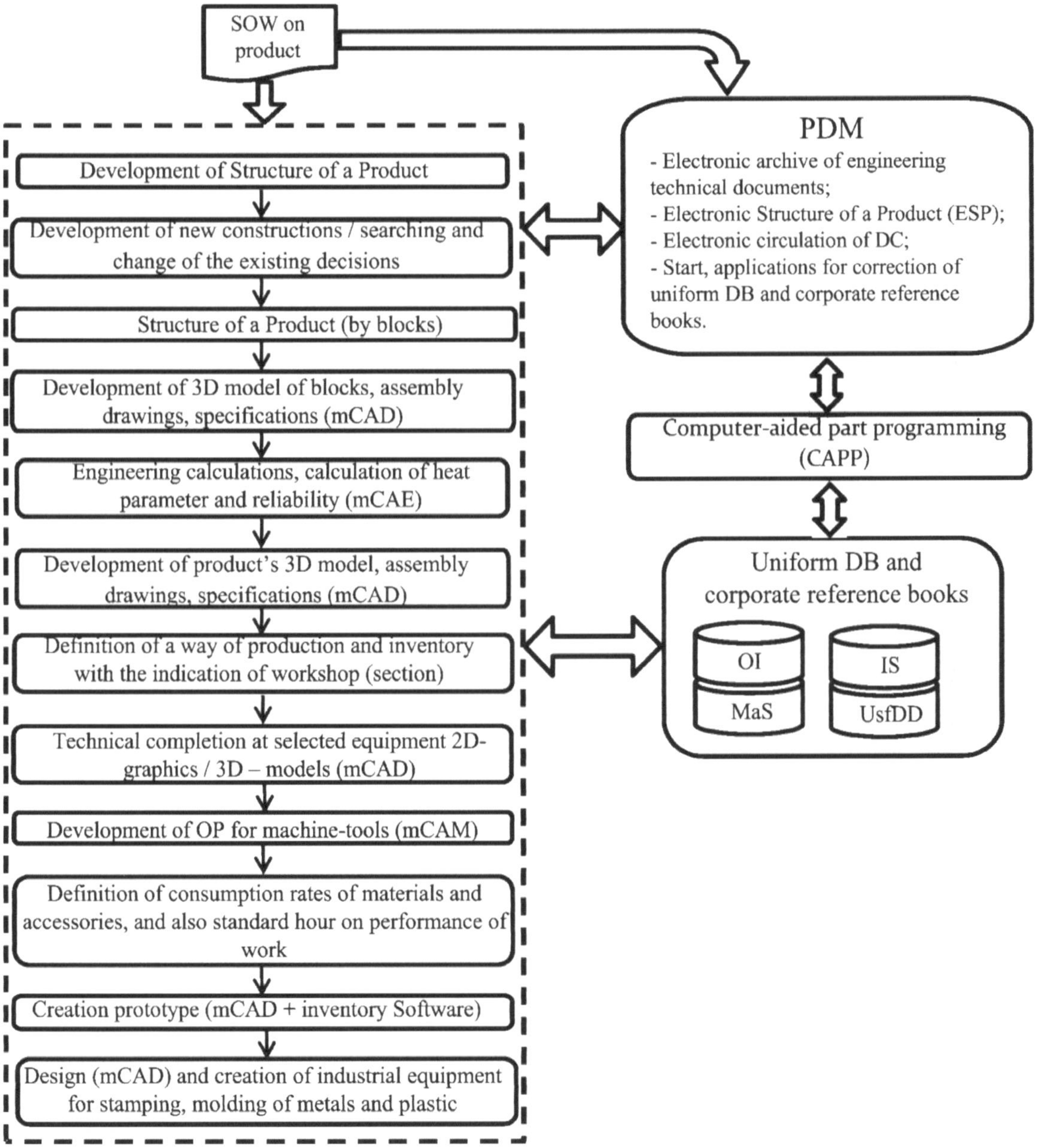

Figure 2. The scheme of processes of design preparation in branch.

3. Theory

The mathematical model of the DTPP system at apy branch is under construction on the basis of theory of probability postulates, in particular, of the theory of a queuing and mathematical statistics [4].

The random process which proceeds in system S is considered as Markov if it proceeds without after-

IOP Conf. Series: Journal of Physics: Conf. Series **858** (2017) 012033 doi:10.1088/1742-6596/858/1/012033

action. In this process future state does not depend on the past, and last state does not influence the future [5].

It is possible to define probabilities of states by the marked state graph

$$p_1(t),\ p_2(t),\ \dots,\ p_n(t)$$

as functions of time. These probabilities satisfy to Kolmogorov's equations. In these differential equations probabilities of states will be unknown functions. Therefore, having solved these equations, it is possible to calculate probabilities of conditions of production as functions of time.

We will take a route list on printed circuit boards, details and clusters for unit of fluency of events λ which switches the services from state x_i to state x_j.

Today in branch the technology of a turn of route lists is not functioning yet. The engineering service is presented by one design-technology department, blanking shops – by third-party contractor, and assembly shops – by one assembly site. In this regard the offered classification of groups for branch is represented in fig. 3.

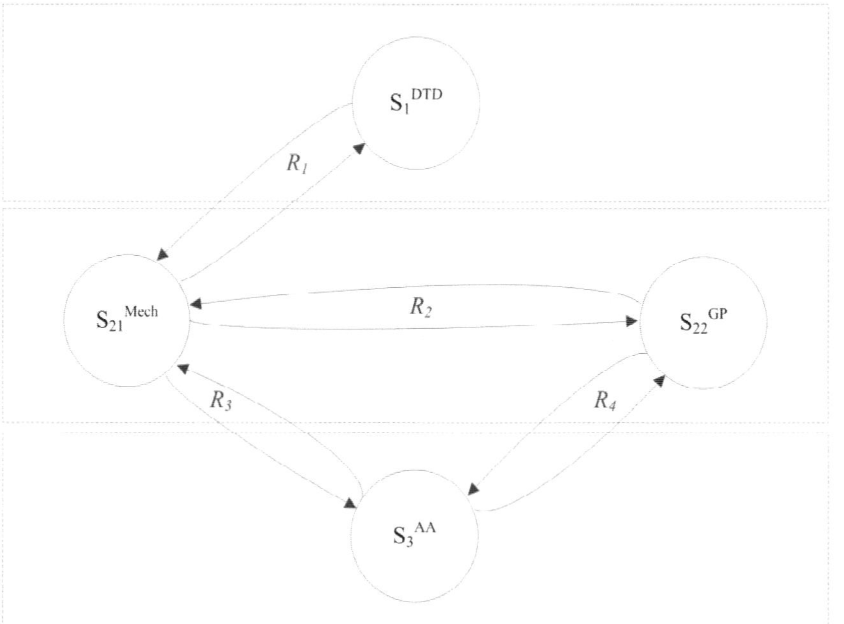

$$R_1\langle=\rangle\begin{cases} z_{1,21}\lambda \\ z_{21,1}\mu \end{cases};\quad R_2\langle=\rangle\begin{cases} m_{21,22}\lambda \\ m_{22,21}\mu \end{cases};\quad R_3\langle=\rangle\begin{cases} m_{21,3}\lambda \\ m_{3,21}\mu \end{cases};\quad R_4\langle=\rangle\begin{cases} m_{22,3}\lambda \\ m_{3,22}\mu \end{cases}$$

Figure 3. The marked state graph of design preparation of production and production at the branch with the description of transitions.

1. Design and Technological Department (DTD). This department is performed design preparation of production and technological study of products. This activity is not performed yet.

2. Third-party contractors. These organizations are manufacture the printboards (PCB), details (Mech), perform galvano-paintings (GP) work.

3. Assembly Area (AA). On this area installation of modules and blocks, assembly, run, adjustment, tests and delivery of blocks and products is performed.

At acceptance of a median number of the events written in one standard primary document or a kit of parts or modules (with the relevant document accompanying them - a route list), for λ, the complete flow of documents, getting to system for their after-treatment, is defined by dependence $x \cdot \lambda$, where x equals z, m в state graph (see figure 4). In our case λ constant value ($\lambda = const$), thus it is about the

elementary stream.
After that:
z – orders for performance of work by third-party contractors;
m – primary documents accompanying details, modules, blocks and products at each production phase - route lists;
μ - stream, constantly serving the applications coming from other departments. The resultant common flow of such documents can be expressed dependence $x \cdot \mu$.For each of the five states of systems S of the count we will find the limiting states probabilities and we will work out Kolmogorov's equations:

$$S_1 : \frac{dp_1}{dt} = -z_{1,21}\lambda p_1 + z_{21,1}\mu p_{21} \tag{1}$$

$$S_{21} : \frac{dp_{21}}{dt} = z_{1,21}\lambda p_1 - z_{21,1}\mu p_{21} - m_{21,22}\lambda p_{21} + m_{22,21}\mu p_{22} - m_{21,3}\lambda p_{21} + m_{3,21}\mu p_3 \tag{2}$$

$$S_{22} : \frac{dp_{22}}{dt} = m_{21,22}\lambda p_{21} - m_{22,21}\mu p_{22} - m_{22,3}\lambda p_{22} + m_{3,22}\mu p_3 \tag{3}$$

$$S_3 : \frac{dp_3}{dt} = m_{21,3}\lambda p_{21} - m_{3,21}\mu p_2 + m_{22,3}\lambda p_{22} - m_{3,22}\mu p_3 \tag{4}$$

For each repartition (design and technological preparation, third-party contractors and an assembly site) we will make equation system of Kolmogorov.
Condition of design-technology preparation:

$$\frac{dp_1}{dt} = z_{21,1}\mu p_{21} - z_{1,21}\lambda p_1 \tag{5}$$

State of procuring production – third-party contractors:

$$\begin{cases} \frac{dp_{21}}{dt} = (z_{1,21}p_1 - (m_{21,22} + m_{21,3})p_{21})\lambda + (m_{22,21}p_{22} - z_{21,1}p_{21} + m_{3,21}p_3)\mu \\ \frac{dp_{22}}{dt} = (m_{21,22}p_{21} - m_{22,3}p_{22})\lambda + (m_{3,22}p_3 - m_{22,21}p_{22})\mu \end{cases} \tag{6}$$

State of an assembly area:

$$\frac{dp_3}{dt} = (m_{21,3}p_{21} + m_{22,3}p_{22})\lambda - (m_{3,21} + m_{3,22})\mu p_3 \tag{7}$$

Integration of the received equation system allows to define state probabilities of design and technological preparation and production as functions of time in each instant [4].

4. Results of experiments

On the basis of monitoring of activity of the divisions involved in production process, authors investigated loading and rhythm of work of services of branch in the course of manufacture of a standard product. Coefficients according to transitions of a state graph of design and technological preparation of production and production were experimentally calculated. For averaging of indexes basic accepted the beginning, the middle and the end of month. Results of researches are given in the Tab. I.

TABLE I
LOADING AND RHYTHM OF WORK OF SERVICES OF BRANCH WITHIN A MONTH

№	Transitions	Coefficients	Indexes within a month		
			Beginning of month	Middle of month	End of month
1	R_1	$z_{1,21}$	10	10	8
2		$z_{21,1}$	1	1	1
3	R_2	$m_{21,22}$	8	10	10
4		$m_{22,21}$	6	8	8
5	R_3	$m_{21,3}$	2	2	2
6		$m_{3,21}$	1	1	1
7	R_4	$m_{22,3}$	2	2	2
8		$m_{3,22}$	1	1	1
9		λ	22	24	22
10		μ	9	11	11

Using the calculated coefficients in equation system (5-7) taking into account that at constant probabilities their derivants are equal 0, and also applying a normalizing condition (8):

$$p_1 + p_{21} + p_{22} + p_{23} + p_3 = 1 \qquad (8)$$

state probabilities of system for the given stages of month were calculated. Results of calculations are presented in the Tab. II. For descriptive reasons they are brought in graphics in Fig 4.

Table II
PROBABILITIES OF CONDITIONS OF SYSTEM WITHIN A MONTH

№	Groups		Probabilities of conditions	Probabilities of conditions of system within a month		
				Beginning of month	Middle of month	End of month
1	DTD		p_1	0,0033	0,0043	0,0065
2	OI	Mech	p_{21}	0,0802	0,0932	0,1035
3		GP	p_{23}	0,2090	0,2197	0,2277
4	AA		p_3	0,745	0,7075	0,6828

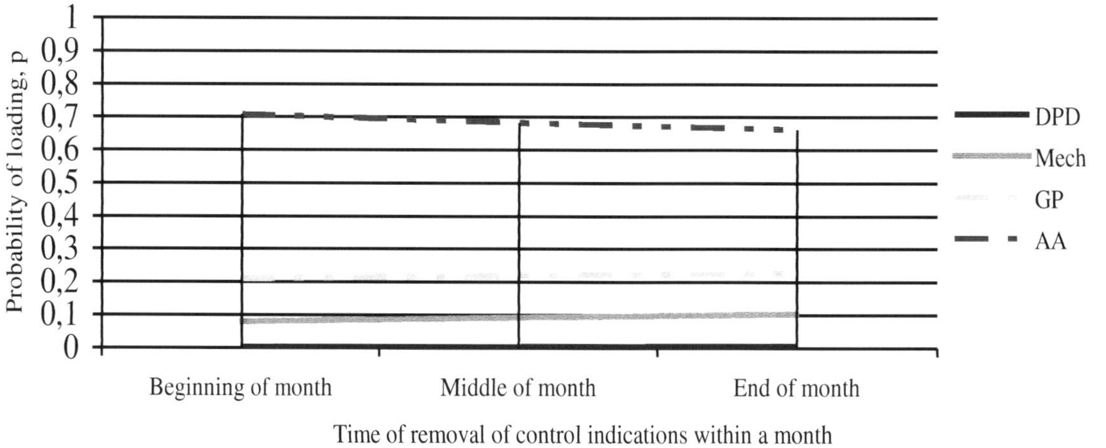

Figure 4. The schedule of probability of loading of divisions of branch within a month

In the course of the analysis of the developed mathematical model, and application of the explained methods and the principles, authors were it is offered to make essential changes to the current business processes of design and technological preparation and production of the enterprise. These changes led to the larger transparence of production consisting in expeditious emergence of actual information about the course of production process to chiefs of divisions of all ranks and persons, responsible for a condition of production, and also to reduction of time and decrease of probability of inaccurate information transfer, details and blocks. It, in turn, positively affected dynamics of implementation of the production schedule, and also prime cost of products. In general, it depend, including, from padding expenses, the bound to losses of time and financial resources on elimination of consequences from irregularly or not in time the made decisions and actions of personnel.

Thereby, proceeding from results of the conducted researches, it becomes apparent that application of the developed DTPP mathematical model and production allows to increase in general loading of services that increases rhythm of their work and, as a result, labor productivity.

5. Discussion of results

From data of research it is visible that probability of loading DTD within a month is step by step increases (from 0.0033 to 0.0065). The probability of loading of the third-party contractor - manufacturer of details (Mech) increases during the month from 0.0802 to 0.1035. The probability of loading of the third-party contractor of galvano-paintings (GP) within a month from value 0.2090 increases to 0.2277. The probability of loading of an assembly site from value 0.7075 decreases to the end of month to 0.6623.

6. Summary and conclusions

The developed mathematical model of system of design and technological preparation and production can be applied to definition of state probabilities of activity of divisions and production as function of time in each instant that will allow to keep account of loading of performance of work in branches of the plant.

FINANCING SOURCE – ON ONE'S OWN ACCOUNT

References

[1] GOST 2.103-68 Uniform System of Design Documentation (USDD). Development stages.
[2] Sim R M 1986 *Recent developments in computer-aided design and engineering*. CAD International Directory, ScienceDirect, p. 11–15. [Electronic resource] - Access mode: http://www.sciencedirect.com/science/article/pii/B9780408255554500056.
[3] Qun Gao 2012 *The Research and Application of the Integration Pattern of PDM and CAD in Aerosun*. (Dept. of Ind. & Commercial Manage., Nanjing Inst. of Ind. Technol., Nanjing, China. Computer Science & Service System (CSSS), International Conference. [Electronic resource] – Access mode: http://ieeexplore.ieee.org/xpl/articleDetails.jsp?arnumber=6394276&newsearch=true&queryText=pdm.
[4] Sosedko V V and Yanishevskaya A G 2012 *Mathematical model of uniform system of design-technology preparation and production at the production enterprise*. Engineering bulletin of Don. Electronic magazine. - No. 4. - Access mode: http://www.ivdon.ru/magazine/archive/n4p2y2012/1404.
[5] Venttsel E S 1972 *Operations research*. (Moscow: Soviet radio). 552 pages.

Morphology, composition and tribological properties of tantalum coatings deposited onto a rubber substrate

Valeriy I Surikov[1], I V Nikolaev[2], D A Polonyankin[1], E P Tselykh[3], E A Rogachev[1], Vadim I Surikov[1]

[1]Omsk State Technical University, 11 Mira Ave., Omsk, 644050, Russia
[2]SB RAS (IT SB RAS), 1 Ac. Lavrentieva Ave., Novosibirsk, 630090, Russia
[3]Federal state unitary enterprise «Federal research productive centre «Progress» (FSUP «FRPC «Progress»), 5 Kordnaya Ave., Omsk, 644018, Russia

E–mail: surval@mail.ru

Abstract. The paper analyzes the features of morphology and composition impact on tribological characteristics of tantalum coatings sputtered onto a rubber substrate. It is shown the influence of sputtering time on the structural changes in the form of clusters and grains that made up the coating. The morphology and composition features correlation with the coatings' wear–resistance characteristics was investigated.

1. Introduction

One of the promising ways of products' creation with high performance characteristics is the direction based on the raw materials modification. There are various approaches and methods of materials' surface modification. One widely used approach is the modification of bulk materials. This method is extensively used for modification of elastomers. The volumetric modification is made by inserting the fillers and modifiers in the matrix. With all the benefits of this method it has disadvantages such as mostly low economy and efficiency. Surface modification is an alternative approach that has become widespread in recent years and consists in modifier's deposition onto the surface of substrate [1]. This method allows obtaining elastomers with improved physical, mechanical and performance properties. It is equally important to solve the same problems at the area of nanostructured coatings deposited onto the surface of rubber. Despite of the small coating thickness it significantly improves the mechanical properties of the products [2] and increases its tribological characteristics.

Processing of the products including the rubber parts without production technology changing should be attributed to advantages of surface modification methods [3]. One of the promising method at this direction is the magnetron sputtering of the nano–dispersed refractory metals onto a rubber substrate. Magnetron sputtering has lots of advantages such as 1) application convenience, 2) high coatings' continuity and 3) possibility of modifiers' deposition at low temperatures. The last of mentioned advantages is particularly important in the case of rubber [4]. As the literature review shown [1, 4] at the present time there are enough investigations devoted to coatings' deposition in application to metal and composite materials modification. With regard to coatings applied to elastomers and particular to rubber products this direction are currently paid insufficient attention. In order to increase the rubber products' efficiency it is important to establish a correlation beetwen morphology and composition of the surface layers with performance characteristics of the products. At the same time it is necessary to establish the effect of sputtering conditions (in particular time and thickness) on the morphology features of the coatings and on the properties of a bulk rubber. The purpose of this paper is devoted to the morphology and composition investigation of the coatings deposited on a rubber substrate by usage tantalum as one of refractory metals in dependence on the sputtering time. The influence of a rubber morphology and composition with tantalum coatings on its abrasive wear and friction coefficient was also developed.

2. Materials and methods

In this study rubber samples were modified by tantalum magnetron sputtering with ADVAVAC VSM–200 setup usage. Samples' preparation, composition and experimental methods are presenter in reference [5, 6]. Investigation of the coatings' morphology was carried out on JEOL JCM–5700 scanning electron microscope equipped with JED–2300 spectrometer. Rubber surface modification was made at the same conditions except the sputtering time (thickness). The rubber samples were preliminarily grinded and polished and then subjected to examination. Tribological properties of rubber were studied on the MI–2 setup (GOST 426–77). The tests were performed with the metallic surface disk used as a counter face under the normal force equal to 26 N (2.6 kgf). The turnovers number of the 11KP structural steel disk during the test for all samples was 5.000.

3. Results and discussion

SEM surface images of the initial rubber and of a rubber modified by tantalum magnetron sputtering are shown on figure 1a and 1b. For comparison on figure 1c is represented a SEM image of the coating after molybdenum sputtering [6]. The figures 1b and 1c show that the morphology of the modified samples substantially different from the initial one (figure 1a). Image analysis suggests a cluster structure of the coatings while maintaining various forms: tantalum clusters unlike the molybdenum clusters have a more elongated shape in one direction. The clusters in both cases have a grain structure. Studying the effect of sputtering time (while maintaining the other parameters) on the morphology, composition and properties of coatings based on refractory metals and possessed the effective protection features for rubber has a great applied interest. Detailed analysis of the time influence on the coatings' morphology showed that during tantalum sputtering on the rubber substrate two stage of grain growth occurs. At the first stage grains with an average grain size of about 3 microns arise after which begins the formation and growth of new smaller grains.

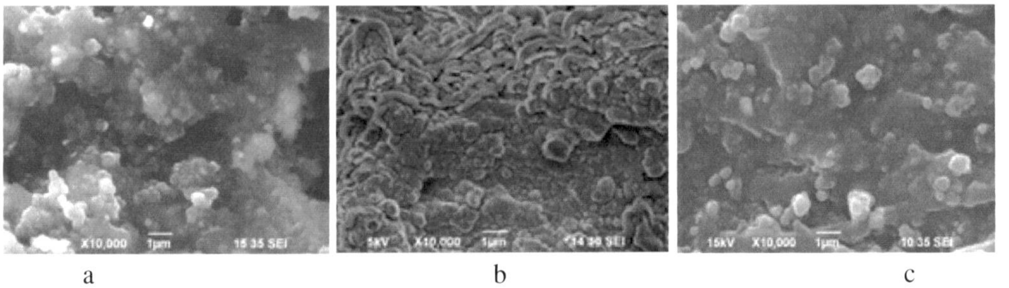

a b c

Figure 1. SEM surface images: a – the initial rubber; b – the rubber modified by tantalum; c – the rubber modified by molybdenum.

After continuous sputtering of more than 170 minutes cracks were appeared on the sample surface and «scaly–form» morphology was formed. Dependence of the average grain size at clusters from the thickness measured by deposition controller is shown on figure 2. The solid line shows the dependence for molybdenum coating, dashed line – for tantalum coating relative to the surface without coating.

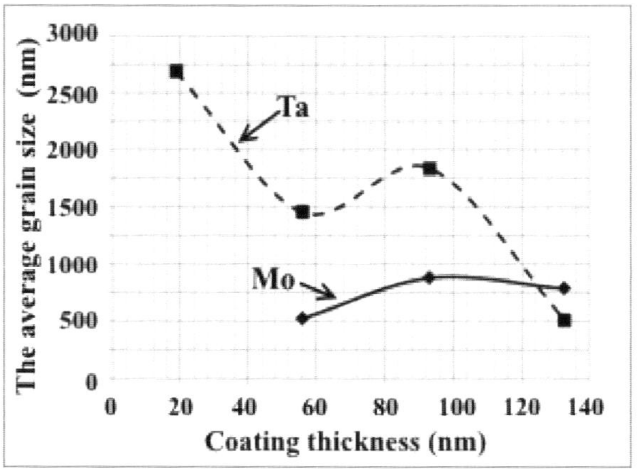

Figure 2. Dependence of the average grain size from coating thickness for tantalum and molybdenum coatings.

As can be seen from the figure 2 the average grain size ranges in a complicated way from about 3 μm (34 minutes sputtering time) to about 0.5 μm (233 minutes sputtering time) with increasing sputtering time. An important role in the coatings' properties formation plays the chemical composition. The chemical composition of the initial rubber and the rubber coated by tantalum for different sputtering times before and after wear resistance test is presented in table 1 and table 2.

Table 1. Chemical composition of the samples before wear resistance test.

Chemical elements and oxides	Mass (%)					
	ST[a]	0	34	100	168	233
C		84.3	13.81	16.52	10.05	5.18
Cl		11.8	2.75	0.00	0.00	0.00
Ta_2O_5		0.00	83.02	83.48	89.95	94.82
SO_3		2.27	0.42	0.00	0.00	0.00
MgO, ZnO		1.60	0.00	0.00	0.00	0.00

[a]ST – sputtering time (minutes).

The results of tribological tests for tantalum coatings deposited onto a rubber substrate for different sputtering time are presented in table 3. From the experimental results we can conclude that the tantalum–based coatings has morphology characteristics defined apparently by peculiarities of plasma forming in the vacuum chamber and by reacting the metal ions with the rubber substrate. Significant influence on the surface morphology formation confirmed by a noticeable difference between the sensor's thickness (tens of nanometers) and the average grain size on the surface of the rubber substrate. Rubber substrate influence on the thickness of coating based on refractory metals shows a rubber cross–section SEM image of the sample with a molybdenum coating presented on figure 3. As can be seen from the figure 3 the coatings' thickness amounts to a few microns.

Table 2. Chemical composition of the samples after wear resistance test.

Chemical	Mass (%)

IOP Conf. Series: Journal of Physics: Conf. Series **858** (2017) 012034 doi:10.1088/1742-6596/858/1/012034

elements and oxides	ST[a] 0	34	100	168	233
C	84.30	37.94	36.04	28.57	19.83
Cl	11.80	9.46	22.71	14.08	4.78
Ta_2O_5	0.00	47.99	31.79	51.54	71.91
SO_3	2.27	0.85	3.75	2.38	3.48
MgO, ZnO	1.60	3.76	5.73	3.43	0.00

[a]ST – sputtering time (minutes).

Sputtering time has a significant impact on the coatings' morphology. Two trends should be allocated in the coatings' structure formation: the first is the grain refinement in clusters formation with increased sputtering time and the second is «scaly–form» structure appearance with infringement of the coatings' integrity for a sufficiently large sputtering time which is more than 170 minutes. Similar trends are not observed in the molybdenum coating's morphology (figure 1c).

Table 3. Abrasive wear and friction coefficient for initial samples and samples modified by tantalum.

Tests' parameters	Sputtering time (minutes)				
	0	34	100	168	233
abrasive wear, $cm^3/kW{\cdot}h$	2.390	0.274	1.124	1.440	2.060
friction coefficient	0.870	0.472	0.517	0.527	0.469

A great impact on rubber performance characteristics including tribological properties must have a chemical composition of the coating in common with the morphology. Analysis of the sample's composition shows that the main component of the coating is apparently tantalum pentoxide (Ta_2O_5) formed by tantalum oxidation by oxygen remaining in the chamber after evacuating.

Figure 3. SEM image of the rubber cross–section with molybdenum coating at a sputtered time of 30 minutes.

The coating also contains carbon inclusions which is confirmed by point analysis. Chemical elements and oxides (C, Cl, SO_3) most likely belong to the near–surface layers of the rubber substrate as

IOP Conf. Series: Journal of Physics: Conf. Series **858** (2017) 012034 doi:10.1088/1742-6596/858/1/012034

indicated by the appearance of these elements and oxides in the samples with a relatively large sputtering time after wear resistance testing. EDS analysis also captures the presence of these elements at low coatings' thicknesses.

Data from the table 3 show that tantalum–based coatings demonstrate a significant improvement of rubber's tribological characteristics. An average friction coefficient is reduced from 1.7 times to 1.8 times at sputtering time of 168 minutes. Tribological characteristics depend on a number of factors. The most important factors are the state of the rubbing surfaces defined by its morphology and composition, surface roughness and adhesion, mechanical properties. The many factors of tribological characteristics make them difficult to analyze. The results obtained in this study allow us to make only a few conclusions about correlation between morphology and abrasion wear of tantalum–based coatings. There is no appreciable necessity for coatings formation with great thickness. Sputtering time can be limited from 30 to 60 minutes. Coatings' continuity is retained at such sputtering time and the grain size is likely to be optimal in terms of surface roughness. Increasing of coating thickness leads to continuity disruption and «scaly–form» structure formation which in turn increases the coatings' wear.

4. Conclusion

As a result it is possible to make a conclusion that the morphology features of tantalum–based coatings on the rubber substrate depend on the sputtering time.

The special features of the tantalum–based coatings' morphology should be attributed the grain's grinding with increasing of sputtering time and the «scaly–form» structure formation with infringement of the coating continuity at a sufficiently high sputtering time (more than 170 minutes).

A coatings formed during tantalum magnetron sputtering consist presumably of tantalum pentoxide with carbon inclusions. The chemical composition of the coatings before and after abrasive wear tests remains invariant. Formation of tantalum–based coatings leads to significant improvements of rubber's tribological characteristics. The friction coefficient is reduced on average in 1.7 times while abrasive wear decreased in 1.8 times (for 168 minutes of sputtering time). The growth of abrasive wear with increasing of sputtering time is explained by morphology features of the coatings. As shown by the abrasive wear test results the optimal sputtering time ranges from 30 to 60 minutes.

For a more detailed research of morphology and composition influence of tantalum–based coatings on tribological properties of rubber products additional studies are needed with attraction of other characteristics such as deformation–strength properties, adhesion and roughness.

Acknowledgments

The work was carried out under the research project № 17053B in Omsk State Technical University (OmSTU).

References

[1] Vityaz P A and Solntsev A K 2011 *Construction technology of nanostructured materials and coatings* (Minsk: Navuka) p 283

[2] Grinberg P B, Poleshchenko K N, Surikov V I and Tarasov E E 2012 *Herald of the Omsk University* **2(64)** 249–252

[3] Makhkamov K K 2016 *J. Frict. Wear+* **37(5)** 500–506

[4] Berlin E V and Seidman L A 2010 *Ion–plasma processes in thin–film technology* (Moscow: Technosphere) p 457

[5] Tselykh E P, Polonyankin D A, Rogachev E A and Surikov V I 2015 *Omsk scientific Herald* **1(137)** 97–100

[6] Surikov V I, Tselykh E P, Polonyankin D A and Rogachev E A 2014 *Dynamics of systems, mechanisms and machines (Dynamics 2014)* 1–6

IOP Conf. Series: Journal of Physics: Conf. Series **858** (2017) 012035 doi:10.1088/1742-6596/858/1/012035

Study on hydraulic direct-acting relief valve

V V Syrkin, P D Balakin, V A Treyer
Omsk State Technical University, 11, Mira ave., Omsk, 644050, Russia

e-mail: SyrkinVV@mail.ru

Abstract. Valves with elastic shut-off-and-regulating elements are increasingly applied in hydraulic systems, working in conditions of excessive vibrations. The objective of this study is to determine the dynamic characteristics of regulators with shut-off-and-regulating elements. The method of phase trajectories establishes the controller's parameters and modes, which provide its work stability. The authors receive dynamic characteristics providing the absence of self-oscillations at the expense of additional damping of the shut-off- and-regulating element. They make a test sample of the pressure regulator with an elastic shut-off-and-regulating element. The article defines the characteristics curve of the pressure regulator (direct operated pressure relief valve) from the initial conditions. The obtained results can be used, when designing hydraulic technological machines. Keywords – oscillation damping, an elastic shut-off-and regulating element.

1. Introduction

Modern technological machines for various purposes often have hydraulic gears and hydraulic tools, which shall comply with the highest requirements to reliability, velocity, stable vibration-free work, constancy of controlled performance, fluid flow, operating accuracy, etc. These requirements are provided by the protection and prevention structure of hydraulic systems, which ensure reliable and steady work of the hydro drive. Protection and prevention components include: safety pressure regulators and relief valves. One of the most important conditions for the work of such devices is dynamic stability. It means that continuous oscillatory processes should not appear in the valve throughout the whole flow range, especially on the main operative range. These phenomena are linked with the presence of elastic medium (working fluid), which may cause malfunction of the whole hydraulic system related to the valve. There are various constructions of devices, such as direct operated pressure relief valve and indirectly working safety valve that are subdivided into many more variant performances including relief valves with elastic shut-off- and-regulating elements. The authors consider that such valves are more prospective [7].

IOP Conf. Series: Journal of Physics: Conf. Series **858** (2017) 012035 doi:10.1088/1742-6596/858/1/012035

2. Materials

Dynamic properties of valves with direct and indirect actions in traditional performance have been studied most thoroughly. In the papers [1, 2, 3, 4] their velocity, transient and frequency characteristics, choice of design factors of structural and functional diagrams are reviewed. At the same time there is practically no information about such studies for devices with elastic shut-off-and-regulating elements. The objective of this study is to determine the dynamic characteristics of regulators with shut-off-and-regulating elements.

3. Theory

Mathematical modeling of dynamic processes of hydraulic pressure regulators typically includes the record keeping of elastic properties of pressure fluid, liquid flow forces, mass of regulator closure, viscous and dry friction during the element motion process, as well as the quadratic law of fluid discharge through the regulator working window [1, 2, 3]. Nonlinear characteristics are subjected in most cases to linearization, differential equations become linear, that allow determining the dynamic characteristics of regulators with sufficient accuracy for engineering analysis.

The regulator under study (Figure 1) contains an enclosure with the internal cylinder 1 with a circular dovetail groove, in which there is a reinforced metal spring 3, an elastic shut-off- and-regulating element 2, ensuring a desired level of hydraulic fluid pressure. There are performed coaxial input and output axial channels I and III enclosed, linked by radial channels (II) through a smaller base of a dovetail groove.

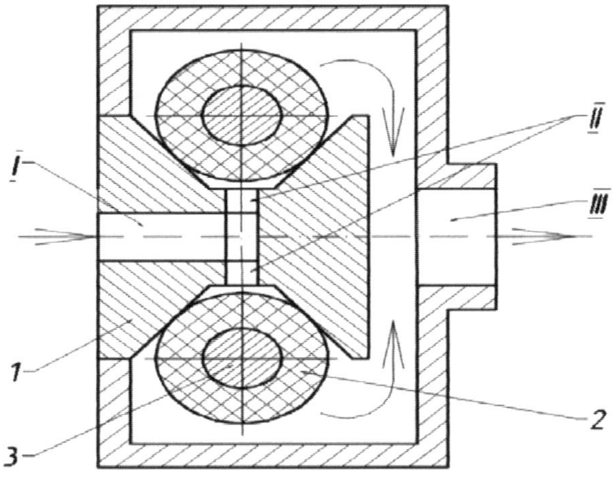

Figure 1. Hydraulic pressure regulator with shut-off and regulating the elastic element.

In the initial position the regulator (valve) is closed, the shut- off element 2 separates the channels (I) and (III). When increasing the pressure in the hydraulic system to which you are connected, the pressure regulator shut-off and regulating element 2 pressure exceeding the level regulator (pressure relief valve), deformed in the radial direction, forming an annular crack between the Conic surfaces grooves and locking element and providing a reset of the working fluid through the output channel III. By reducing the pressure shut-off element under the action of the elastic force element (springs) returns to its original position, cutting through the input and output channels III, I..

IOP Conf. Series: Journal of Physics: Conf. Series **858** (2017) 012035 doi:10.1088/1742-6596/858/1/012035

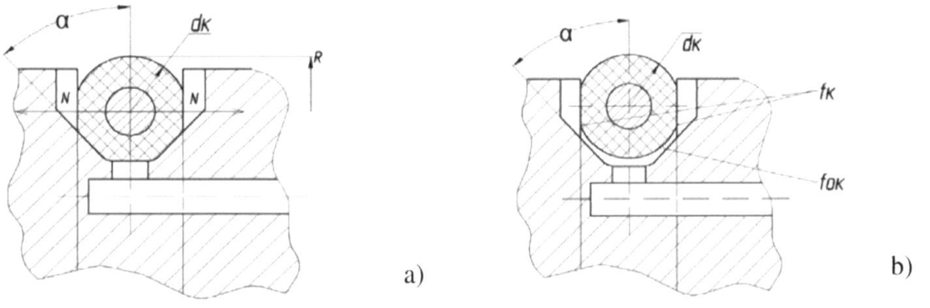

Figure 2. Position locking-regulating element a), working b).

For damping vibrations locking-regulating element in the design of the valves can be provided with vertical walls of the Groove in which the locking-regulating element (Figure 2) for sealing surfaces which creates a constant contact pressure excessive based on elastic material properties (E) and compression, when ε<30% of the pressure determined according to [1]

$$p_k = 1.25 \cdot \varepsilon \cdot E \cdot 10^{-3}\ MPa.$$

Contact area f_k locking-regulating element with vertical walls is $f_k = 2 \cdot \pi \cdot d \cdot \lambda$ defined as the grooves in which the width of the contact with the wall element is determined by the degree of deformation of this element when mounting for rubber hardness H_p = 55...75 according to Shore and it can be defined by the empirical formula $\lambda_k = 0.03 \cdot d_k \cdot \varepsilon$, mm [1].

The normal force is defined as $N_k = p_k f_k$ and the corresponding friction, taking into account the non-linear nature can be determined as $F_{mp} = f_{mp} N_k$ signsr, where r is the radial displacement locking-regulating element s-operator of differentiation, ftr-coefficient of friction between the vertical wall and locking-regulating element.

Mathematical model of the regulator with preserved fluid volume can be represented by the equations of equilibrium of the forces on the locking- regulating element and balance costs in the working area of the regulator (pressure relief valve). This assumes that the pressure at all points of pressure line is changed at the same time, the discharge coefficient of the working fluid through the working window regulator and its constant temperature, hydraulic resistance of drain line slightly, hydrodynamic forces are not counted because of small fluid flow speeds.

Applying the principle of D'alembert and the law of conservation of mass for some fixed hydraulic circuit, we get a system of differential equations that describe the dynamic processes occurring in the hydraulic circuit, which includes a pressure regulator.

$$Ms^2r + \beta sr + c(r_0 + r) + F_{mp} = pf,\ (1)$$

where m is the mass of the regulatory element; r – move an element; β – coefficient of viscous friction; c – the stiffness of the spring; r_0 is the pretension of the spring; f – square, which operates fluid pressure; p = inlet pressure regulator; s – operator of differentiation.

Balance the cost equation for fluid pressure regulator can be represented as follows

$$Q = Q_p + Q_{сж} + Q_{п},,\ (2)$$

where Q is the flow rate for preserved volume, $Q = f_1\ su$; $Q_p = B_p\ f_{ok}\ sqrt(\Delta p)$ – liquid flow rate passing through the window; $Q_{сж} = kWsp$ – fluid consumption caused by its compression; $Q_n = f\ sr$ – fluid comsumption formed by the offset of the regulating element; f_1 - area of preserved volume which is bounded by the edges of the cylinder volume; B_p - constant of working window regulator; Δp - pressure difference in the regulator; k - coefficient of compressibility of fluid; W is the amount of input cavity of the regulator; u - move of preserved volume.

Now the system of differential equation takes the following form:

$$ms^2r = pf - \beta sr - c(r_0 + r) - F_{mp},;$$

$$f_1 su = B_p f_{ok}\ sqrt(\Delta p) + kWsp + fsr. \ (3)$$

The dynamic stability of the pressure regulator may be the presence or absence of autooscillations in a wide range of frequencies, from acoustic resonance phenomena to exciting fluctuations caused by fluid compressibility or slackness has been established pump drive [2]. So, the volatility of the regulator can be caused by coincidence own frequency regulator with frequency pulsation of pump pressure.

The volatility caused by Governor compressibility of fluids, can be called as follows. On the application of a pressure pulse to the regulatory body in the direction of closing pressure in pressure cavity increases. Released when compressing the volume to fill the newly received liquid. After removing the pulse pressure regulator will begin to move in the opposite direction, that is accompanied by a decrease in pressure in the pressure valve cavity. Through the disc annular gap along with the main regulator flow must pass and the consumption that is released when expansion of the liquid, resulting in additional forces, aimed in the direction of movement of the shutter (regulatory body). If it is smaller, the damping force adjuster after some time return to the established State. Otherwise, the controller will enter the mode self- excited oscillations with constant or increasing amplitude.

For linear systems, defining the boundaries of a steady- state is trivial in nature. For nonlinear systems require further analysis, particularly in terms of how to define those boundaries. It is necessary to determine the feasibility of linearization varying nonlinearity identified when solving particular tasks, and also obtained when this error.

Taking into account non-linearities this model it is advisable to apply the method of phase trajectories [5], which allows you to explore nonlinear vibrational systems without having to build a whole picture phase trajectories. When you use this method, built phase trajectory of polucikla (in the case of symmetric phase trajectories) or (in the case of asymmetrical phase trajectories) traffic system, which defines a function turned [5], on which to construct the diagram Kenigsa-Lamereâ. On the type of chart, you can judge the system limit cycles or subcarinal regimes [6]. The initial values of the system parameters and disturbing impact of maximum are chosen. In the absence of limit cycles can argue that they will be absent in any other primary values and revolting influences.

Study of dynamic stability regulator conducted Runge-Kutta numerical method. Solution algorithm of mathematical model of pressure regulator (pressure relief valve direct action) provides for the transition process and phase trajectory in polar coordinates. For the study were taken the following initial conditions:

$$p_0 = 2MPa;\ m = 0.1\cdot 10^{-3}\ kg;\ f = 0.5\cdot 10^{-4}\ m;$$

$$f_{mp}=0.075;\ d = 0.8{\cdot}10^{-2}\ m;\ b\ = 50N{\cdot}s/\ m;\ c=103N/m;$$

$$r = 3{\cdot}10^{-3}\ N/m;\ f = 0.16\cdot 10^{-2}\ m^2;\ m = 0.4;$$

$$B = 0.18{\cdot}10^{-3}m^3/Ns;\ k\ = 0.714\cdot 10^{-3}\ 1/\ MPa;$$

$$W = 2{\cdot}10^{-4}\ m^3.$$

Phase trajectory obtained by numerical method, compared with constructed graphically (Figure 3). The resulting difference should not exceed 5%, which can be considered as quite acceptable.

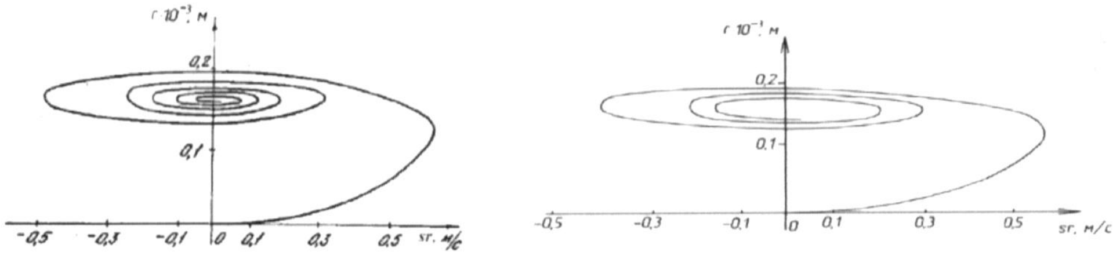

a) b)

Figure 3. Phase trajectory of locking-regulating valve element (shutter) received (a) numerical and graphical (b) methods.

The study found that the dynamic stability of safety valves direct action largely depends on: the mass closure (shutter), the stiffness of the spring, volume pressure line, dry and viscous friction.

Built phase trajectory of shut-off and regulating element without (Figure 4 a) and taking into account the forces of friction (Figure 4 b) shows that in the first case there has been a limit cycle, while the second is a fading transition.

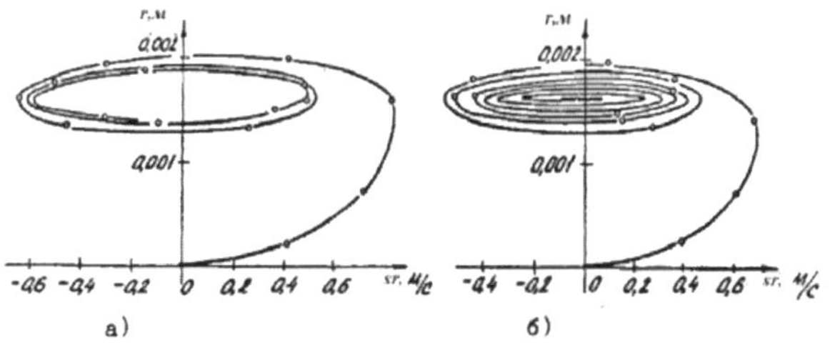

Figure 4. Phase regulator shutter motion path excluding (a) and with (b) the force of friction bolt.

In the research process as changing initial settings were adopted: the stiffness of the spring, damping factor, consumption and the starting offset of the shutter. Relevant phase trajectories are presented in Figure 5, 6. Based on the results of the calculations are constructed diagrams Kenigsa- Lamereâ that can be used to judge the nature of transients in hydraulic circuit with pressure regulator (direct-acting relief valve with an elastic locking-regulating element).

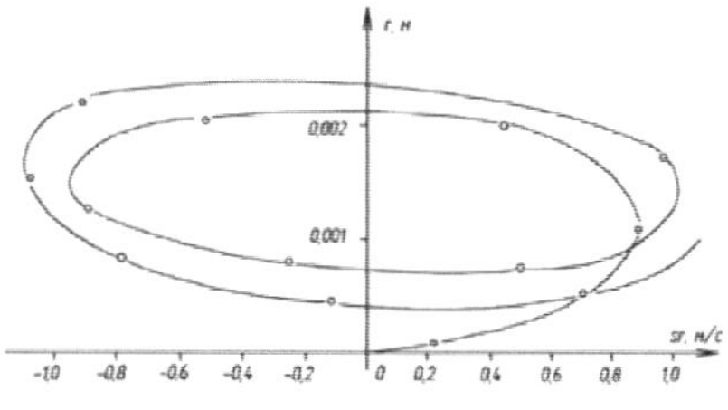

a)

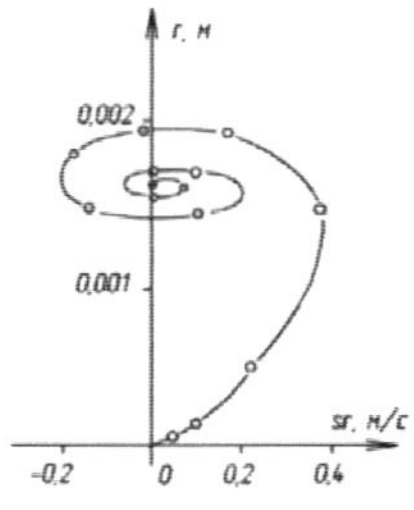

b)

Figure 5. Phase trajectories of the valve movement when changing damping coefficient of the regulator.

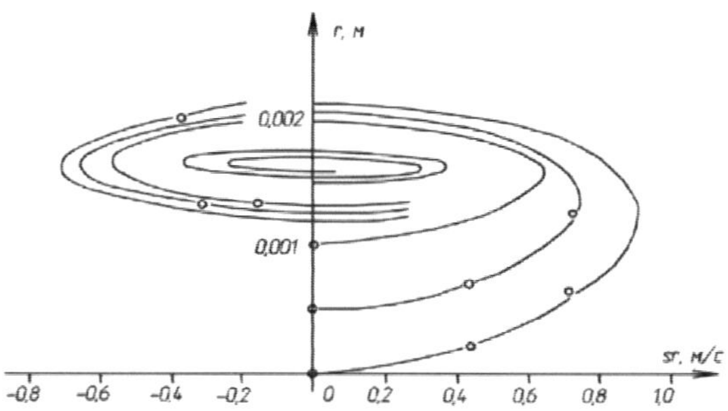

Figure 6. Phase trajectories of the regulator valve when changing the initial displacement of the r_0 valve.

4. Results

Dynamic characteristics are obtained to ensure the absence of self-oscillations due to the additional damping of the shut-off- and-regulating element. The prototype of the elastic pressure regulator with the shut-off-and-regulating element is constructed. Characteristics dependences of the pressure regulator (pressure relief valve with direct action) from the initial conditions are defined.

Figures 7, 9 present the diagrams, obtained by changing the initial values of the spring stiffness. The diagram first return function, is under the bisecting line, indicating the absence of limit cycles in the system and the presence of the damping transition process [5]

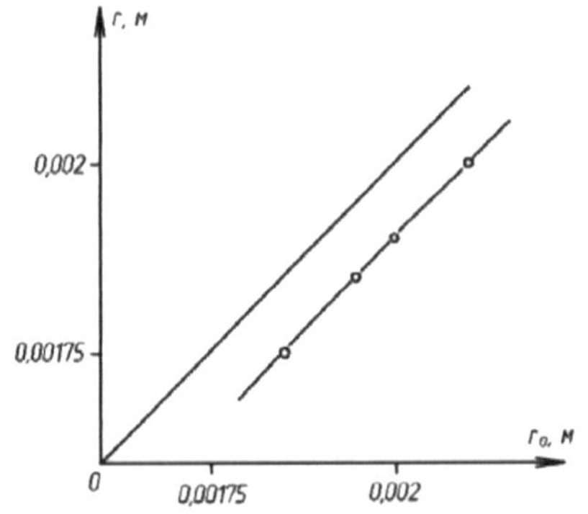

Figure 7. Kenigs-Lamereâ's diagram obtained by changing the spring stiffness.

Figures 8, 10 present the diagrams, obtained by changing the initial values of the damping coefficient and pressure. In both cases, the first return functions cross the bisector line; it shows that there is a limit cycle in the system, i.e. a damping transition process.

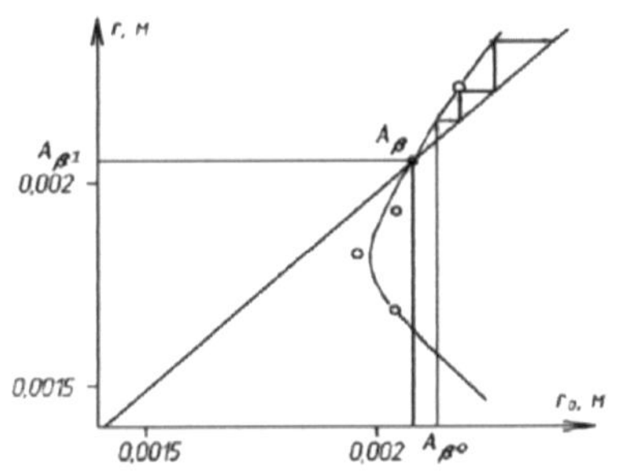

Figure 8. Kenigs-Lamereâ's diagram obtained by modification of the regulator damping coefficient β.

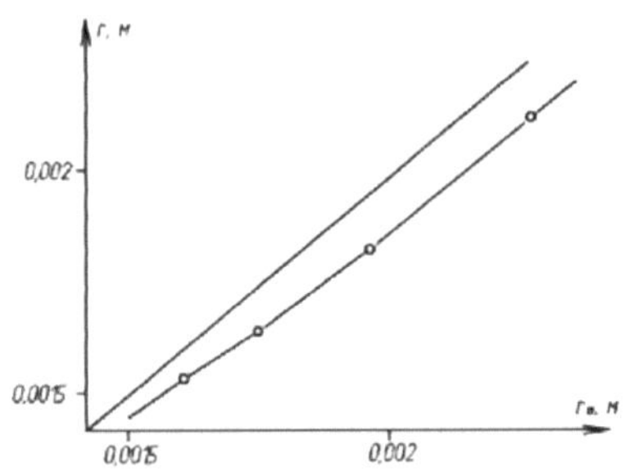

Figure 9. Kenigs-Lamereâ's diagram obtained by the modification of the input flow rate Q into the regulator.

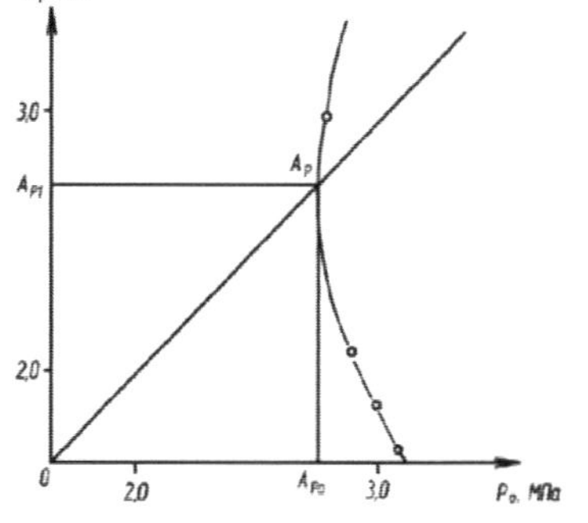

Figure 10. Kenigs-Lamereâ's diagram obtained by modification of the initial pressure p_0.

5. Discussion of results

Projecting points of intersection A_p an A_β on the relevant axes of self-excited oscillations amplitudes $A_{\beta0}$ and $A_{\beta1}$; Ap_0 and Ap_1 accordingly. The period of self-oscillations is determined by the phase trajectory in time of the single cycle of self-oscillations in a certain integration step.

After comparing the values of self-oscillations parameters of the researched system, which were obtained experimentally, (A_a = 1.8 MPa, f_a = 350 Hz) and in response to the numerical solution of the mathematical model (A_a = = 1.9MPa; f'_a = 330 Hz) we establish that the proposed mathematical model describes adequately the dynamic processes occurring in the hydraulic system with hydraulic regulator of the proposed design.

6. Summary and conclusions

The analysis of the results shows that the main parameters affecting the dynamic stability of the pressure regulator is the damping factor and the initial pressure. Finding the best values of these parameters was carried out through the analysis of transient characteristics of the valve when the impact on it of the standart disturbance on consumption. It has been established that valve mode is unstable in the case when the damping coefficient is < 40 Ns/m and the initial pressure is > 3 MPa. When you change other parameters limiting unstable modes were not observed. The obtained results can be used when designing hydraulics of the technological machines.

References

[1] Bashta T M, Khaindrava Z K 1974 Effect Of Vibrations On The Friction And Hermetic Sealing Of Plunger Pairs In Hydraulic Assemblies *Russ. Eng. J* **7** 3-5

[2] But'ko V S and Pogodaev F G 1972 Resistance of hydraulic bleed-off valves with the exhaust chamber Hydrodrive and hydraulic and pneumatic control systems of machines **8** pp 85–88

[3] Sitnikov B T, Matveev N B 1972 Calculation and research of safety and bleed-off valves. (Moscow, Mechanical Engeneering)

[4] Tyagunov F F 2004 Pressure hydraulic valves with indirect action. Moscow

[5] Andronow A A, Chaikin C E 1949 Theory Of Oscillations *Princeton University Press* 358

[6] Dragoslav D. Šiljak 1968 Nonlinear systems: the parameter analysis and design. Wiley

[7] Syrkin V V, Treier V A 2016 Dynamics of an indirect hydraulic pressure regulator with an elastic element *Russian Engineering Research* **36(4)** pp 270-272

IOP Conf. Series: Journal of Physics: Conf. Series **858** (2017) 012036 doi:10.1088/1742-6596/858/1/012036

Dynamics of the process boom machine working equipment under the real law of the hydraulic distributor electric spool control

V N Tarasov, I V Boyarkina

Siberian State Motor Road University, 5, Mira ave., Omsk, 644050, Russia

E-mail: tarasov_vladimir07@mail.ru

Abstract. Analytical calculation methods of dynamic processes of the self-propelled boom hydraulic machines working equipment are more preferable in comparison with numerical methods. The analytical research method of dynamic processes of the boom hydraulic machines working equipment by means of differential equations of acceleration and braking of the working equipment is proposed. The real control law of a hydraulic distributor electric spool is considered containing the linear law of the electric spool activation and stepped law of the electric spool deactivation. Dependences of dynamic processes of the working equipment on reduced mass, stiffness of hydraulic power cylinder, viscous drag coefficient, piston acceleration, pressure in hydraulic cylinders, inertia force are obtained. Definite recommendations relative to the reduction of dynamic loads, appearing during the working equipment control are considered as the research result. The nature and rate of parameter variations of the speed and piston acceleration dynamic process depend on the law of the ports opening and closure of the hydraulic distributor electric spool. Dynamic loads in the working equipment are decreased during a smooth linear activation of the hydraulic distributor electric spool.

1. Introduction

The improvement and complication of engineering systems stimulate the development and improvement of the differential equation solution methods. During the dynamic processes research the differential high-order equations are substituted by equivalent systems of differential equations of first and second order, as the solution methods of these equations are well-proven. Papers [1, 2, 3] are devoted to the solution methods improvement of the differential equation systems of first and second order.

In paper [1] the authors M. Abas, A. Libošvárová improve the solution of the linear differential equations systems with constant coefficients in the field of mathematical modeling of dynamic systems. New numerical solutions of nonhomogeneous linear differential equations with several variables are developed by the authors Y. Kida, T. Kida in paper [2]. In paper [3] the authors M. Greitans, E. Hermanis, A. Greitane consider linear differential equations of first and second order. Second-order equations contain constant coefficients, in paper [3] first-order equations contain time-varying coefficients.

In paper [4] the authors G. Reissig, H. Boche, P.I. Barton improve the initial conditions determination methods for the linear differential equations systems of first order. In paper [5] the authors X. Lin, K. Lu, H. Wei, Y. Bai examine the problems of solvability of time-varying linear differential equations of second order.

In paper [6] the authors M. H. Eghlidi, K. Mehrany, B. Rashidian propose the general solution of the liner differential equations by means of the differential method.

In paper [7] the authors Y. Zhang, J. Sun examine the stability of the impulse linear differential equations with delay.

In paper [8] the authors J. Zhou, Q. Wu discuss the problems of exponential stability of pulse delays in the linear differential equations.

For the mathematical description of engineering systems the differential equations of second order are used in the form of differential equations of free, damped or forced oscillations of mechanical systems. More complex dynamic processes are described by the linear differential equations of third and higher orders.

The boom bucket machines working equipment dynamics is described in this article by the differential equation of third order, which is represented by differential equation of hydraulic power cylinder [9]

$$m\dddot{s} + \vartheta\ddot{s} + C\dot{s} = Kx. \qquad (1)$$

Equation (1) is a differential equation of hydraulic power cylinder, where constant coefficients have a definite physical content: m – mass of the working equipment reduced to the piston ϑ – viscous resistance coefficient; C – stiffness coefficient; s – movement of the hydraulic cylinder piston of the working equipment; x – ports opening size of the hydraulic distributor electric spool.

The right side of equation (1) may be different and represents the ports opening size of the hydraulic distributor electric spool.

The derivation of the hydraulic power cylinder differential equation is performed by V.N. Tarasov in paper [9]. According to physics the equation of third order is equivalent to the system of two equations: second-order equation (oscillatory link) and first-order equation in the form of an integrating factor.

The numerical solution of differential equation (1) is performed in paper [10] using the software package MATLAB, where third-order differential equation is represented in the form of the differential equations system of first order

$$\dot{S}_1 = S_2;\ \dot{S}_2 = S_3;\ \dot{S}_3 = -2nS_3 - \omega^2 x_2 + d(t).$$

The numerical solution has the indicated peculiarities, that's why this article proposes the direct analytical solution of differential equation (1). The direct analytical solution of the hydraulic power cylinder differential equation is considered as third-order differential equation with the right side in the form of the ports opening linear size of the hydraulic distributor electric spool.

Process boom bucket machines possess multifunctional working equipment for usage in mining industry construction, agriculture and other spheres.

Figure 1 shows as an example a front-end hydraulic loader, which hydraulic power cylinders of the working equipment together with the hydraulic distributor and hydraulic pump receive masses reduced to the piston, possess elastic characteristics and damping properties.

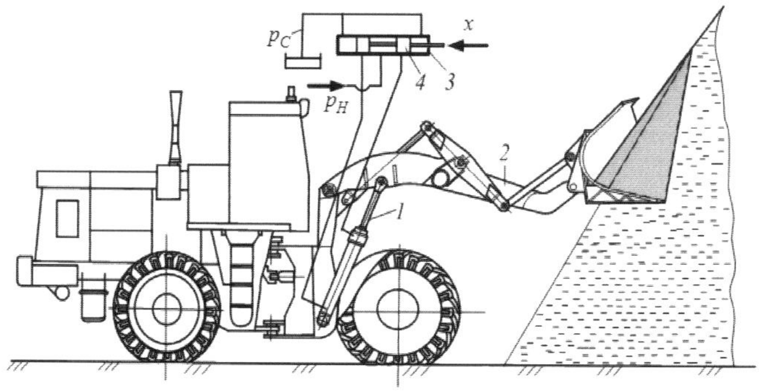

Figure 1. Front-end loader: *1* – hydraulic power control cylinders; *2* – working equipment; *3* – hydraulic distributor of working equipment; *4* – electric spool; p_H, p_C – working fluid pressure during delivery and discharge to the tank, respectively.

Modern process bucket machines are in demand and form size ranges which include small-size and capacity machines, heavy machines and super heavy models of machines.
Small-size and medium-size machines are sufficiently studied and well-tried to a greater extent.

2. Task definition

The article sets the task to develop the analytical solution of third-order differential equation for the dynamics research of the hydraulic machines working equipment transient processes.
Regarding high-capacity machines and super models the dynamic characteristics and performance features of dynamic parameters among which the most important ones are represented by inertia masses of the working equipment reduced to the hydraulic cylinder piston, stiffness of the hydraulic cylinder, viscous resistance, pressure in the hydraulic system etc. are not sufficiently studied.
The dignity of modern hydraulic power cylinders lies in a quick acceleration of the working equipment during the hydraulic distributor electric spool activation and rigid fixation of working equipment positions in the technological process.

3. Theory

The linear differential equation of third order contains the right side in the form of the coordinate *x*, characterizing the ports opening size of the hydraulic distributor electric spool [9, 10, 11]. At present the solution of the differential equations is usually performed by numerical methods [10]. But the proposed analytical solution is simpler and more effective.
The dynamic transient processes of the parameters variation: movements of the piston *s*, piston speed $\dot{s}$, piston acceleration $\ddot{s}$ depend on the hydraulic distributor electric spool activation modes $x=f(t)$.
Figure 2 shows the basic kinds of the distributor control laws: stepped (step-wise); linear; real laws of the hydraulic distributor electric spool control.

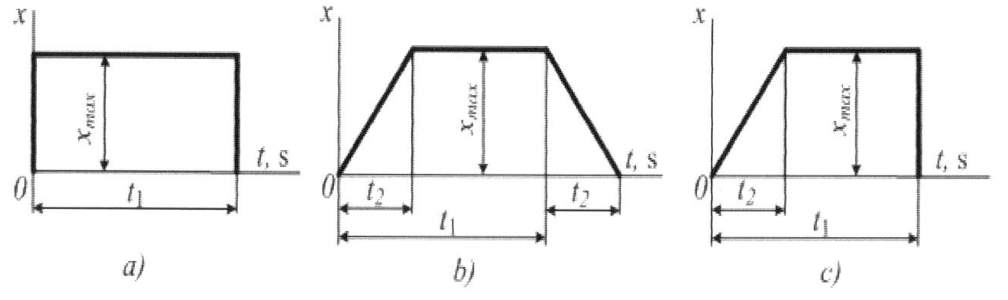

Figure 2. Control laws of the hydraulic distributor electric spool:
a) stepped law $t = t_1$; *b)* linear law $t = t_2$; *c)* real law.

The linear diagram of the hydraulic distributor electric spool control (Figure 2,*b)* is as follows:
electric spool activation mode is as follows

$$x = x_{\max}\frac{t}{t_2} \quad \text{at } 0 \le t \le t_2; \tag{2}$$

electric spool activated status mode

$$x = x_{\max} \quad \text{at } t_2 < t \le t_1; \tag{3}$$

electric spool deactivation mode

$$x = x_{\max}\left(1 - \frac{t}{t_2}\right) \quad \text{at } t_1 \le t \le (t_1 + t_2); \tag{4}$$

locked status mode of the hydraulic power cylinder chambers

$$x = 0 \quad \text{at } t > (t_1 + t_2). \tag{5}$$

Figure 2,*c* shows that the linear real activation mode of the electric spool is complemented with the real stepped deactivation of the hydraulic distributor electric spool.
These four considered statuses of the hydraulic distributor electric spool characterize sequential cases of the working equipment dynamic transient processes appearance and their attenuation. By means of replacing $\dddot{s} = \ddot{V}$, $\ddot{s} = \dot{V}$, $\dot{s} = V$ differential equation (1) of third order is reduced to differential equation of second order with the right side [9, 11]

$$\ddot{V} + 2n\dot{V} + \omega^2 V = \frac{K}{m}x, \tag{6}$$

where $2n = \vartheta/m$; n – resistance coefficient; ω – natural vibration frequency, $\omega = \sqrt{C/m}$.
The dynamic study of the process machines working equipment under the linear law of the electric spool activation is fulfilled using differential equation

$$\ddot{V} + 2n\dot{V} + \omega^2 V = \frac{K}{m}x_{\max}\frac{t}{t_2}, \tag{7}$$

where $K = \dfrac{CV_{con}t_2}{x_{\max}}$; $n = \beta\omega$.

The solution of differential equation (7) is considered as the sum of two solutions $V = V_1 + V_2$, where V_1 – solution of the differential equation (7) without the right side; V_2 – particular solution of general equation (7), $V_2 = C_3 + C_4 t$.
The analytical solution of differential equation (7) in case of the linear activation of the electric spool is as follows

$$V = e^{-nt}\left(C_1\cos\omega_1 t + C_2\sin\omega_1 t\right) + C_3 + C_4 t; \tag{8}$$

$$\dot{V} = e^{-nt}\left[(C_2\omega_1 - C_1 n)\cos\omega_1 t - (C_2 n + C_1\omega_1)\sin\omega_1 t\right] + C_4, \tag{9}$$

where ω_1 – circular frequency of damping oscillations, $\omega_1 = \sqrt{\omega^2 - n^2}$.
The initial conditions of differential equation (7) are as follows: at t=0 $V_0 = 0$; $\dot{V}_0 = 0$. Constants of integration in obtained expressions (8), (9) are determined proceeding from the initial conditions as per expressions:

IOP Conf. Series: Journal of Physics: Conf. Series **858** (2017) 012036 doi:10.1088/1742-6596/858/1/012036

$$C_1 = 2\frac{\beta V_{con}}{\omega t_2};\ C_2 = \frac{V_{con}}{\omega_1 t_2}(2\beta^2 - 1);\ C_3 = -\frac{2\beta V_{con}}{\omega t_2};\quad C_4 = \frac{V_{con}}{t_2}, \tag{10}$$

where β – damping coefficient of the hydraulic power cylinder, $\beta = n/\omega$; V_{con} – steady-state value of the piston speed after the oscillating process damping.

The electric spool activation operation is performed during the period of time $t = t_2$.

During this period the speed and acceleration are varied according to formulas (8), (9) during the time period of the electric spool ports opening.

After the electric spool activation the right side of the differential equation becomes constant and is as follows

$$\ddot{V} + 2n\dot{V} + \omega^2 V = \frac{K}{m}x_{\max}. \tag{11}$$

The initial conditions in differential equation (11), characterizing the previous process conditions are as follows: at t=0 $V = V_0$; $\dot{V} = \dot{V}_0$. The process of the working equipment movement to a new position is performed as a damping oscillation process of the speed transition to a steady-state value with completely open ports of the electric spool during the period of time $t = t_1 - t_2$ of the working equipment movement.

After obtaining a required movement of the working equipment the electric spool linear deactivation operation is performed using the differential equation with the right side

$$\ddot{V} + 2n\dot{V} + \omega^2 V = \frac{Kx_{\max}}{m}\left(1 - \frac{t}{t_2}\right). \tag{12}$$

The constant values of differential equation (12) are as follows

$$\omega^2 = \frac{C}{m};\ K = \frac{CV_{con}}{x_{\max}};\ n = \beta\omega.$$

The initial conditions of equation (12) are as follows: at t=0 $V = V_0$; $\dot{V} = \dot{V}_0$. We take the electric spool deactivation time as $t = t_2$.

The solution of differential equation (12) is as follows

$$V = e^{-nt}\left(C_1 \cos\omega_1 t + C_2 \sin\omega_1 t\right) + C_3 + C_4 t; \tag{13}$$

$$\dot{V} = e^{-nt}\left[\left(C_2\omega_1 - C_1 n\right)\cos\omega_1 t - \left(C_2 n + C_1\omega_1\right)\sin\omega_1 t\right] + C_4. \tag{14}$$

The obtained constants of integration correspond to the adopted initial conditions

$$C_1 = V_0 - V_{con}(1 + 2\frac{\beta}{t_2\omega});\ C_2 = \frac{\dot{V}_0 + C_1 n - C_4}{\omega_1};\ C_3 = V_{con}(1 + 2\frac{\beta}{t_2\omega});\ C_4 = -\frac{V_{con}}{t_2}. \tag{15}$$

The working equipment transient process with the electric spool linear deactivation according to equations (13), (14) is performed during the time period $t = t_2$.

At the moment of the complete closure of the electric spool ports the process of the working equipment oscillations damping is started which is simulated by the differential equation with zero right side:

$$\ddot{V} + 2n\dot{V} + \omega^2 V = 0. \tag{16}$$

IOP Conf. Series: Journal of Physics: Conf. Series **858** (2017) 012036 doi:10.1088/1742-6596/858/1/012036

The solution of differential equation (16) with the deactivated electric spool is as follows

$$V = e^{-nt}\left(C_1 \cos \omega_1 t + C_2 \sin \omega_1 t\right); \tag{17}$$

$$\dot{V} = e^{-nt}\left[\left(C_2\omega_1 - C_1 n\right)\cos \omega_1 t - \left(C_2 n + C_1\omega_1\right)\sin \omega_1 t\right], \tag{18}$$

where C_1, C_2 – constants of integration which are as follows

$$C_1 = V_0; \quad C_2 = \frac{\dot{V}_0 + nV_0}{\omega_1}. \tag{19}$$

Analytical solutions of differentials equations of the working equipment movement are recorded for the first time and they have allowed to fulfill the researches of the hydraulic drive parameters influence on the dynamics of transient processes with the linear control of the hydraulic distributor electric spool.

4. Research results

The nature and rate of parameter variations of the speed and piston acceleration dynamic process depend on the law of the ports opening and closure of the hydraulic distributor electric spool.
Figure 3 shows the transient processes research results of the hydraulic working equipment of a front-end loader with the loading capacity of 3 tons under the linear control law of the hydraulic distributor electric spool by means of varying the activation and deactivation time of the electric spool t_2 = var.
Figure 3,*a* shows that with the electric spool activation time t_2 = 0.05s the transient processes of the acceleration *a* and speed V_1 possess the oscillatory nature with overshoot.

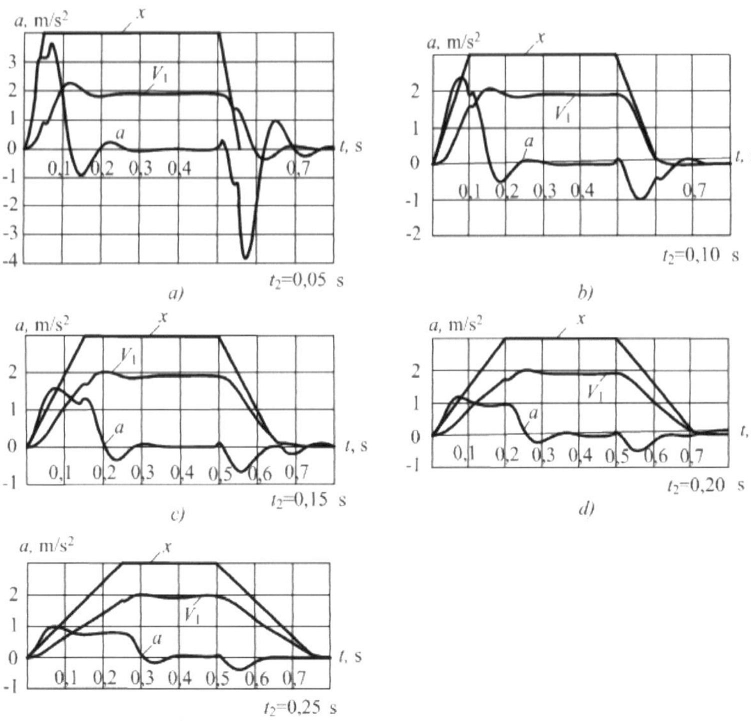

Figure 3. Transient processes of the hydraulic cylinder piston during the variation of the electric spool activation time for the loader with the loading capacity of 3 tons.

5. Discussion of results

The electric spool activation time increase leads to the reduction of oscillability and parameters overshoot. At a maximum value of the electric spool activation time $t_2 = 0.25$s the acceleration a reduced to a=0.8 m/s^2, at the same time the speed overshoot V_1 practically disappeared (Figure 3,*e*).

That's why the electric spool activation time increase, i.e. a smooth nonstepped adjustment is the reserve of the transient processes dynamics adjustment smoothness increase of the bucket machines working equipment with hydraulic power cylinders.

That's why an important conclusion arises that the electric spool activation time increase, i.e. a smooth nonstepped adjustment is the reserve of the transient processes dynamics adjustment smoothness increase of the bucket machines working equipment with hydraulic power cylinders. Another important conclusion proceeding from the submitted results of the research represents the conclusion that with the electric spool activation time $t_2 > 0.2$s the adjustment time t_p practically coincides with the electric spool linear activation time $t_p = t_2$.

This means that in case when $t_p = t_2$, after the electric spool activation the acceleration $a \to 0$, and the speed V_1 acquires the steady-state value $V_1 = V_{con}$ practically without oscillations.

Figure 4 shows dependences of the acceleration a of the reduced mass on the opening time of the electric spool ports t_2 for the boom machine with the lifting capacity of Q=3 t.

Curve *1* corresponds to the piston speed V_1=0.1 m/s, curve *2* – to the piston speed V_1=0.2 m/s.

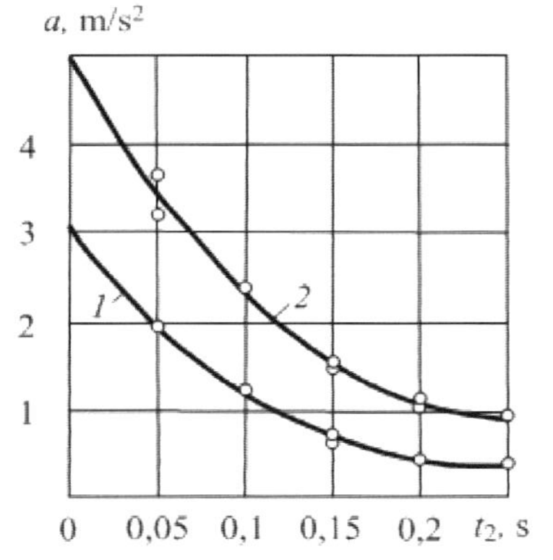

Figure 4. Dependence of acceleration on the electric spool activation time.

The inertia force at the piston is determined by the formula

$$\Phi_d = ma.$$

The dynamic additional pressure excited in hydraulic cylinders by the force Φ_d, is determined by the formula

$$p_d = \frac{\Phi_d}{n_c(\pi D_c^2/4)},$$

where n_c – number of hydraulic power cylinders, n_c=2; D_c – diameter of hydraulic power cylinder, D_c=0.125 m.

The indicated pressure p_d is a pulse pressure, which operates during the time period $t = 0.1$s (see Figure 3,*a*).

For the loader with the lifting capacity of Q=3 t, reduced mass of which is equal to m=80270 kg,

IOP Conf. Series: Journal of Physics: Conf. Series **858** (2017) 012036 doi:10.1088/1742-6596/858/1/012036

acceleration $a_{\max}$ =3 m/s^2 (see Figure 4), inertia force and pressure have the real values Φ_d =240810 N, p_d =9.811 MPa.

6. Summary and conclusion

1. The dynamics of the modern boom machines working equipment, simulated by third-order linear differential equation with the right side, depends on the electric spool ports opening size and nature.
2. It's proposed to describe the dynamics of the boom bucket machines working equipment using the analytical method by means of the differential equation depression of the working equipment hydraulic power cylinders.
3. Control processes of the working equipment acceleration and deceleration dynamics are adjusted by the right side of the differential equation. The damped oscillating dynamic process of the working equipment is simulated by the differential equation with zero right side.

References

[1] M. Abas, and A. Libošvárová, "Solutions of systems of linear differential equations with constant coefficients in education of automation," Interactive Collaborative Learning (ICL), 2013, International Conference. pp. 776 – 777.
[2] Y. Kida, and T. Kida, "Development of a new numerical solution of inhomogeneous linear partial differential equations with many independent variables," Information Theory and its Applications (ISITA), 2010 International Symposium on Year: 2010. pp. 627 – 632.
[3] M. Greitans, E. Hermanis, and A. Greitane, "Construction of growth models using linear differential equations," EUROCON 2009, EUROCON '09. IEEE Year: 2009. pp. 112 – 117.
[4] G. Reissig, H. Boche, and P. I. Barton, "On inconsistent initial conditions for linear time-invariant differential-algebraic equations," IEEE Transactions on Circuits and Systems I: Fundamental Theory and Applications Year: 2002, Volume: 49, Issue: 11. pp. 1646 – 1648.
[5] X. Lin, K. Lu, H. Wei, and Y. Bai, "On the solvability of time varying linear differential equations of second order," Multimedia Technology (ICMT), 2011, International Conference, pp. 2515 – 2518.
[6] M. H. Eghlidi, K. Mehrany, and B. Rashidian, "General solution of linear differential equations by using differential transfer matrix method," Proceedings of the 2005 European Conference on Circuit Theory and Design, 2005. Volume: 3, p. III/113 - III/116. vol. 3.
[7] Y. Zhang, and J. Sun, "Stability of impulsive linear differential equations with time delay," IEEE Transactions on Circuits and Systems II: Express Briefs Year: 2005, Volume: 52, Issue: 10. pp. 701 – 705.
[8] J. Zhou, and Q. Wu, "Exponential Stability of Impulsive Delayed Linear Differential Equations," IEEE Transactions on Circuits and Systems II: Express Briefs Year: 2009, Volume: 56, Issue: 9. pp. 744 – 748.
[9] V.N. Tarasov. Dynamics of control systems of working processes of earth-moving machines (monography) / V.N. Tarasov. – West Siberian Book Publisher, Omsk, 1975. – 182 pages.
[10] V.N. Tarasov. Theory of impact in the construction and mechanical engineering industries/ V.N. Tarasov, I.V. Boyarkina and others – M.: ASV Publishers, 2006. – 336 pages.
[11] I.V. Boyarkina. Technological mechanics of single-bucket front-end loaders: monography/ I.V. Boyarkina. – Omsk; SibADI, 2011. – 336 pages.

IOP Conf. Series: Journal of Physics: Conf. Series **858** (2017) 012037 doi:10.1088/1742-6596/858/1/012037

Software correction of geometric errors for multi-axis systems

V Teleshevskii [1] **and V Sokolov** [1]

[1] MSTU «STANKIN», 1, Vadkovsky per., Moscow, 127994, Russia

E-mail: Vasokolov-stankin@yandex.ru

Abstract. Accuracy control for software-controlled multi-axis systems (i.e. machining centers and CMMs) is an important problem for contemporary industry. The paper concerns a method of systematic geometric errors correction for such systems. Practically this method increases machining and measuring accuracy without mechanical adjustments of the machine. The goal of accuracy improvement was achieved by means of completing the following tasks: measuring components of volumetric error by means of multi-function laser interferometer; visualizing volumetric error distribution within the machine's workspace (error mapping); developing of error correction strategy based on the mapping; developing of postprocessor for CNC programs in order to get minimal errors within the machine's workspace. The paper includes results of volumetric errors correction for three-axis CNC-controlled precision machine-tools. The results prove adequacy of models and efficiency of the method.

1. Introduction (Heading 1)

Software-controlled multi-axis systems, both technological (various machine-tools and robotic systems) and measuring (coordinate-measuring machines and devices) are the basis of modern industry [1,2]. Present-day systems of this kind may be comprised of as many as 5 or 6 axes and even more.

Parts with complex three-dimensional surfaces are widely used in modern industry. This requires not only high positioning accuracy for machine's axes, but also high volumetric accuracy - ability of a machine to precisely create a complex three-dimensional shapes by means of a cutting edge or gaging tip. The problem of volumetric accuracy improvement is often considered in recent papers [1-3].

2. Problem Description

The volumetric accuracy is defined by volumetric error (similarly to measurement error as difference between measure and datum value) – a **vector** from **nominal** (program defined) and **actual position** of a cutting edge or gaging tip within a certain point of machine's workspace. Volumetric error for CNC-controlled multi-axis systems is defined as follows [3]. A real Cartesian coordinate system of a machine is concerned with three axes X_M, Y_M, Z_M, with a movement of ΔX, ΔY, ΔZ from point A to point B (Figure 1 a). This movement in an ideal coordinate system X_I,Y_I,Z_I moves not from point A to point B_I with coordinates ΔX_I, ΔY_I, ΔZ_I, but from point A to point B_M. This difference is presented by error vector B_M - B_I with components EX, EY, EZ.

This error is caused by the following. The moving part of a machine, e.g. table, not only moves linearly in the desired axis direction. It also has a number of unwanted degrees of freedom: two linear traversal degrees and three rotary degrees of freedom that can't be completely excluded. Furthermore, the three axes of a machine have squareness errors. Thus 3 axes with 6 degrees of freedom each and 3

IOP Conf. Series: Journal of Physics: Conf. Series **858** (2017) 012037 doi:10.1088/1742-6596/858/1/012037

straightness errors make up 21 volumetric error components. These components form volumetric errors vector field within the machine's workspace (Figure 1.b). The components are listed below. The angular errors are marked with Greek letter ε, linear errors are marked with Greek letter δ, squareness errors are marked with Greek letter - α [3].

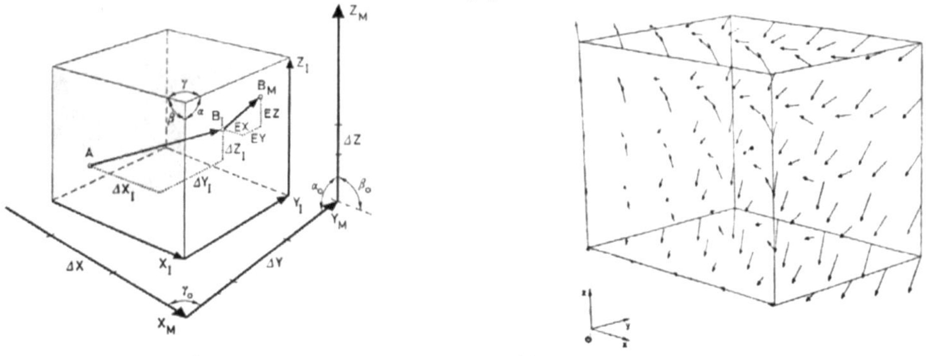

a) b)

Figure 1. Volumetric accuracy of machines (a); Errors field (b).

1) X positioning (scale) error - $\delta_{xx}(X)$
2) Y straightness of X Yslraibhlncss of X - $\delta_{yx}(X)$
3) Z straightness of X - $\delta_{zx}(X)$
4) Roll of X - $\varepsilon_{xx}(X)$
5) Pitch of X - $\varepsilon_{yx}(X)$
6) Yaw of X - $\varepsilon_{zx}(X)$
7) Y positioning (scale) error - $\delta_{yy}(Y)$
8) X straightness of Y - $\delta_{xy}(Y)$
9) Z straightness of Y - $\delta_{zy}(Y)$
10) Roll of Y - $\varepsilon_{yy}(Y)$
11) Pitch of Y - $\varepsilon_{xy}(Y)$
12) Yaw of Y - $\varepsilon_{zy}(Y)$
13) Z positioning (scale) error - $\delta_{zz}(Z)$
14) X straightness of Z - $\delta_{xz}(Z)$
15) Y straightness of Z - $\delta_{yz}(Z)$
16) Roll of Z - $\varepsilon_{zz}(Z)$
17) Pitch of Z - $\varepsilon_{xz}(Z)$
18) Yaw of Z - $\varepsilon_{yz}(Z)$
19) X,Y squareness - α_{xy}
20) Y,Z squareness - α_{yz}
21) X,Z squareness - α_{zx}

In order to effectively correct the volumetric errors an error map – error distribution within the machine's workspace – is required. Next, a correction strategy will be developed based on the error mapping [4]. Thus, the problem of the paper is to develop a method of measurement and correction of errors.

3. Theory

A modern machine-tool or CMM comprises of a software-controlled linear or rotary axes. Movement errors and axes displacement sum up into volumetric errors field [4].

The three-dimensional workspace of a machine is the machining or measurement space. This space is being deformed due to the linear and angular motion errors, thus the workpiece is being processed or

IOP Conf. Series: Journal of Physics: Conf. Series **858** (2017) 012037 doi:10.1088/1742-6596/858/1/012037

measured with errors. The problem arises to determine the errors of a part (e.g. after machining) or part's image (e.g. 3d-model after measuring) based on axes errors measurement results. The axes errors should be measured and volumetric error map should be visualized in order to solve this problem.
The authors use Rigid body kinematics method and Denavit-Hartenberg notation in order to describe motion of machine's parts [3].
The volumetric geometric error in machine's workspace points is calculated by means of solving a matrix equation (1):

$$XYZ = R_X\left[R_Y\left(R_Z^{-1}T + Z - Y\right) - X\right] \quad (1)$$

where $\mathbf{X},\mathbf{Y},\mathbf{Z}$ – linear and squareness error vectors; R_X, R_Y, R_Z – rotation matrices; T– tool size correction vector.
Equation (1) expands into equations for error vector components for any workspace point:
X axis error component:

$$\begin{aligned}\Delta X &= \delta_{XX}(X) + \delta_{XZ}(Z) + \delta_{XY}(Y) + Y[\varepsilon_{ZY}(Y) + \varepsilon_{ZX}(X)] + Y\alpha_{YX} - Z[\varepsilon_{YY}(Y) + \varepsilon_{YX}(X)] - \\ &- Z\alpha_{ZX} + X_T - Y_T[\varepsilon_{ZX}(X) + \varepsilon_{ZY}(Y)] - Z_T[\varepsilon_{YZ}(Z) + \varepsilon_{YY}(Y) + \varepsilon_{YX}(X)]\end{aligned} \quad (2)$$

Y axis error component:

$$\begin{aligned}\Delta Y &= \delta_{YY}(Y) + \delta_{YX}(X) + \delta_{YZ}(Z) - X\varepsilon_{ZX}(X) - Z[\varepsilon_{XY}(Y) + \varepsilon_{XX}(X)] - Z\alpha_{ZY} + \\ &+ X_T[\in_{ZY}(Y) + \in_{ZX}(X)] + Y_T - Z_T[\varepsilon_{XY}(Y) + \varepsilon_{XX}(X) + \varepsilon_{XZ}(Z)]\end{aligned} \quad (3)$$

Z axis error component:

$$\begin{aligned}\Delta Z &= \delta_{ZZ}(Z) + \delta_{ZX}(X) + \delta_{ZY}(Y) - X\varepsilon_{YX}(X) - Y[\varepsilon_{XY}(Y) + \varepsilon_{XX}(X)] + \\ &+ X_T[\varepsilon_{YX}(X) + \varepsilon_{YZ}(Z) + \varepsilon_{YY}(Y)] + Y_T[\varepsilon_{XY}(Y) + \varepsilon_{XZ}(Z) + \varepsilon_{XX}(X)] + Z_T\end{aligned} \quad (4)$$

Where X, Y, Z – coordinates of a workspace point;
X_T, Y_T, Z_T – vector T components;
Other values were described in part 2.
The absolute value of geometric error vector for a workspace point is calculated as (5).

$$\Delta = \sqrt{\Delta X^2 + \Delta Y^2 + \Delta Z^2} \quad (5)$$

Thus a large volume of observation is required in order to obtain information on volumetric geometric errors. 21 error components should be measured for a 3-axes system, and over 30 components are required for a 5-axes system. Thus the volumetric accuracy control task can only be solved by means of high-performance multi-functional noncontact measuring systems.
Using laser interferometers makes solution of this kind of task possible [4]. Large volume of observation and computer-based information processing gives new quality to cnc-controlled systems. It is now possible to develop an error correction strategy based on laser measurements and automatically correct the errors by means of CNC system. Next, a **post-processor can be introduced between a CAM system and CNC system**. The postprocessor is based on error distribution information and introduces corrections into CNC program. This provides a possibility to correct volumetric errors by means of software without mechanical adjustments of the machine [1].
The authors have developed a software for error distribution visualization. The software functions as follows:

1) The software calculates absolute value of volumetric error vectors based on interference measurements results according to (1-5) for a number of workspace points;
2) Each absolute value is associated with a color accordng to color scale, i.e. minimal to maximal error green to red or black to white;

IOP Conf. Series: Journal of Physics: Conf. Series **858** (2017) 012037 doi:10.1088/1742-6596/858/1/012037

3) The workspace is output on a screen with its points colored according to the color scale (Figure 2).

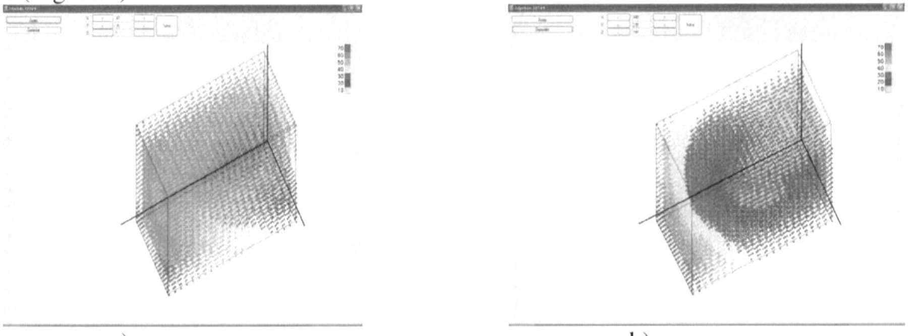

a) b)

Figure 2. Error distribution visualization software depicting errors before (a) and after (b) correction. Colorscale in μm.

The authors have developed a CNC program postprocessor introducing the error correction. Its sequence of operation is as follows:
The volumetric error components and absolute values are calculated for all the considered workspace points according to (1-4);

1) The zero of machine's coordinate system is displaced by value of each axes error components for a k workspace point. Thus the error within the k point is completely compensated and equals zero.
2) The error components within all other points are changed by same displacement value, thus the error can either increase or decrease for different workspace points.
3) The steps 2-3 are being repeated for each i point of workspace. As a result of this search a workspace point called «**zero point**» is found, so that the sum of absolute alues for all th workspace points is minimal (6):

$$\sum_i \Delta_i \rightarrow \min \tag{6}$$

Condition (6) provides minimal volumetric error within maximal workspace zone.

4. Results of experiments

The measuring system required for solving the considered problem should be capable of data storing and transmitting and compatible with data processing software and error mapping visualization software.
A Renishaw XL-80 was used for data collection, being capable of performing noncontact measurements within wide range, with high precision and on high speed of moving parts.
A Kondia A-10 was used as error correction object (Figure 3 a). The type of the machine is a widespread kind including stacked horizontal slides and vertical spindle (Figure 3 b).

IOP Conf. Series: Journal of Physics: Conf. Series **858** (2017) 012037 doi:10.1088/1742-6596/858/1/012037

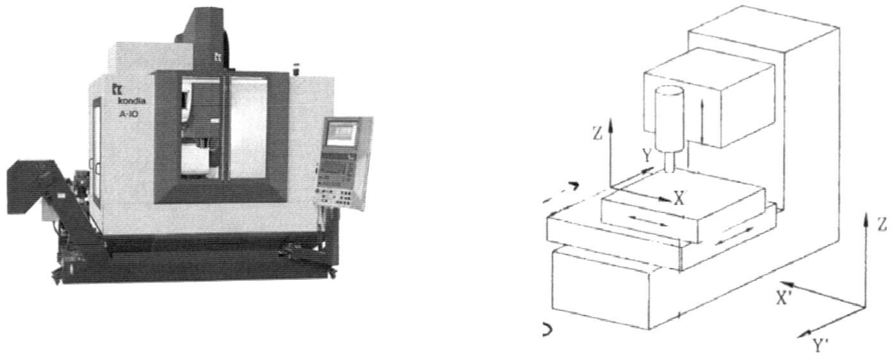

a) b)

Figure 3. Kondia A-10 exterior (a) and schematic (b).

The authors have measured the 21 error components of the object. Fig. 4-6 represent schematics, pictures and results of measurements of selected error components.
The experiment has determined that a point "A" with coordinates X=440, Y=230, Z=180 complies with (6). The Figure 2 b represents volumetric error distribution after correction using this point. The "green" area of low error (under 20 μm) at Fig. 2b is much larger compared to the Figure 2a shown. Before correction this area included 8% of workspace points; after the correction the "green" area included more than 50% of workspace points.
After computer simulation of correction the machining center's CNC system was corrected according to the simulation results. Next, the 21 error components were measured again according to part IV in order to validate the simulation results. The actual error values differed from simulation results by no more than 3%.

5. Discussion

The volumetric error mapping makes the following possible:

- Visualization of volumetric geometric machining or measurement errors within the machine's workspace;
- Determination of workspace areas with minimal or required error values;
- Determination of machined or measured surfaces deformation;
- Calculation of correcting software adjustments of CNC system providing minimal errors within the workspace.

a) b) c)

Figure 4. Schematics (a), picture (b) and results (μm) (c) of Y axis positioning errors measurements.

IOP Conf. Series: Journal of Physics: Conf. Series **858** (2017) 012037 doi:10.1088/1742-6596/858/1/012037

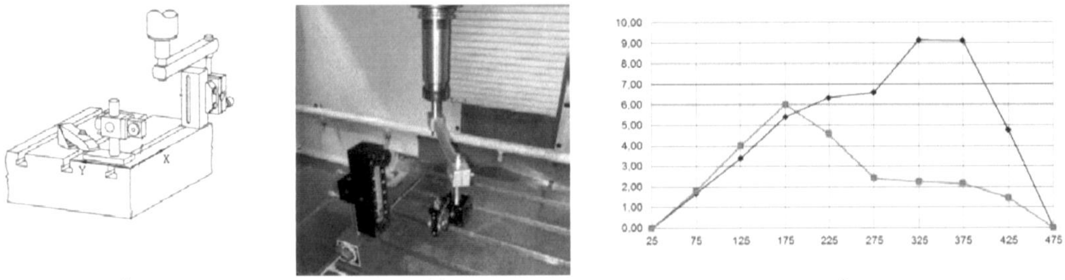

a) b) c)

Figure 5. Schematics (a), picture (b) and results (μm) (c) of Y axis straightness errors measurements.

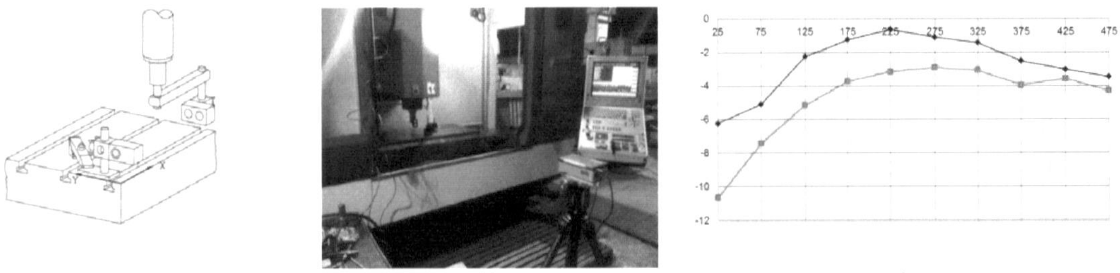

a) b) c)

Figure 6. Schematics (a), picture (b) and results (μm/mm) (c) of Y axis angular errors measurements.

6. Conclusion

- Advances in measuring systems and CNC-controlled machinery provides new opportunities for volumetric geometric errors improvement.
- Modern laser measuring devices, being noncontact, multifunctional and efficient, make it possible to acquire large array of measurement data on the tested machine by means of fast and precise measurement of multiple error functions. All sizes and quality classes of machining and measuring equipment can be rapidly tested. Thus previously unsolvable problem of volumetric error correction can now be solved.
- The algorithm and software used for this paper provide high speed of calculations (i.e. it processes 12000 points in 40 seconds).
- The "zero point" error correction strategy is introduced, providing minimal errors within maximal workspace zone.
- The authors described variational search algorithm. The algorithm finds the "zero point" by means of processing the deformed workspace in several dozens of seconds.
- Using CAM and CNC systems with a postprocessor corrects errors of considered machine based on calculated geometric error distribution.

The article describes a method of significant improvement of multi-axis machines volumetric accuracy by means of only software procedures and without mechanical adjustments of the machine.

The method's effectiveness is confirmed by real measurements within physical workspace of a multi-axis machine.

7. Acknowledgment

The paper was sponsored by The ministery of Science and Education of Russian Federation, research program № 1883.

References

[1] V Teleshevskii, V Sokolov 2012 Laser correction of geometric errors of multi-axis programmed-controlled systems *Measurement Techniques* vol. 55 **5** pp 535-41

[2] S Kononogov, V Lysenko and S Zolotarevsky 2008 Concept of geometric surface parameters measurements traceability *Pribory* **3** pp 1-13

[3] H Schwenke et al. 2008 Geometric error management and compensation of machines – an update *Ann. CIRP* vol. 57 pp 660–75

[4] V Sokolov, K Basalaev 2014 Laser measurements based for volumetric accuracy improvement of multi-axis systems *Physics Procedia* vol. 56 pp 1297-1304

IOP Conf. Series: Journal of Physics: Conf. Series **858** (2017) 012038 doi:10.1088/1742-6596/858/1/012038

Control of the offset printing image quality indices

L G Varepo[1], A Yu Brazhnikov[1], A A Volinsky[2], I V Nagornova[3], A P Kondratov[3]

[1] Omsk State Technical University, 11, Mira ave., Omsk, 644050, Russia
[2] University of South Florida, Tampa, FL 33620, USA
[3] Moscow Polytechnic University, 38, Bolshaya Semenovskaya str., Moscow, 107023, Russia

E-mail: larisavarepo@yandex.ru

Abstract. The problem of assessing printing quality is urgent nowadays and requires informative adaptation and development methods. The study of "printing ink – substrate" boundary, taking into account a great variety of printing ink, is of high scientific and applied interest. The paper presents research results of the printouts made on the coated paper by the offset printing method. Visualization of the components distribution of offset ink in substrate surface and volumetric layers of the substrate is presented. Based on the scanning electron microscopy and energy dispersive spectra analysis of the printouts cross-sections, colorimetric parameters and the character of ink layers distribution in pores of the paper, coated layers of two coloured inks at different applications during printing, are compared. The tendency of bigger distortion of the required colour at higher depth of the ink penetration is observed. The obtained data can be used for the values of pre-distortions, predefined in the pre-press, so that the printing ink can be chosen depending on the required reproduction accuracy.

1. Introduction

The quality indicators of the offset printout significantly depend on the stage of the ink fixation on the printing substrate during printing, a complicated processes, which depends on the physicochemical characteristics of the interacting systems.

The analysis of the printing main elements allowed to identify the following factors influencing the product quality: printing material and ink, dampening solution and equipment configuration options. The problem of printing quality assessment is quite complicated and urgent, requiring informative methods of adaptation and development.

In references [1, 2], computers are utilized for assessing offset printing quality, including dots deviation control and imitation of ink film thickness distribution for various images. Printability of functional inks on multilayer curtain coated substrates was investigated in reference [3], where pore size was the determining factor for penetration of the nano-sized ink.

Electron microscopy can be used for both printing material analysis [4, 5], particularly, the form and the pores location in the paper volume, and for monitoring ink distribution in the printing substrate structure [6, 7]. Besides, it can be used for analyzing image quality and for assessing other factors, which characterize interactions between printing ink and substrate surface [8].

IOP Conf. Series: Journal of Physics: Conf. Series **858** (2017) 012038 doi:10.1088/1742-6596/858/1/012038

The ink components were visualized in reference [9] by focused ion beam to determine which of the ink components entered the coating layer. This work agrees with other published findings, which show that the pigments and hard resins remain on the surface of the coating, whereas mineral and vegetable oils are mobile and penetrate into the coating layer. Ester solvent and dissolved wax also penetrate into the coating layer. Liquid spreading on solid surfaces and penetration into porous matrices (powders and coated papers) are also investigated in reference [10] both experimentally and theoretically.

2. Problem statement

Investigating the "printing ink–substrate" boundary, taking into account a great variety of printing inks, is of high scientific and applied interest. In papers [5, 6, 11] there is some data showing electron microscopy images of the "printing ink–board" boundary. However, the factors influencing the printout process and curing on the substrate surface, have not been considered.

The aim of the research is the study and monitor the influence of the offset printing ink emulsifying efficiency on the process of its absorption and curing on the substrate surface. Besides, it is practically important to control the degree of ink emulsification under printing conditions, taking into account simple reliability indices of the printout quality.

3. Experiment methodology

Printouts printed with two offset inks by leading manufacturers were studied: Litoflora FTX FlintGroup (**I1**) and Toyo (**I2**) in a standard manner of 100% covering black (**K**)-Cyan (**C**) –Magenta (**M**). Coated TitanGloss paper was used as a substrate. The printouts were printed under comparable conditions on the four-color ManRoland704 printing machine without thermal control of the printing apparatus with alcohol damping.

To obtain and keep the acidity within the pH=4.8-5.3 range, the dampening solution concentrate Hydrofast 307 GS was used. Chemically pure isopropyl alcohol was used as an additive, while acetone content was 0.03vol.%. Feed rate of the dampening by the duct roller was 23%. The degree of the printing ink emulsifying efficiency (D_e) at the interaction with dampening solution was defined according to the reference [12].

Quantitative assessment of the depth of ink absorption into the substrate structure was defined by the "printing ink–substrate" boundary, according to the cross-section images obtained with high-resolution field emission scanning electron microscope JSM7500F (JEOL). Spatial resolution was better than 1 nm. To decrease charging and prevent destruction of the non-conductive samples under the electron beam, the sample surface was coated with platinum layer (about 10 nm thick) in argon atmosphere using the AutoFineCoater JFC-1600 magnetron sputtering system.

For element composition and distribution on the surface and in the volume of paper layers, energy dispersive spectrometer (EDS, Oxford X-Max80) was used with the SATW window at 10 kV accelerating voltage, $2 \cdot 10^{-10}$ A probe current, 40 s acquisition time and minimal counting rate. Si was used as the optimization element.

For the printout quality indices, optical density and colour difference ΔE in the system of Lab colour space with regard to the colour coordinates were considered, given in the file. Optical density was defined by the conventional method using the GretagMacbeth SpectroEye spectrophotometer.

4. Experimental results

Experimental results are presented in Table 1 and in Figure 1–2 in the form of SEM images of the printout cross sections, printed by the following offset printing ink: Litoflora FTX FlintGroup (**I1**) and Toyo (**I2**) in a standard manner of 100% covering black (**K**) –Cyan (**C**) –Magenta (**M**), with superimposed EDS mapping of the ink layers elements distribution, along with the X-ray signal intensity at different adsorption depth into the paper pores.

Based on the obtained SEM images and EDS analysis of the printouts cross-sections, colorimetric values and the character of the ink layer distribution on the surface and in the pores of the paper coated layer, two colored ink series are compared during the printing process, specifically: **K / KC / KCM**.

IOP Conf. Series: Journal of Physics: Conf. Series **858** (2017) 012038 doi:10.1088/1742-6596/858/1/012038

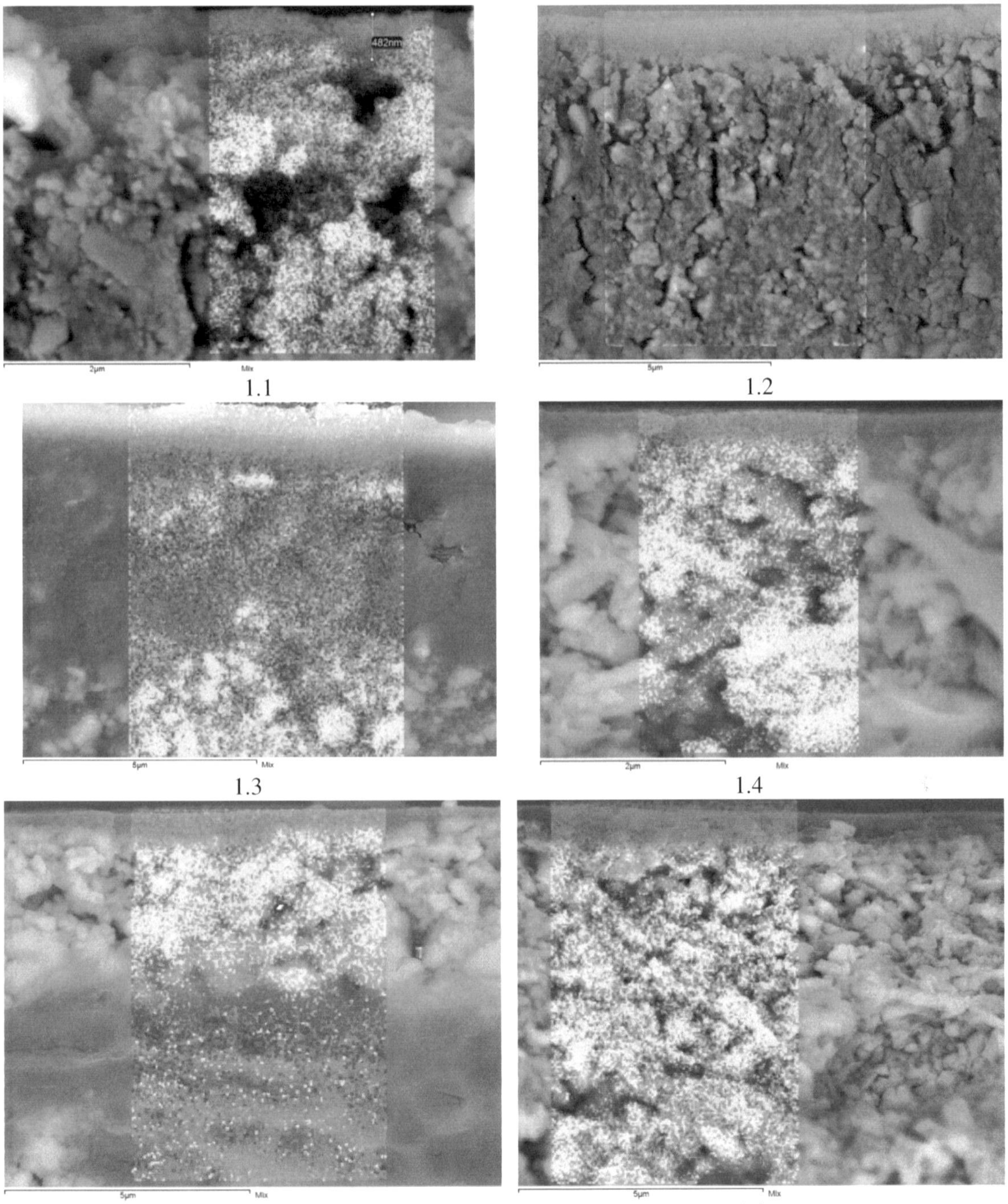

Figure 1. Beginning. SEM images of printouts cross-section: 1.1 –**I1 K** (20000x); 1.2 – **I1 K/C** (10000x); 1.3 – **I1 K/C/M** (10000x); 1.4 – **I2 K** (20000x); 1.5 – **I2 K/C** (10000x); 1.6 – **I2 K/C/M** (10000x).

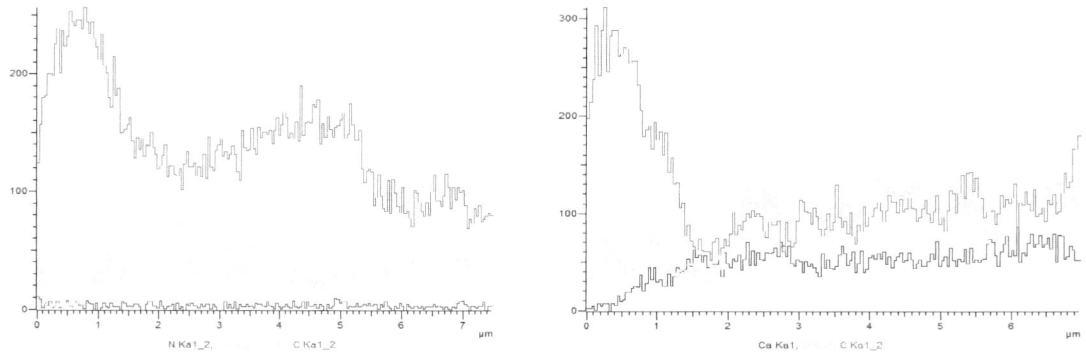

1 2

Figure 2. EDS mapping of elements distribution and X-ray signal intensity as a function of penetration depth: 1 – **I1 K/C/M** (10000x); 2 - **I2 K/C/M** (10000x).

5. Results and discussion

The thickness of the ink layer varies according to the zones, depending on the total area of the printing elements in the given zone and is defined at the "printing ink–substrate" boundary. The thickness and the optical density of black (**K**) ink layers are equal in both cases, while the values of the colour difference are moderate, and the ink layer penetration into the coated layer surface is almost invisible (Table 1, Figure 1, positions 1.1 and 1.4).

Table 1. Ink layers parameters.

Ink	D_e %	Optical density	ΔE	The layer thickness on the surface, µm	Adsorption depth, µm
		K / KC / KCM	*K / KC / KCM*	*K / KC / KCM*	*K / KC / KCM*
I1	65	1.7 / 1.4 / 1.5± 0.1	2.2 / 1.7 / 4.0±0.2	0.4±0.1 / 0.8±0.3 / 0.9±0.2	0.3±0.1 / 1.5±0.2 / 4.5±0.3
I2	54	1.7 / 1.4 / 1.4± 0.1	1.8 / 1.0 / 0.7±0.1	0.4±0.1 / 0.5±0.1 / 0.6±0.1	0.4±0.1 / 0.4±0.1 / 0.6±0.2

After the application of the second ink (**C**, wet printing), the difference becomes more essential (Table1, Fig. 1, positions 1.2 and 1.5) and obtains significant values after the application of the third ink (**M**, Table 1, Figure 1, positions 1.3 and 1.6). It can be seen in elements distribution curves (Figure 2, positions 1 and 2) that the **I1** depth of penetration into the pores of the coated layer is about 4.5 µm, and the thickness of the ink layer directly on the paper surface is approximately 1 µm (determined by the boundary of increasing Ca signal, which constitutes the coated paper layer basis). For the **I2** these values are approximately 0.6 µm. It should be noted that the thickness of the ink layers in all **I2** applications and the depth of penetration into the paper have approximately equal values.

Such ink behavior can be explained by varying degree of emulsification while printing. On the one hand, the printing elements contact the dampening form roller during each plate cylinder rotation cycle and bear a layer of dampening solution on their surface, which migrates into the inking unit during subsequent contacts. On the other hand, form printing roller transfers the dampening solution into the inking unit, while in contact with the dampened whitespace areas of the printing plate. Higher emulsifying value of the **I1** ink (Table 1) means that there is possibly more dampening solution in the transferable ink layer. Herewith, optical density of the ink layers both for **I1,** and for **I2** have comparable values, and are within the tolerance limits permitted by the ISO 13655 standard.

The values of the colour differences are significant: for **I1** **ΔE=4±0.2** is the boundary and the maximum possible value when printing with the so-called psychological fidelity (differences are visible to the naked eye); **ΔE=0.7±0.1** for **I2** is an exact reproduction of the specified object colour.

IOP Conf. Series: Journal of Physics: Conf. Series **858** (2017) 012038 doi:10.1088/1742-6596/858/1/012038

6. Conclusions

In conclusion, precision of the given colour reproduction on the printout and the character of different ink adsorption in surface pores and volumetric paper structure layers were reasonably compared.

The tendency of the larger distortion of the required colour can be observed at all versions of ink application for the case of the maximum penetration of the printing ink components into the structure of the coated material.

Based on the collected pre-distortion data values defined in the pre-press process, the choice of ink for the specified printing system can be made with regard to the required accuracy of the multicolored image reproduction.

References

[1] Lundstrom J and Verikas A 2010 Detecting halftone dots for offset print quality assessment using soft computing *Fuzzy Systems (FUZZ),* IEEE Conf. Publications pp 1–7

[2] Yu J and Wei P 2011 Modeling the axial oil distribution of offset printing *Electrical and Control Engineering (ICECE),* IEEE Conf. Publications pp 4968–4971

[3] Bollström R, Tobjörk D, Dolietis P, Salminen P, Preston J, Österbacka R and Toivakka M 2013 Printability of functional inks on multilayer curtain coated paper *Chemical Engineering and Processing: Process Intensification, Advances in Coating and Drying of Thin Films* vol 68 pp 13–20

[4] Kugge Ch, Craig V S J and Daicic J 2004 A scanning electron microscope study of the surface structure of mineral pigments, lattices and thickeners used for paper coating on non-absorbent substrates *Colloids and Surfaces A: Physicochemical and Engineering Aspects* **238(1-3)** 1-114

[5] Ozaki Y and Kimura M 2000 Visualization of printing of ink vehicle on paper surfaces by a SEM technique *Appita J.* **3** 216-219

[6] Preston J S, Elton N J, Husband J C, Dalton J S, Heard P J and Allen G C 2002 Investigation into the distribution of ink components on printed coated paper *Colloids and Surfaces A: Physicochemical and Engineering Aspects* **205(3)** 183–198

[7] Verikas A, Lundström J and Gelzinis A 2011 Advances in computational intelligence-based print quality assessment and control in offset colour printing *Expert Systems with Applications* **38(10)** 13441–13447

[8] Pietrov D, Gereige I and Gourgon C 2013 Automatic detection of NIL defects using microscopy and image processing *Microelectronic Engineering* **112** 163–167

[9] Koivula H, Preston J S, Heard P J and Toivakka M 2008 Visualisation of the distribution of offset ink components printed onto coated paper *Colloids and Surfaces A: Physicochemical and Engineering Aspects* **317(1–3)** 557–567

[10] Rosenholm J 2015 Liquid spreading on solid surfaces and penetration into porous matrices: coated and uncoated papers *Advances in Colloid and Interface Science* **220** 8–53

[11] Varepo L G, Nagornova I V, and Trapeznikova O V 2015 Application of electron microscopy to quality control of surface ink layers *Procedia Engineering* **113** 357–361

[12] Surland A 1983 Factors determining the efficiency of lithographic inks *TAGA Proc.* pp 191–233

IOP Conf. Series: Journal of Physics: Conf. Series **858** (2017) 012039 doi:10.1088/1742-6596/858/1/012039

Computer visualization of the results for calculating the Ink «dusting»

L G Varepo[1], O V Trapeznikova[1], A V Panichkin[2], I V Nagornova[3], V I Bobrov[3]

[1] Omsk State Technical University, 11, Mira ave., Omsk, 644050, Russia
[2] S L Sobolev Omsk branch of Institute of mathematics of the Siberian Branch of the Russian Academy of Sciences, 13, Pevtsova st., Omsk, 644043, Russia
[3] Moscow Polytechnic University, 38, Bolshaya Semenovskaya str., Moscow, 107023, Russia

E-mail:panich@ofim.oscsbras.ru

Abstract. The paper presents the results of the numerical modeling of the quantitative assessment of the offset printing ink "dusting" index at the outlet of the printing contact zone while transferring it to the substrates having various surface characteristics. The modelling was carried out with the help of finite-difference methods. The paper considers the results of practical implementation of the software calculating the printing ink "dusting". The result of the ink filaments splitting into many fine particles and the intensive spraying these particles into the surrounding space by the centrifugal force reduces the print quality. Creating new concepts of reducing the level of "dusting" in offset printing and its quantitative assessment is important. A new approach to solving the problem of reducing the "dusting", which is characterized by the influence on the ink in the engagement zone of the idle pulse of varying power, is suggested. To quantify the ink transferred into noodles and participating in the formation of ink "dusting", process numerical modeling techniques using solution finite difference methods are applied. Graphic visualization of the given solution results on the basis of the graphic modeling is presented. The practical implementation of this method improves the quality of the final printing product, and allows to predict the quantitative assessment of the "dusting" indices directly in the process of preparing the order for printing and to optimize the selection of the printing system components.

1. Introduction

Dusting is a result of the ink filaments splitting into many fine particles and the intensive spraying these particles into the surrounding space by the centrifugal force. The disadvantages of this process are the reduced quality of printing and high environmental safety risk.

Finding the solutions for managing the flows of the viscous fluid relatively to the rotating cylinders [1-3] is still urgent nowadays.

An ink preset model for offset press based on least squares support vector machine optimized by chaotic bee colony (combining with the least squares support vector machine) was proposed in [4] to solve the adverse effects for the lateral ink flow due to the movement of the vibrator rollers.

According to [5] it can calculate ink distribution on the paper by the use of computer simulation, hence a matrix that reflects ink flow characteristics of the inking system under the current image coverage can be established.

In [6] emphasis is placed on seeking the optimal control of time-averaged forces on the cylinder surface using the rotation rate as a control parameter. However, ink "dusting" parameters are not discussed in this study.

Together with the previously known methods of "dusting" preventing, i.e. the group of electro physical methods and the group of methods based on using various chemical additives, some new approaches are suggested by scientists [7, 8].

Here for the quantitative analysis of the image, obtained with the help of so-called "dusting" eliminator, is used. Such "dusting" eliminator is made of a paper sheet, installed around the upper element of the roll. The image was scanned and analyzed by measuring the ink "dusting" drops distribution and its total area, using special software at the end of the experiment.

Reducing the ink "dusting" is achieved by "splitting" the filament until its forming by positioning a wire or a blade into the gap between the rolls in parallel with them.

The results, obtained in [7, 8] have shown, that the blade reduces the dusting level by 92% in average and is more efficient at higher speeds.

The wire, in contrast, acts better at low speeds, while at higher speeds the wire vibration causes an increase in the force, so it cannot be installed close to the point of rupture at high printing speeds.

Despite these positive trends in "dusting" reducing, it should be noted that there is the difficulty of defining the "dusting", because the increase in speed or cut-off of the distributing shaft vibration have a significant impact on the dusting index. The image obtained by the eliminator depends on the accuracy of its mounting and fixation, its shape, size, etc.

The analysis of the periodic literature on the subject has revealed that the creation of new concepts concerning the reducing of the "dusting" level in offset printing and its quantitative evaluation is relevant currently; that is why it is the aim of the present study.

2. Statement of the problem

Based on the analysis of papers [1, 2, 7, 8] it was revealed that using of additional devices into the printing engagement zone has some limitations taking into account that:

- structure complication and, consequently, increase in the price of the printing equipment;
- direct dependency of the image obtained by using eliminator on the accuracy of its installation and fixing, its shape, size, etc.

The direction of the development of new approaches to solving the problem of reducing ink "dusting" and quantification of this index, due to high complexity of its definition, is now of both scientific and practical interest and is the purpose of the given study.

3. Theory

In the proposed technical solution, by using mathematical modeling and mathematical statistics methods, in the controlled area of the ink transfer upon offset printed imaging the area with separate ink noodles having the highest "dusting" probability is determined; while in a controlled area of the printed image the area a the control zone, having the highest "dusting" probability is determined.

To quantify the printing ink transferred into the ink noodles and participating in the formation of the ink "dusting", the methods of numerical and graphical process modeling [9-11] using the solution finite difference methods are applied.

For calculating the fluid flows between the cylinders the following formula was used (1) [11]:

$$x_c = \frac{\frac{\tau^2}{H\rho}\left[(P_1-P_0)-\frac{E\Delta r}{H\,2}\right]}{\left(1+\frac{E\tau^2}{H^2\rho}\right)} \pm \frac{\sqrt{\frac{\tau^4}{H^2\rho^2}\left[(P_1-P_0)-\frac{E\Delta r}{H\,2}\right]^2+\left[1+\frac{E\tau^2}{H^2\rho}\right]^2\tau^2\cdot\dot{x}_{c,1}^2}}{1+\frac{E\tau^2}{H^2\rho}} \quad (1)$$

where Δr is the value of the boundary deformation at a time pointt_1 H is the initial thickness of the deformable layer, ρ is the fluid density, E is the Young's modulus of rubber or a substrate, P_0 is the external atmospheric pressure, P_1 is the pressure on the deformable material surface.

The ink "dusting" index was determined using the developed numerical modeling algorithm as the difference between the following parameters (indices): the total number of ink transferred from the ink-offsetting surface onto the coated material and remaining on the ink-offsetting surface after transfer and the ink splitting in the printing engagement zone.

At the outlet of the printing engagement zone the "dusting" reduction, carried out by cross-cutting the ink noodles by using thermal effect of infrared laser beam was controlled.

4. Experimental results

In this setting numerical and graphical modeling of transferring the ink between the contacting surfaces and the control of the "dusting" index was carried out up to a certain time value at the stages of printing contact and at the outlet of the printing engagement zone.

The input data for the modeling includes the following: the cylinders diameters in the printing unit (offset and printing) $R_1 = R_2 = 0.30$ m, $P = P_{atm} = 10^5$ N/m^{2}is the ambient pressure; kinematic viscosity $\nu = 0.012$ m^2/s, density $\rho = 10^3$kg/m^3, surface tension coefficient $c_n = 0.03$ N/m.

Numerical solutions on the grid 80×80 are given at the angular rate equal to 10 rad/s for different time points. The experiment was carried out with substrates having surface roughness indices R_a – 0.3 ± 0.02 (substrate 1) and 2.8 ± 0.3 (substrate 2) correspondingly.

The modeling was carried out with regard to the contacting surfaces deformation. Elasticity modules are: of rubber $E_1=2.9\cdot10^7$ N/m^2, of the substrate $E_2=4.8\cdot10^6$ N/m^2.
Pulse strength of the pulse guided into the engagement zone was varied from 0 to 7.0 W.d.

5. Results and discussion

The results of the numerical modeling for the "dusting" assessment are given in Figure 1.

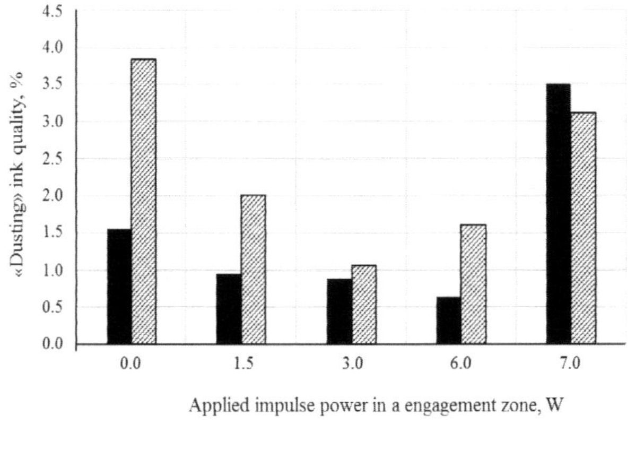

Figure 1. The results of numerical modeling of the laser power influence on the "dusting" index: ■ - substrate roughness 0.3± 0.02; ▨ - substrate roughness 2.8± 0.3.

IOP Conf. Series: Journal of Physics: Conf. Series **858** (2017) 012039 doi:10.1088/1742-6596/858/1/012039

The results of the graphic modeling of the "dusting" control are given in Figure 2 (where 1 is standard offset printing, 2 is printing with introducing the additional pulse into the printing engagement zone for time period $t = 0.38 \times 10^{-5}$ s) for substrate 1.

Figure 3 presents the process of full ink layer splitting while transferring onto the substrate 1 (1 – without additional pulse effect, 2 – under additional pulse effect).

The transfer of the printing ink onto the substrate in the printing engagement zone (the very moment of the ink splitting centers forming) is shown.

The positive effect of the practical application of the developed method for reducing ink "dusting" is visually demonstrated in Figure 3 (position 2).

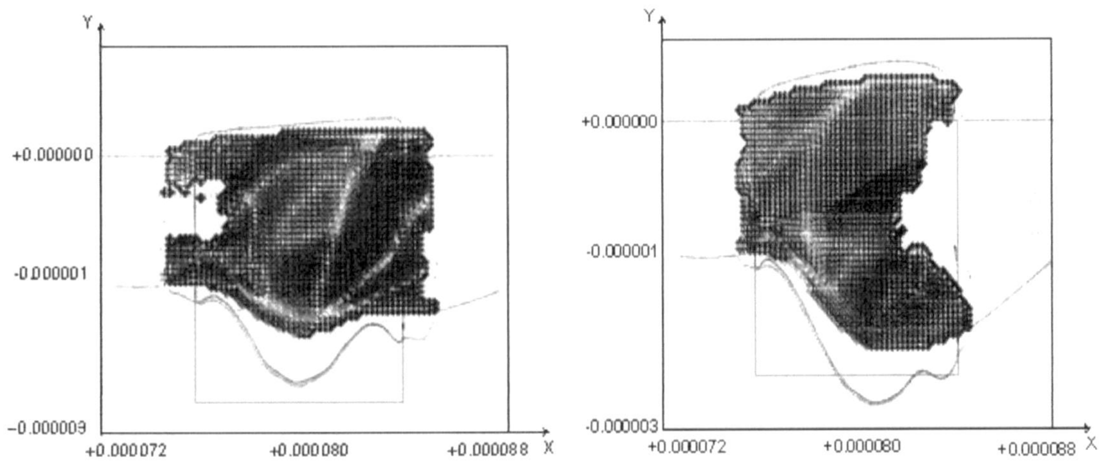

1 2

Figure 2. The results of graphic visualization.

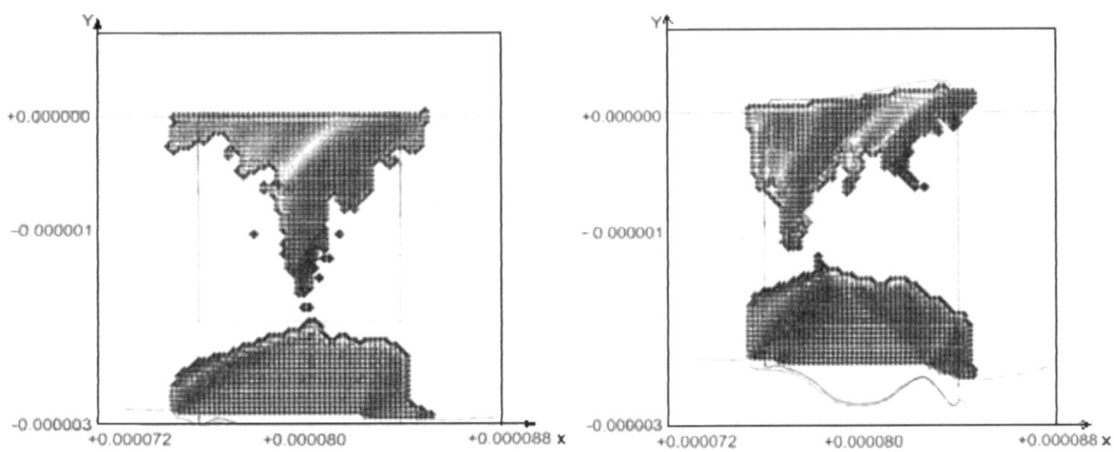

1 2

Figure 3. The results of graphic visualization.

As it can be observed, the quantity of the ink, transferred into the "dusting", which is shown in the form of separate graphic pixels, is considerably reduced.

The analysis of the research results has shown that the strength of additional pulse, under which the effect of reducing the ink "dusting" when it is splitting at the outlet of the printing engagement zone occurs, depends on the characteristics of microgeometry of the ink-carrying surface.

For substrates with low surface roughness and low porosity the ink layer splitting occurs in practice between the ink-carrying and the ink-absorbing surfaces.

Increasing the surface roughness and the substrate porosity increases the ink "dusting".

The amount of ink, transferred into the "dusting" for substrate 2 is considerably higher than for substrate 1, which can be explained by the composition of the test samples.

It is found out that, taking into account the guide pulse effect on the ink in the contact zone, the amount of ink transferred into the "dusting" with the pulse power increase on the estimated cell up to 1.50 W is reduced by 2.4 times (from 1.55 to 0.64%), the process of ink splitting goes faster (Figure 1).

With the further increase in the pulse power, on the other hand, there is an growth in this index, which is probably caused by the rising in cavitation.

It should be noted that the amount of ink on the offset cylinder after the splitting with the additional pulse into the engagement zone increases slightly with increasing the pulse power due to the "dusting" reducing.

6. Conclusions

The practical importance of the suggested approach on the basis of the selected criteria for the specific components of the printing system is the following.

The proposed solution allows controlling and predicting the quantitative assessment of the "dusting" index directly in the process of preparing the order for printing as well as optimizing the selection of the printing system components, and reduces the labor intensity due to the use of information processing technical means.

Besides, the implementation of this approach allows to control the influence of surface and structure indices on the changes in the ink layer at different stages of the printing process in the into engagement zone.

It also contributes to the environmental protection component of the printing unit of the printing press, and improves the quality of the final product of the multicolor transfer of images to a substrate.

References

[1] Ou Y-R 1992 Decision and Control relative to a rotating cylinder Decision and Control, IEEE Conf. Publications vol 4 pp 3399–3404

[2] Yu J andWei P 2011 Modeling the axial oil distribution of offset printing *Electrical and Control Engineering (ICECE),* IEEE Conf. Publications pp 4968–4971

[3] Tai C H, Zhao Y and Liew K M 2005 Parallel computation of unsteady incompressible viscous flows around moving rigid bodies using an immersed object method with overlapping grids, J. of Computational Phys. **207(1)** 151–172

[4] Yu J and Lin J 2012 The ink preset algorithm based on the model optimized by chaotic bee colony 5th Int. Congress on Image and Signal Processing pp 547–551

[5] Yu J and Tian P 2010 Study on print ink presetting through singular value decomposition *3rd Int. Congress on Image and Signal Processing (CISP)* vol 5 pp 2405–2408

[6] Ou Y-R 1992 Active flow control relative to a rotating cylinder Proc. of the 31st IEEE Conf. on Decision and Control vol 4 pp 3404–3399

[7] Vlachopoulos G, Claypole T and Bould D 2010 Ink "dusting" formation in roller trains *IARIGAI 2010 Proceedings: Advances in Printing and Media Technology* vol 37 pp 227–234

[8] Claypole J, Williams P R and Deganello D 2012 Control of breakup of ink filaments in offset printing *IARIGAI 2012 Proceedings: Advances in Printing and Media Technology* vol 39 pp 207–211

[9] Panichkin A V and Varepo L G 2014 The simulation of the viscous fluid splitting at the outlet of the engagement zone between cylinder surfaces and substrate Dynamics of Systems, Mechanisms and Machines, IEEE Conf. Publications pp 1–4

[10] Panichkin A V and Varepo L G 2014 The Numerical Calculation of a viscous Incompressible Fluid transfer onto Poros Surface between Rotating Cylinders Int. Multidisciplinary Microscopy Congress, Springer Proc. in Physics vol 154 pp 79–83

[11] Varepo L G, Panichkin A V and Trapeznikova O V 2016 The numerical calculation of the viscous incompressible fluid transfer between contacting surfaces *IOP Conf. Series: Materials Science and Engineering* vol 124 pp 1–6

Method for Evaluating Information to Solve Problems of Control, Monitoring and Diagnostics

V A Vasil`ev, N V Dobrynina

Penza State University, 40, Krasnayastr., Penza, 440026, Russia

E-mail: opto@bk.ru

Abstract. The article describes a method for evaluating information to solve problems of control, monitoring and diagnostics. It is necessary for reducing the dimensionality of informational indicators of situations, bringing them to relative units, for calculating generalized information indicators on their basis, ranking them by characteristic levels, for calculating the efficiency criterion of a system functioning in real time. The design of information evaluation system has been developed on its basis that allows analyzing, processing and assessing information about the object. Such object can be a complex technical, economic and social system. The method and the based system thereof can find a wide application in the field of analysis, processing and evaluation of information on the functioning of the systems, regardless of their purpose, goals, tasks and complexity. For example, they can be used to assess the innovation capacities of industrial enterprises and management decisions.

1. Introduction

Decisionsupport systems (DSS) are useful to improve the efficiency of complex systems management and make the informed management decisions, based on the latest information technologies. They allow one to convert and analyze the accumulated information and develop the information control action [1–3].

Usually, the indicator system of a complex technical, economic, or social object of study is a complex hierarchical structure with many private indicators, which, depending on the management tasks, may include various criteria, reflecting its status [4–9].

It is advisable for management tasks to have a generalized criterion of the scorecard (for example, reflecting the level of innovative capacity of the system to solve the problem of assessing the functioning of research and production enterprise), a comprehensive assessment of information system performance (including financial condition, intellectual support, quality personnel, manufacturing flow, corporate organization, culture of innovation, innovation receptivity in solving the problem of estimation of innovative potential of the enterprise) to characterize the overall information about the actual state of the system (technical, economic and social).

The actual problem is the creation of systems conversion, analysis, management and information processing, enabling a comprehensive assessment of various kinds of information collected from various sources and created in the process of functioning of complex technical, economic and social systems [10–16].

IOP Conf. Series: Journal of Physics: Conf. Series **858** (2017) 012040 doi:10.1088/1742-6596/858/1/012040

2. The problem statement and objectives

To solve the problems of efficiency increase of systems analysis functioning, processing and evaluation of information it requires the development of a new, improved method and a system that allows an integrated criteria-based assessment of heterogeneous information, on the basis of which one would produce management, control and diagnostics of better quality.

In particular, scientific and practical interest is the creation of assessing information system about the innovative potential of mechanical engineering enterprises, which allows the assessment of an innovative potential level of the enterprise in real time, to analyze and make decisions on enterprise management, while providing the enhanced functionality of the system, improving the efficiency of information processing about the innovation potential and to provide reasonable criteria-based evaluation of its components.

3. Description of the method and the system based thereof

A new method for evaluating information for solving problems of control, monitoring and diagnostics consists in reducing the dimension of information indicators of situations; conversion of figures in relative units; calculation of generalized information metrics; ranking them by characteristic levels; calculating an efficiency criterion of functioning of the system in real time. This method has been developed and the system has been designed on the basis thereof, for the analysis and information processing on the functioning of the systems. For example, they can be used to assess the innovation capacities of industrial enterprises and management decisions.

The task is to develop a new, improved method and a system enabling the integrated criteria-based assessment of heterogeneous information, on the basis of which better management, control and diagnosis could be implemented, solved by:

- a new workflow, which allows reducing the dimension of information indicators of situations that lead them to relative units, calculating generalized information metrics on their basis, ranking them by the characteristic levels, calculating the performance criteria of the system functioning for the current situation;
- reducing the number of decision-makers (DM) to one, eliminating the need for control over the timeliness of decision made by many decision-makers, eliminating the need of calculating weighting coefficients according to the i-th DM, the elimination of calculations to determine the likelihood of timely and correct decision, enhancing the visibility of information display about the results of the assessment of the current state of the system (by ranking the levels of information on performance situations);
- reducing time and amount of the required computational resources and memory, by avoiding the dimensionality (redundancy) of information and indicators of a number of calculations, improving the objectivity of information assessment.

The evaluating information method is aimed at assessing the effectiveness of decisions and functioning of the system.

The necessary source data is stored in a storage device (SD) in the form of arrays of variables and constant values of information performance situations.

The following parameters are used as the initial information: the decision matrix A of dimension m x h, where h is a number of solutions; the array B of the current situation of dimension m. A linguistic variable may be quantitative or qualitative characteristic of an argument that affects the choice of the solution. For example, if the argument is a certain dimension in some physical quantity, the linguistic variable can have values like “small”, “medium”, and “large”.

Performance management indicator W of systems with the automated decision making is evaluated by

$$W = P_{cd} \times P_{t.i}, \tag{1}$$

where $P_{cd}$is probability of correct decision; $P_{t.i}$is probability of timely implementation of correct and timely decision (the correct and timely implementation of the decision refers to the timely delivery of properly executed solution to the performers on communication lines).

$$P_{cd} = \frac{1}{M}\sum_{j=1}^{M} R_j, \; j = \overline{1,M}, \tag{2}$$

where $M$is a number of considered situations; $R_j$is correct decision for the *j-th* situation;

$$R_j = \sum_{j=1}^{M} R_{ij}, \; j = \overline{1,M} \; . \tag{3}$$

Here

$$R_i = \begin{cases} 1, if a solution is found \\ 0, if a solution is not found \end{cases} ; \tag{4}$$

$$P_{t.i} = \frac{1}{Q\times M}\sum_{j=1}^{M} \frac{C_j}{R_j}, \; j = \overline{1,M}, \tag{5}$$

where C_j is a number of timely implemented decisions for the j-th situation;

$$C_j = \sum_{q=1}^{Q} C_{ij}, \; q = \overline{1,Q} \; , \tag{6}$$

where $Q$is a number of communication lines corresponding to the number of performers;

$$C_{qj} = \begin{cases} 1, if t_{qj} \le t_{add.i} \\ 0, if t_{qj} > t_{add.i} \end{cases} , \tag{7}$$

where $t_{qj}$is time spent on the implementation of decisions of the performers in the *j-th* situation; $t_{add.i}$is time allowed for implementation of the decision.

After recording information in the storage device, the dimensionality of the data is reduced using the factor and correlation analysis, leading information indicators$V_j$for each *j-th* situation to relative units using the scales according to

$$V_j = \begin{cases} [0;20), v_j \in (-\infty; l_{2j}) \\ [20;40), \; v_j \in [l_{2j}; \, l_{2j} + 20u_j) \\ [40;60), \; v_j \in [l_{2j} + \, 20u_j, l_{2j} + 40u_j), j = \overline{1,M}, \\ [60;80), \; v_j \in [l_{2j} * 40u_j, l_{4j}) \\ [80;100], \; v_j \in [l_{4j}; \infty) \end{cases} \tag{8}$$

where v_j is an information indicator for the *j-th* situation; l_{2j} is a standard value of an information indicator corresponding to the second level (low value); l_{4j} is a standard value of an information indicator corresponding to the fourth level (highest value). Here is a step interval of the casting scale

$$u_j = \frac{l_{2j} - l_{4j}}{3} \times \frac{1}{20}, j = \overline{1,M}. \tag{9}$$

The next step of the system is the identification of the generalized indicators of information situations for P_i groups with defined indicators of the system state, and calculation of the characteristic level for each group by

$$X_i = \begin{cases} first level, P_i \in [0;20) \\ second level, \; P_i \in [20;40) \\ third level, \; P_i \in [40;60) \\ fourth level, \; P_i \in [60;80) \\ fifth level, \; P_i \in [80;100] \end{cases} , \tag{10}$$

where X_i is the characteristic group-level indicators of the system state, here

$$P_i = \frac{\sum_{j=1}^{M} V_j}{M}, \tag{11}$$

IOP Conf. Series: Journal of Physics: Conf. Series **858** (2017) 012040 doi:10.1088/1742-6596/858/1/012040

$$K_{effect} = \left(\sum_{i=2}^{n} \frac{1}{2} P_i P_{i-1} \sin\left(\frac{360}{n}\right) \Big/ \frac{1}{2} {P_{great}}^2 n \sin\left(\frac{360}{n}\right) \right) \times 100\% =$$

$$= \left(\sum_{i=2}^{n} P_i P_{i-1} \Big/ {P_{great}}^2 n \right) \times 100\% \,, \qquad (12)$$

where P_{great} is the maximum possible value of the information indicator of the situation on the scale of these values (P_{great} =100); n is a number of generalized information indicators.

The next step is to analyze the compliance options and situations, then output information about the situation on the display unit screen.

After determination the probability of good decision-making and the probability of timely implementation of decisions, we determine the value of the management efficiency indicator by (1), output on the screen of the display unit, and analyze the received information.

Logic systems for information evaluating on the basis of the proposed method are presented in Fig. 1 and Fig. 2, where signs marked ○ are the inputs of the system, and ◻ is the relationship between the elements.

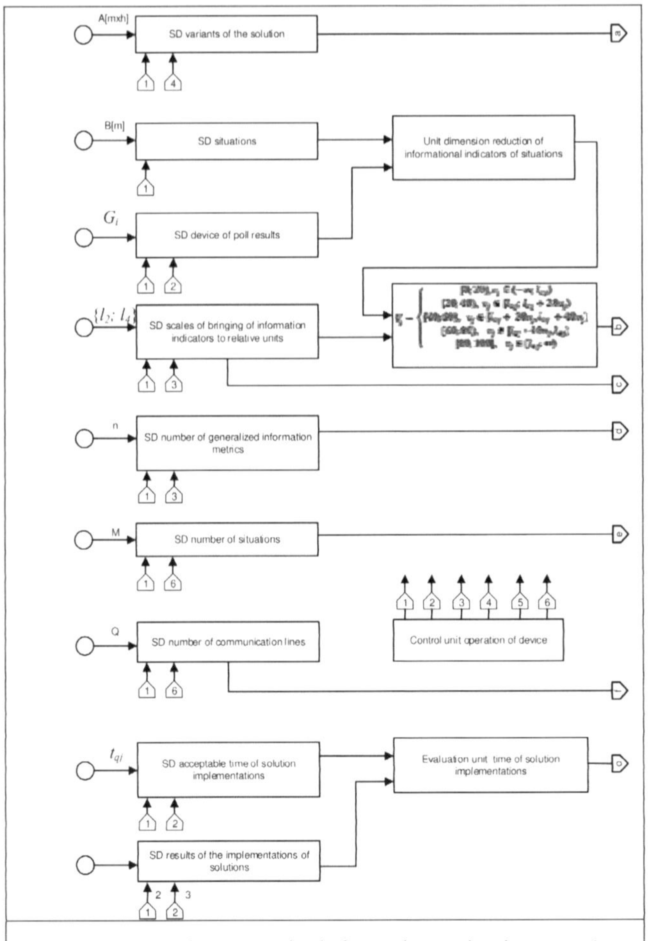

Figure 1. Logic system for informationevaluating, part 1.

IOP Conf. Series: Journal of Physics: Conf. Series **858** (2017) 012040 doi:10.1088/1742-6596/858/1/012040

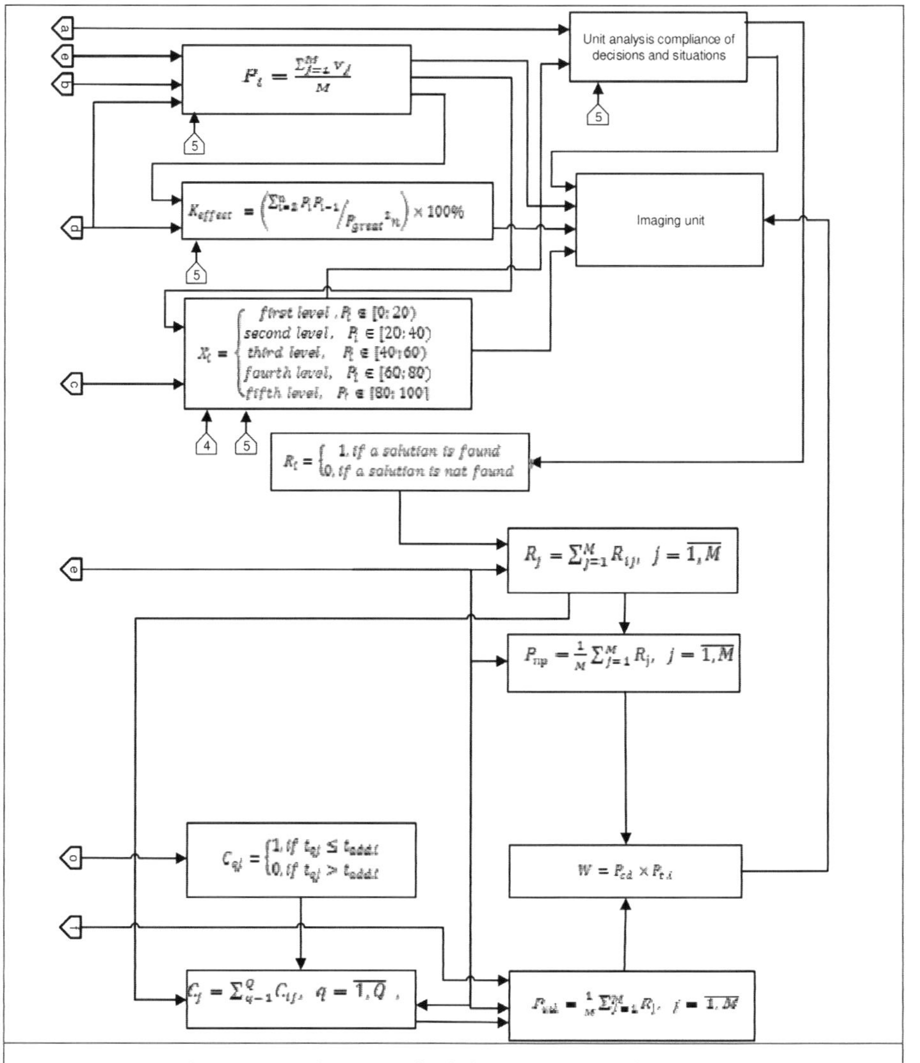

Figure 2. Logic system for informationevaluating, part 2.

The essence of the proposed approach to the assessment of information is to produce associations between qualitative and quantitative results of the evaluation components included in the model evaluate the information and solutions possible under the current situation.

4. Results and Prospects of Their Use

Below there are the estimation results of innovative potential of research and production instrument-making enterprise using the developed evaluation information method.

Table 1 provides numerical values of information indices in relative units (the maximum possible value is of 100 relative units) and the numbers of levels of groups of information indicators (1 to 5) obtained using the described method of evaluating information.

IOP Conf. Series: Journal of Physics: Conf. Series **858** (2017) 012040 doi:10.1088/1742-6596/858/1/012040

Table 1. Values and levels of information indices groups of innovation potential of the enterprise

Information IndicesGroup	Value	Levels
Financial potential	50	*thirdlevel*
Intellectual potential	61	*fourthlevel*
Organizational and managerial potential	40	*second level*
Marketing potential	42	*second level*
Information and methodological support	70	*fourthlevel*
External innovation climate	66	*fourthlevel*
Logistical potential	45	*second level*
Innovation culture	50	*thirdlevel*

By the numerical values of the groups and the numbers of levels of information indices of innovative potential, shown in Table 1, it is possible to draw a conclusion about the adequacy or inadequacy of a specific index group, based on the specified requirements.

Fig. 3 shows a radar chart with the values of information indicators groups depicted in Table 1. As you can see from the chart, the selected object for research – research-and-production instrument-making enterprise – had a relatively low organizational and management capacity during information evaluation, and it was being increased. The average level was characterized by the marketing potential, experience of realization of innovative projects, the potential of material-technical base and innovative culture of the enterprise. The intellectual capacity, information and methodological support, external innovation climate had a high level.

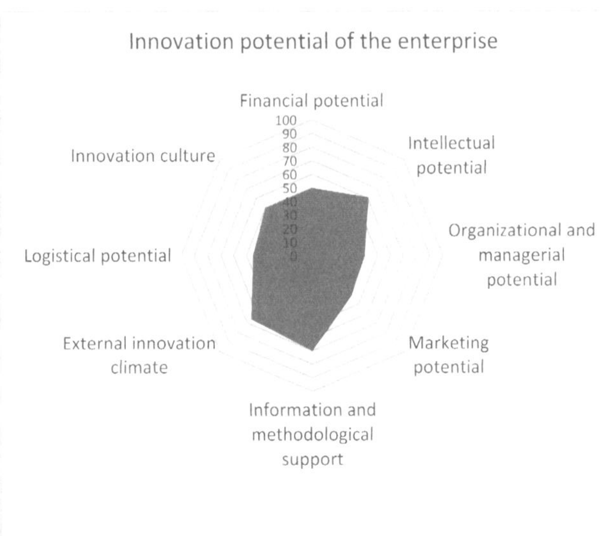

Figure 3. A radar chart for the system efficiency.

The efficiency indicator of the investigated object (the system company) functioning, calculated by (12), has a value of

$$K_{effect} = \left(\frac{50 \times 61 + 61 \times 40 \ldots 45 \times 50}{100^2 \times n}\right) \times 100\% =$$

$$= \left(\frac{3050 + 2440 + \cdots + 2970 + 2250}{70000}\right) \times 100\% = 28.5\%$$

The example confirms the efficiency of the proposed method and the possibility of its use to evaluate the information about the object, which can be any technical, economic and social system.

The developed method can be used to evaluate the information in the tasks of management, control and diagnostics of technical, economic and social systems. In particular, it can be used to assess the level of innovative potential of the enterprise in real time, analysis and decision-making in the enterprise management. While it provides the enhanced functionality of the system, it improves the efficiency of information processing about the innovation potential, and allows a proper evaluation of criteria-based components of the innovative potential.

5. Conclusion

The example confirms the efficiency of the proposed method and the possibility of its use to evaluate the information about the object, which can be any technical, economic and social system.

The advantage of the proposed method and the system for information evaluating based on thereofis its application to assess both quantitative and qualitative information. Heterogeneous information is evaluated and presented in the form of numerical values of the information indices of groups and numbers of the group levels. And the performance score of the studied object as a whole is determined by the values of information indices groups.

The method and the system for evaluating information to solvethe problems of management, control and diagnostics can be used in the analysis, processing and evaluation of information on the functioning of technical, economic and social systems, regardless of their purpose, goals, tasks and complexity. For example, they can be used to monitor and diagnose complex equipment to evaluate information about the innovative potential of the instrument-making enterprise.

References

[1] A. A. Burba, V. A. Makarov, and S. P. Khripunov, RF Patent No. 2207621, Rev. Fig., **18**, 2003

[2] A. A. Burba, V. A. Makarov, and B. D. Tret'yakov, RF Patent No. 2158955, Rev. Fig., **31**, 2000

[3] V. A. Selivanov and V. V. Selifanov, RF Patent No. 2326442, Rev. Fig., **16**, 2008

[4] V. A. Vasil'ev, "Dynamic management of financial and material resources of the enterprise," Proc. Conf. Information Technologies in Science, Education, Telecommunication, Business and Protection of Natural Resources, Ukraine, Crimea, Yalta-Gurzuf (20-30 May 1999), Zaporozhye, 1999, p. 185

[5] V. A. Vasil'ev, "Information resources management in small business," Proc. Conf. Information Technologies in Science, Education, Telecommunications and Business, Ukraine, Crimea, Yalta-Gurzuf (May 18-28 2000), Zaporozhye, 2000, pp. 123–124

[6] N. V. Dobrynina, "Application of statistical and econometric analysis to identify systematic relations of the fuzzy set evaluation model of enterprise's innovative potential," Devices and Systems. Management, Control, Diagnostics, 2015, **8**, pp. 53–58

[7] N. V. Dobrynina, "Assessment of innovation capacities of industrial enterprises with application of hybrid expert systems," Devices and Systems. Management, Control, Diagnostics, 2015, **9**, pp. 49–53

[8] N. V. Dobrynina, "Evaluation of innovative potential of enterprise based on fuzzy-set

descriptions," Devices and Systems. Management, Control, Diagnostics, 2015, **10**, pp. 51–54
[9] V. A. Vasil'ev,"Information resource of recording solid-state structures," Measurement Techniques, vol. 47, **7**. USA, New York: Springer, 2002, pp. 706–709
[10] V. A. Vasil'ev, "Homeostasis and homeokinesis in improving information efficiency of data converters," Metrology, **5**. Moscow, 2003, pp. 3–21
[11] V. A. Vasil'ev, "Decision making in the design of data converters," Automation and Modern Technology, **11**. Moscow, 2003
[12] V. A. Vasil'ev, "Information-structural analysis and synthesis systems transform information from heterogeneous structures," Devices and Systems. Management, Control, Diagnostics, 2004, **1**, pp. 24–27
[13] E. M. Belozubov, V. A. Vasil'ev, and N. V. Gromkov,"Modeling of converters of information systems for measuring, monitoring and control," Proc. III Int. Conf. Analytical and Numerical Methods for Modelling Natural and Social Problems, Penza (15-16 October 2008), Penza: Volga House of Knowledge, 2008. pp. 229–232
[14] V. A. Vasil'ev, "Principles of construction of models of devices and systems," Devices and Systems. Management, Control, Diagnostics, 2003, No**. 6**. pp. 40–45
[15] V. A. Vasil'ev, and N. V. Gromkov,"System approach to improvement of measurement transducers," Devices and Systems. Management, Control, Diagnostics, 2010, **4**, pp. 33–38
[16] V. A. Vasil'ev, "Synergistic aspects in the activity of enterprises,"Automation and Modern Technology, **3**. Moscow, 2008, pp. 44–46

Analysis of electromagnetic valve with spring at the joint of plunger and plug

N V Zakharenkov[1], **V V Zakharenkov**

[1] Omsk State Technical University, 30, Mira ave., Omsk, 644050, Russia

E-mail: znickbar@mail.ru

Abstract. Widespread use of electromagnetic valves in the engineering fields makes urgent the problems which are aimed to improve efficiency by optimization of their design. The paper aims to examine of efficiency of the electromagnetic valves with an accumulative spring at the joint in the kinematic chain with a plunger and a plug. To achieve the primary goal, the two problems are formulated and considered: analysis of the electromagnetic valve by considering its operation as separate phases; experimental investigation of the valves and comparison of electromagnetic characteristics of the valves having an accumulative spring with the same characteristics of classical electromagnetic valves. The design and principle of valves with an accumulative spring operation are considered. The dynamic equations of electromagnetic valves with an accumulative spring, which are took into account a pressure drop while the valve is opened, are stated. The equation for accumulative spring stiffness that is proportional to the given resistant force created by the pressure of fluid is written. The results of experimental investigation of the valves with an accumulative spring show that current in a coil is linear for all values of operating environment pressure. The power demand can be reduced by using an accumulative spring.

1. Introduction

A classical electromagnetic valve can be defined as a device in which movement of a plunger occurs due to electromagnetic forces and wherein the plunger is rigidly connected with a plug. Such valves are widely used in many engineering fields. In recent years, the electromagnetic valves are being used more and more in combustion engines instead of cam gears [1, 2].

Performance and operational reliability of the industrial systems with control valves are greatly affected by characteristics of these electromagnetic valves [1, 3].

Geometrical dimensions and energy consumption of an electromagnetic valve actuator are important characteristics in terms of the operational reliability and the performance both the valve itself and a control object. Therefore, optimization of the electromagnetic actuator of the direct acting valves for geometrical dimensions and energy consumption is an important problem.

here are a lot of research papers in the field of electromagnetic valves optimization that are floating around applications of electromagnetic valves in different technical devices [1, 3, 4]. Moreover, the valve actuator, which is generally an electromagnet, is the aim of optimization. Today the growing interest in application of the electromagnetic actuator in mechanotronics [5, 6] makes urgent the investigation of new approaches to design of electromagnetic actuators that have a lower weight of used material per unit work under given solenoid heating conditions [6].

Determination of the relation between electrical characteristics and geometrical parameters is an

IOP Conf. Series: Journal of Physics: Conf. Series **858** (2017) 012041 doi:10.1088/1742-6596/858/1/012041

important problem both in terms of finding engineering calculation techniques [4, 7, 8] and in terms of optimization of a design. There is approach to optimization providing a low power consumption without changing dimensions of the valve. It is based on application of accumulative springs in the kinematic chain with a plunger and a plug [9]. Such valves we will call electromagnetic valves with an accumulative spring.

2. Problem statement

The paper aims to analyze the an electromagnetic valve having an accumulative spring in the kinematic chain with the joint of a plunger and a plug. To achieve the primary goal, the following problems are formulated and considered:

1. Analysis of the electromagnetic valve by considering its operation as separate phases;
2. Experimental investigation of the valves. Comparison of electromagnetic characteristics of the valves having an accumulative spring with the same characteristics of classical electromagnetic valves.

2.1. Design analysis.

The basic scheme of classical electromagnetic valve and valve having accumulative spring consists of two parts (see Fig. 1) without regard to a design and over-all dimensions.

The first part is a valve. A valve body 5 has an inlet channel 10 and outlet channel 9 with a seat 8 which is overlapped by a plug 7. The plug 7 is provided with a return spring 6 and connected with a rod 4 which is linked to a plunger 3.

The second part is an electromagnet by which the valve is shifted to open position. The electromagnet comprises the magnetic core 1, a coil 2 and the plunger 3.

The electromagnetic valve having an accumulative spring has been changed in connection of the rod 4 and the plunger 3. The connection has linear elastic element – a spring. Generally, the accumulative spring can be in the form of any spring element efficiently storing deformation energy.

Fig. 1 shows the electromagnetic valve that differs from classical electromagnetic valve in containing of an accumulative spring 13, a bucket 11 and a tail stock 12 linked to the rod 4.

The accumulative spring 13 is placed in the bucket 11 and set so that it encloses the rod 4. In addition, it rests at one end against the tail stock 12. The bucket 11 and the plunger 3 are connected rigidly.

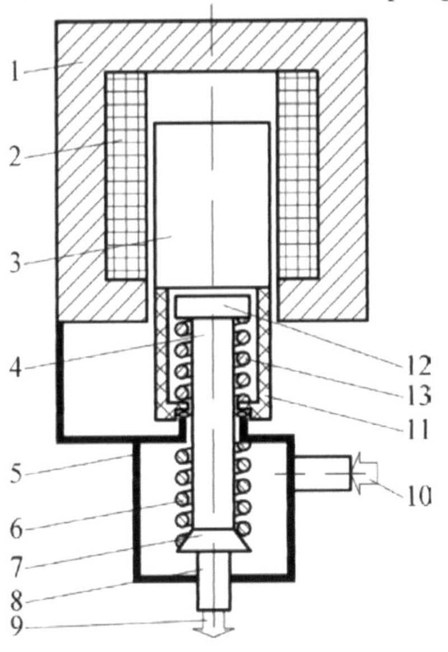

Figure 1. Basic scheme of electromagnetic valve having accumulative spring: 1 - magnetic core; 2 - coil; 3 - plunger; 4 - rod; 5 - valve body; 6 - return spring; 7 - plug; 8 - seat; 9 - outlet channel; 10 - inlet channel; 11 bucket; 12 - tail stock; 13 - accumulative spring.

Operation of classical electromagnetic valve is to supply current to the coil 2 for the valve to be opened, i.e. for the plug 7 to be moved (out) from the seat 8. Thus, electromagnetic force arises. The plunger 3 passes it to the plug 7 through the rod 4. Separation of the plug 7 from the seat 8 takes place when the electromagnetic force exceeds an external load (counter force) due to pressure of an overlapped area and of a return spring. Further, the plunger 3, the rod 4, and the plug 7 move to the magnetic core 1 by a value of plunger full stroke xmax.. When current from the coil 2 is switched off, the plug 7 is set at the initial position and overlaps the seat 8 under influence of the return spring 6. Let us agree that the classical valve opening process occurs per one phase.

The opening process of the electromagnetic valve with an accumulative spring is accomplished in a few phases. Now we define these phases for design shown in Fig. 1.

With current supplied to the coil, electromagnetic force arises at the first phase. The force moves the plunger and the bucket by a value of the air gap δ_s between them (value of accumulative spring strain). At the same time, the electromagnet overcomes the elastic spring force of an accumulative spring. The accumulative spring accumulates potential energy of deformation under the action of electromagnetic force. The first phase ended at that point.

A plunger stroke attains value δ_s at the second phase when the maximum degree of accumulative spring deformation is reached. It results in significant growth of electromagnetic force that is sufficient to separate the plug 7 from the seat 8 and to open the valve. It is assumed that accumulative spring is in a compressed state. Distance between the plug and the seat, before the valve is fully opened, is less than given by a value of accumulative spring deformation. The second phase is ended at that point.

At the third phase movement of the plug into the given position relatively to the seat is due to the process where potential energy of accumulative spring deformation converts to kinetic energy of the plug.

When current from the coil is switched off, the plunger returns to initial position under influence of the return spring and the plug overlaps the seat.

2.2. Experimental procedure.

Valves analysis is carried out for various designs of the electromagnetic valves by energizing valves at a different pressure of operating environment. Direct acting, normally closed valves – the classical electromagnetic valves with different valve flow area (II, III) and valves having an accumulative spring (I) – with technical characteristics presented in Table 1 were considered.

Electromagnets of the valves I and II are intended for valves having a seat with a diameter of 0.8 mm and pressure $24 \cdot 10^5$ Pa. The valve electromagnet III is for valve having seat with a diameter of 1 mm and pressure $54 \cdot 10^5$ Pa. The flow area of valves I, II and III are modified to a seat diameter of 9 mm, i.e. the valve design in all tests is identical with a seat diameter of 9 mm. Such seat diameter is not a nominal valve size for its electromagnet for all three valves.

The characteristics of the valve I and the valve II are the same. The connection of the plunger and the plug in the valve I has an accumulative spring, which is shown in Fig.1.

Table 1. Specifications of test specimens (electromagnetic valves)

Electromagnet mass, kg	Coil resistance, ohm	Plunger diameter, mm	Diameter of valve flow area, mm	Plunger stroke, mm	DC voltage, V
		With an accumulative spring I			
0.2	75	14	9	6	27
		Classical II			
0.2	75	14	9	6	27
		Classical III			
1.2	50	22	9	6	27

In the first set of tests, a current level of electromagnetic valve I is compared with a current level of

the valve II at the moment when the valves are opening and pressure of operating environment is from $2 \cdot 10^5$ Pa to 10^6 Pa.
In the second set of tests the valves I and III are compared.
Pressure of operating environment and valve response time are basic parameters.
Experimental equipment amounts to a testing workbench that has a compressor, a pneumatic pipeline, a manometer, fittings, the testing valves, wires, a DC source, an ammeter, and switches for current to be supplied from the DC source to a coil of the testing valves.

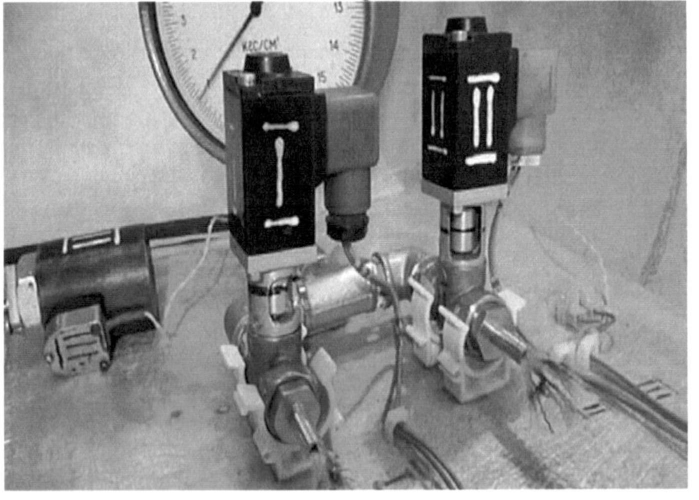

Figure 2. Experimental equipment.

Valve response time (from the moment of current supplying to a coil to the moment of the plug separation) was measured through observations from the time of pressing a current supply button to the time of the beginning of the plug movement.

3. Theory

3.1. Plunger dynamic equation.
Let us consider dynamic of plunger for design in Fig. 1. There are following forces at the functioning plunger:

- inertial force which is produced by moving plunger mass;
- reaction force of operating environment that is proportional to a seat area S and to a pressure difference ΔP between inlet and outlet channels;
- elastic spring forces of an accumulative spring and a return spring;
- dry friction force;
- viscous friction force.

Direct acting valves are characterized by the following equation

$$m\ddot{x} = F - P_1S_1 + P_2S_2 - k_{rs}x - F_{0rs} - \alpha\dot{x} - \beta \operatorname{sgn} \dot{x} \qquad (1)$$

where x – a plunger stroke; m – mass of moving parts; F – moving force; $P_{1(2)}$ – inlet and outlet channels the pressure of fluid; $S_{1(2)}$ – inlet and outlet areas; $\alpha\dot{x}$ – viscous friction force; $\beta \operatorname{sgn} \dot{x}$ – dry friction force; k_{rs} – an elastic coefficient of a return spring; F_{0rs} – force created by initial set of a return spring.
Equation (1) of electromagnetic valve having an accumulative spring is written as follows taking into account pressure drop while the valve is being opened:

$$m\ddot{x} = F_e + (-P_1S_1 + P_2S_2)\rho(x) - k_{rs}x - F_{0rs} - k_{as}x - F_{0as} - \alpha\dot{x} - \beta \operatorname{sgn} \dot{x},$$

where F_e – electromagnetic force; k_{as} – an elastic coefficient of an accumulative spring, F_{0as} – force

IOP Conf. Series: Journal of Physics: Conf. Series **858** (2017) 012041 doi:10.1088/1742-6596/858/1/012041

created by initial set of an accumulative spring; $\rho(x)$ – a function which takes into account pressure drop while the valve is being opened. When the function is equal to 1 – maximum value of the function – no pressure drop occurs and the system is in a state of equilibrium.
Following formula describes electromagnetic force [6]

$$F_e = \frac{SN^2\mu_r^2\mu_0 i(t)^2}{2(l_2 + l_{eq} + (\mu_r - 1)x)^2},$$

For the operating time of an electromagnet in different phases to be calculated we will use equation (2)

$$mdv = [F_e + (-P_1S_1 + P_2S_2)\rho(x) - k_{rs}x - F_{0rs} - k_{as}x - F_{0as} - \alpha\dot{x} - \beta\,\mathrm{sgn}\,\dot{x}]dt, \tag{2}$$

where v – linear velocity.
Let us consider dynamic of electromagnet at the first phase for the classical electromagnetic valve and the valve having an accumulative spring after current supplying. Before electromagnet being activated, the plunger is at the standstill position due to resistant force caused by an environment pressure and a return spring. Moreover, velocity and acceleration components, dry friction force, and viscous friction force are equal to zero, while $\rho(x) = 1$. Using equation (2) for classical electromagnetic valve, we obtain equation of statics equilibrium:

$$F_{e1} + (-P_1S_1 + P_2S_2) - F_{0rs} = 0, \tag{3}$$

where $F_{e1} = \dfrac{SN^2\mu_r^2\mu_0 I^2}{2(l_2 + l_{eq} + (\mu_r - 1)x_{\max})^2}$; $x_{\max}$ – maximum value of a plunger stroke.
Equation of statics equilibrium for electromagnetic valve having an accumulative spring can be written as:

$$(-P_1S_1 + P_2S_2) - F_{0rs} >> k_{as}x + F_{0as}, \tag{4}$$

where $F_{e2} = \dfrac{SN^2\mu_r^2\mu_0 I^2}{2(l_2 + l_{eq} + (\mu_r - 1))(x_{\max} - \delta_s))^2}$.

Formula (3) means that motion of plunger starts from overcoming forces caused by the pressure of fluid and a return spring, while formula (4) shows that motion of plunger starts from overcoming forces created by an accumulative spring.
3.2. Accumulative spring stiffness.
Accumulative spring stiffness for given resistant force created by the pressure of fluid and for maximum degree of accumulative spring deformation δ_s can be defined from (3) and (4) can be expressed as:

$$(-P_1S_1 + P_2S_2) - F_{0rs} - k_{as}x - F_{0as} = 0.$$

For $x = \delta_s$, i.e. if valve is activated, accumulative spring stiffness can be written as:

$$k_{as} = \frac{(-P_1S_1 + P_2S_2) - F_{0rs} - F_{0as}}{\delta_s}.$$

4. The results of experiments
The first set of tests shows that the current in a coil for the classical electromagnetic valve II changes from 0.65 A to 1.6 A, however, it is not for all pressure values. When pressure was $1 \cdot 10^6$ Pa the valve was not opened. The current in the valve I with an accumulative spring, which has operated for all pressure values, is equal to 0.4 A for any pressure.

IOP Conf. Series: Journal of Physics: Conf. Series **858** (2017) 012041 doi:10.1088/1742-6596/858/1/012041

The valve response time of the valve II for pressure $2\cdot10^5$ Pa is 0.36 s, for all other values of pressure the valve response time more than 8 s. The valve response time of the valve I changes from 0.16 s to 0.8 s in accordance with growth of pressure of operating environment.
The value of current in a coil for the classical electromagnetic valve III was defined at the second set of tests. The current changes from 0.3 A to 0.65 A. The valve response time of the valve III changes from 0.56 s to 1.44 s. Although, the valve response time was defined not for any pressure of operating environment.
A current change for the classical electromagnetic valves is linear for all values of operating environment pressure. Although, for the electromagnetic valves with an accumulative spring it is constant and equal to 0.4 A for all values of pressure. The valve response time increases with growth of pressure.

5. Discussion of the results

In the theoretical part of the article, the model of plunger movement for the valves having an accumulative spring is presented. The dynamic equation of the valves with an accumulative spring is different from classical ones by having phases in their operating. Separation of the phases is caused by nonlinearity of the valve operation that, in turn, is produced by moving mass changes. At the first phase it is a plunger mass, at the second one – masses of the plunger and the rod with the plug.
The refined model takes into account pressure drop when the valve is opened. It is necessary, since one must make allowance for a state of the system between the first and the second phases. Definition of an accumulative spring stiffness, considering that valve is operated, allows to find a number of springs for application in the valves depending on a plunger stroke, coil parameters, and value of operating environment pressure.
The considered above dynamic equations of electromagnetic valves with spring at the joint of the plunger and the plug allow to determine the time of all phases in operation of the electromagnetic valves with an accumulative spring. The main interest for further research is formulation of electromagnet energy equation for each phase of operation. In addition, further research suggests determination of correlation between electrical parameters and geometrical valve dimensions.
The obtained experimental data are shown in Table II, they allow to evaluate results of electromagnetic valves comparison: the classical valves and the valves with an accumulative spring.

Table 2. Experimental results (Coil current, A)

Operating environment pressure, Pa	Electromagnetic valve With an accumulative spring I	Classical II	Classical III
$2\cdot10^5$	0.4 / 0.16	0.65 / 0.36	0.3 / 0.56
$3\cdot10^5$	0.4 / 0.32	0.85 / 8.00	0.37 / –
$4\cdot10^5$	0.4 / –	0.95 / –	0.45 / –
$5\cdot10^5$	0.4 / 0.72	1.0 / 15.6	0.47 / –
$6\cdot10^5$	0.4 / 0.76	1.14 / –	0.5 / 1.12
$7\cdot10^5$	0.4 / 0.76	1.58 / 14.88	0.55 / –
$8\cdot10^5$	0.4 / –	1.60 / –	0.60 / –
$1\cdot10^6$	0.4 / 0.8	Not opened	0.65 / 1.44

IOP Conf. Series: Journal of Physics: Conf. Series **858** (2017) 012041 doi:10.1088/1742-6596/858/1/012041

6. Conclusions

The performed analysis of known direct acting valves with spring element shows that, up to the present moment, the theory of optimal design of such electromagnetic valves has not yet been developed. In addition to that, there is no engineering techniques of calculation of them.

According to the results, it is clear that obtained value of current for the valve I indicates that this current value is required to overcome elastic spring force of an accumulative spring. Therefore, it is reasonable to conclude that the power consumption can be reduced by using an accumulative spring.

For further investigation the evaluation should be performed for a physical model that will take into consideration various designs of valves. After that, the numerical modeling should be used for comparison of obtained results of physical modeling.

The next step of experimental investigation is close examination of search for basic parameters, and also, is obtaining initial information for the designing of testing devices. Later, it is expected to carry out a numerical experiment.

In view of the aforementioned, new problem statement can be formulated – the development of calculation methods and theory of optimal design of electromagnetic valves with an accumulative spring.

References

[1] Ohdachi Y, Kawase Y, Murakami Y, Inaguma Y 1991*IEEE Transactions on Magnetics***27**5226–5228
[2] Lee S H, Yi H C, Han K, Kim J H 2015*Energies***8**pp 13222–13230
[3] Mahajan D P, Narayanaswamy R, Bavisetti S 2014 *IEEE 23rd International Symposium on Industrial Electronics (ISIE). Istanbul*pp 1245–1249
[4] Paden B A, Snyder S T, Paden B E, Ricci M R 2015*IEEE/ASME Transactions on Mechatronics***20**pp2654–2665
[5] Melbert J, Uhlenbrock R 2003*Vehicular Technology Conference***5**pp3202–3206
[6] Gomis-Bellmunt O, Campanile L F 2010Design Rules for Actuators in Active Mechanical Systems. Springer-Verlag London Limited205
[7] Baek-Ju S, Eun-Woong L 2005 *International Conference on Power Electronics and Drives Systems***2**pp 1140–1145
[8] Schinnerl B, Gerling D 2008 *18th International Conference on, Vilamoura*pp 1–5
[9] Kotelnikov VI, Zakharenkov VV 2010Patent RF, no. 95379

Institute of Physics
Dirac House, Temple Back
Bristol BS1 6BE UK

ISSN: 1742-6588
ISBN 978-1-5108-4246-5

CURRAN ASSOCIATES INC.
proceedings.com